Lecture Notes in Networks and Systems

Volume 78

The series "Lecture Notes in Networks and Systems" publishes the latest developments in Networks and Systems—quickly, informally and with high quality. Original research reported in proceedings and post-proceedings represents the core of LNNS.

Volumes published in LNNS embrace all aspects and subfields of, as well as new challenges in, Networks and Systems.

The series contains proceedings and edited volumes in systems and networks, spanning the areas of Cyber-Physical Systems, Autonomous Systems, Sensor Networks, Control Systems, Energy Systems, Automotive Systems, Biological Systems, Vehicular Networking and Connected Vehicles, Aerospace Systems, Automation, Manufacturing, Smart Grids, Nonlinear Systems, Power Systems, Robotics, Social Systems, Economic Systems and other. Of particular value to both the contributors and the readership are the short publication timeframe and the world-wide distribution and exposure which enable both a wide and rapid dissemination of research output.

The series covers the theory, applications, and perspectives on the state of the art and future developments relevant to systems and networks, decision making, control, complex processes and related areas, as embedded in the fields of interdisciplinary and applied sciences, engineering, computer science, physics, economics, social, and life sciences, as well as the paradigms and methodologies behind them.

** Indexing: The books of this series are submitted to ISI Proceedings, SCOPUS, Google Scholar and Springerlink **

More information about this series at http://www.springer.com/series/15179

Tatiana Antipova
Editor

Integrated Science in Digital Age

ICIS 2019

 Springer

Editor
Tatiana Antipova
Institute of Certified Specialists (ICS)
Perm, Russia

ISSN 2367-3370   ISSN 2367-3389   (electronic)
Lecture Notes in Networks and Systems
ISBN 978-3-030-22492-9   ISBN 978-3-030-22493-6   (eBook)
https://doi.org/10.1007/978-3-030-22493-6

This Springer imprint is published by the registered company Springer Nature Switzerland AG
The registered company address is: Gewerbestrasse 11, 6330 Cham, Switzerland

Preface

This book is based on selected papers accepted for the presentation and discussion at the 2019 International Conference on Integrated Science in Digital Age (ICIS 2019). This conference had the support of the Institute of Certified Specialists (ICS), Russia, and Springer. It took place at Batumi, Georgia, May 10–12, 2019.

The International Conference on Integrated Science in Digital Age (ICIS) is a network of scholars interested in natural and social sciences original research results. The network was established to promote cooperation between scholars of different countries. An important characteristic feature of the conference should be the short publication time and worldwide distribution. This conference enables fast dissemination, so the conference participants can publish their papers in print and electronic format, which is then made available worldwide and accessible by numerous researchers.

The Scientific Committee of ICIS 2019 was composed of a multidisciplinary group of 110 experts from 38 countries around the world; 33 invited reviewers, united by science, have had the responsibility for papers' evaluating in a "double-blind review" process. ICIS 2019 received about 50 contributions from 6 countries. All the selected papers went through a round of the peer-review process. The main topics of the included papers are following: Agriculture in Digital Age; Artificial Intelligence Research; Business & Finance in Digital Age; Digital Accounting & Auditing; Digital Economics; Educational Sciences in Digital Age; Health Management Informatics; Materials Science in Digital Age; Information Management System; Social Science in Digital Age.

The papers accepted for the presentation and discussion at the conference are publishing by Springer (this book) and will be submitted for indexing by ISI, SCOPUS, among others. We acknowledge all of those that contributed to the staging of ICIS 2019 (authors, committees, reviewers, organizers, and sponsors). We deeply appreciate their involvement and support that was crucial for the success of ICIS 2019.

May 2019   Tatiana Antipova

ICIS'19 Committee

General Chair

Tatiana Antipova   Institute of Certified Specialists, Russia

Scientific Committee

Abel Suing   Universidad Tecnica Particular de Loja, Ecuador
Aijaz Shaikh   University of Jyväskylä, Finland
Aijun Chen   Central South University, China
Alan Sangster   University of Aberdeen, UK
Alejandro Medina Santiago   INAOE, Mexico
Aleksejs Jurenoks   Riga Technical University, Latvia
Alex Pedersen   Aarhus Universitet, Denmark
Alexandra Arellano   Universidad Nacional, Colombia
Altinay Fahriye   Near East University, Cyprus
Andre Carlos Busanelli   University of Sao Paulo (USP), Brasilia
  de Aquino
Ankur Bist   KIET Ghaziabad, India
Archi Addo   ERAU, USA
Arnold D. Kaluzny   UNC Gillings School of Global Public Health,
                                      USA
Arooj Arshad   The University of Lahore, Pakistan
Arun Kumar   University College Dublin, Ireland
Bruno Veloso   INESC TEC, Portugal
Carlos Alexandre Silva   Federal Institute of Minas Gerais, Brazil
Carlos Carreto   Polytechnic Institute of Guarda, Portugal
Carlos Ortiz   Universidad Técnica Particular de Loja, Ecuador
Celio Goncalo Marques   Instituto Politecnico de Tomar, Portugal
Cristovam Guerreiro Diniz   Instituto Federal do Pará, Brazil

David Krantz	Arizona State University, USA
Diana Andone	Politehnica University of Timisoara, Romania
Dolores Rojas	Universidad Tecnica Particular de Loja, Ecuador
Edgar Chavez	Universidad Autonoma de Baja California, Mexico
Elena Fleaca	University Politehnica of Bucharest, Romania
Elhem Younes	Universite Paris, France
Elizabeth Carvalho	Universidade Aberta, Portugal
Erick Cristobal Onate Gonzalez	FCB, Universidad Autónoma de Nuevo León, México
Erick Toledo Gómez	Universidad Autónoma de Ciudad Juárez, Mexico
Fairouz Al Dhmour	Mutah University, Jordan
Francesco Caviglia	Aarnus University, Denmark
Francisco Kelsen de Oliveira	IFPE, Brazil
George Danko	University of Nevada, USA
Giuseppe Grossi	Nord University, Norway
Helmer Munoz	Universidad del Sinu, Colombia
Henrique Manuel Pires Gil	Age.Comm-Instituto Politecnico de Castelo Branco, Portugal
Howard Frank	Florida International University, USA
Hu-Chen Liu	Shanghai University, China
Indrawati Yuhertiana	Universitas Pembangunan Nasional Veteran Jawa Timur, Indonesia
Irina Shikina	Central Research Institute for Organization and Informatization of Health Care, Russia
Ivaldir H. de Farias Junior	Federal University of Pernambuco, Brazil
Janneth Chicaiza	Universidad Técnica Particular de Loja, Ecuador
Janusz Grabara	Czestochowa University, Poland
Jelena Jovanovic	University of Nis, Serbia
John Dumay	Macquarie University, Australia
Julia Belyasova	Center for Social Assistance to Local and Foreign Population, Belgium
Juriy Voskanyan	Medical Academy of Continuing Professional Education, Russia
Kalina Trenevska Blagoeva	University St. Cyril and Methodius, Macedonia
Kangkana Bora	Institute of Advanced Study in Science and Technology, India
Khalid Al Marri	British University in Dubai, UAE
Kim Bul	Coventry University, UK
Laura Alcaide Munoz	University of Granada, Spain
Leandro Rodríguez-Hernández	Universidad Autónoma de Ciudad Juárez, Mexico
Lidice Haz	Universidad de Guayaquil, Ecuador
Linda Kidwell	Nova Southeastern University, USA

Lipi B. Mahanta	Institute of Advanced Study in Science and Technology, India
Luca Bartocci	University of Perugia, Italy
Lucas Oliveira Gomes Ferreira	Federal Court of Accounts, Brasilia
Lucas Tomczyk	Pedagogical University of Cracow, Poland
Lukas Kralik	Tomas Bata University, Czech Republic
Luz Sussy Bayona Ore	Universidad Nacional Mayor de San Marcos, Peru
Lydia Nahla Driff	USTHB, Algeria
Mangal Patil	Bharati Vidyapeeth's College of Engineering, India
Manuel do Carmo	Universidade Europeia, Portugal
Marc Behrendt	Ohio University, USA
Marc Dreyer	Futopedia Consulting, Switzerland
Maria Cristina Marcelino Bento	UNIFATEA, Brazil
Marina Gurskaya	Kuban State University, Russia
Maristela Holanda	University of Brasilia, Brazil
Matthew Kenzie	University of Cambridge, UK
Michael Esew	Ahmadu Bello University Zaria, Nigeria
Mikhail Kuter	Kuban State University, Russia
Naci Karkin	Pamukkale University, Turkey
Nemias Saboya Rios	Universidad Peruana Union, Peru
Nicolae Ilias	University of Petrosani, Romania
Nizar Alsharari	University of Sharjah, UAE
Ochoa Humberto	Universidad Autónoma de Ciudad Juárez, Mexico
Oleg Golosov	"Fors" Group of Companies, Russia
Olga Khlynova	Russian Academy of Science, Russia
Omar Jawabreh	University of Jordan, Jordan
Omar Leonel Loaiza Jara	Universidad Peruana Union, Peru
Oscar Hernandez	Universidad Autonoma Metropolitana, Mexico
Pakize Erdogmus	Duzce University, Turkey
Panduranga Rao M. V.	BTL Institute of Technology, India
Patricia Cano Olivos	UPAEP, Mexico
Peter Nabende	Makerere University, Uganda
Ramon Bouzas-Lorenzo	University of Santiago de Compostela, Spain
Reza Mohammadpour	University of Tehran, Iran
Rodolfo Gustinelli	UTC, Brazil
Rodrigo Bortoletto	Sao Paulo Federal Institute of Education, Science and Technology, Brazil
Roland Moraru	University of Petrosani, Romania
Rui Pedro Marques	University of Aveiro, Portugal
Sadik Hasan	University of Dhaka, Bangladesh

Contents

Social Science in Digital Age

Agriculture in Digital Age

The Impact of Motivation on Labor Resources Reproduction

Alexander Zakharov$^{(\boxtimes)}$ ⓘ, Vasily Kozlov$^{(\boxtimes)}$, Anatoly Shamin$^{(\boxtimes)}$ ⓘ,
and Yulia Sysoeva$^{(\boxtimes)}$ ⓘ

Nizhny Novgorod Region, Nizhny Novgorod State Engineering and Economics
University, 606340 Knyaginino, Russia
79081692376@yandex.ru, Kozlov.kovado@yandex.ru,
ngiei-126@mail.ru, uushka@mail.ru

Abstract. The process of reproduction of labor resources is paid attention in all economically developed countries of the world. The article presents a new approach to improving the efficiency of motivation in the modern economy.

In the digital economy, new demands are being placed on labor resources. At the same time, the problem of replacing a person with digital means appears. Based on a study of the demographic situation in rural areas and the work of scientists dealing with labor resources, the need to overcome «the digital gap» and improve the labor resources quality has been identified.

The authors noted the need to consider motivation as the cumulative effect of the government bodies of different levels efforts and human behavior in the current economic conditions. The accumulative or cumulative effect of motivation is substantiated.

The country leaders have planned a breakthrough in the agriculture efficiency associated with the digitization of the economy and the use of the latest achievements of genetics, robotics, and innovations development. And it ultimately depends on the competencies and motivation of people. The authors concluded on the basis of the proposed approach to the consideration of motivation.

Keywords: Reproduction · Agriculture · Forecast · Labor resources quality · Competencies · Cumulative effect

1   Introduction

Despite the current level of technology development that could be used in agriculture, labor resources can be a deterrent to the agriculture development and its digitalization. Therefore, solving the problem of the effectiveness of their intensive reproduction can be the key to improving agriculture efficiency as a whole. Even with increasing attention from the government, increasing investment, provided that agriculture is equipped with all the latest achievements of science, technology, it is directly dependent on workforce, which should apply and use them. Labor resources are the weakest link. All the achievements of science and technology man controls. Therefore, the most important point is to improve labor resources quality (their level of education, skills, competencies).

© Springer Nature Switzerland AG 2020
T. Antipova (Ed.): ICIS 2019, LNNS 78, pp. 3–10, 2020.
https://doi.org/10.1007/978-3-030-22493-6_1

At the same time, the term «labor resources» is an abstract concept applied in economics in relation to an industry, a region, or a country. At the organization level, we can talk, first of all, about people, man. So «workforce» ultimately is human resources. The main thing is the person, with his knowledge, competencies, experience, physical and moral condition, having the ability to accumulate and develop his labor and cultural potential, etc.

2 Methodology

In the digital economy conditions development, the problem of rising unemployment is brewing due to release of a large amount of labor, which is noted by a number of Russian and foreign authors [1, p. 106; 2, p. 22; 3, p. 36; 4, p. 319; 18; 19, pp. 2–3]. The transition to digital technologies and high-tech means of production is gradually changing the structure of staffing needs. New professions are emerging with a new set of competencies, production processes are becoming «uncrowded», some professions are becoming disappearing. In agriculture, as in other areas of the economy, digitalization process is not static. This can lead to the decrease in the agricultural workers number.

Some studies show impossibility to unequivocally conclude that technology development will lead to a reduction in jobs number [18, p. 9]. According to the forecasts of socio-economic development, a gradual increase is expected not only in labor force, but also in jobs, and unemployment rate will decline by 2024 [5, 6]. But still, there are concerns. It depends on negative indicators of labor force growth, according to their forecast for 2018–2020 (Table 1). In addition, in the long term after 2024, these fears seem even more real.

Table 1. Forecast of labor resources balance for 2018–2020, thousand people.

Indicator	2017	Forecast			Rate of increase, %
		2018	2019	2020	
The number of labor resources in the Russian Federation	90372,8	89694,7	89151,7	88674,2	98,1
Including agriculture, hunting and forestry	5360,0	5340,0	5300,0	5300,0	98,9
The number of labor resources in the Nizhny Novgorod Region	1946600	1930000	1921500	1914500	98,4
Including agriculture, hunting and forestry	69100	69050	69000	68900	99,7

Human resources are a national priority according to the May Presidential Decree. Therefore, that is why by improving reproduction efficiency, it is possible to solve assigned economic problems, including implementation of the digital economy program [9].

The revision of the attitude towards human resources leads to the conversion of this concept into human capital. Therefore, it is necessary to reproduce not human resources, but human capital.

A very important condition, while still maintaining the urgency of the problem of rural personnel shortage, is to support human resources. In region of the Russian Federation, such measures are being taken to attract specialists to villages and improve their living conditions [10–12].

However, at the current economic level of development and in digitalization conditions, the extensive path of reproduction process does not make sense. Therefore, the problem of human resources reproduction should be solved not through a simple increase in their number, but by improving quality. It is the improvement of human resources quality that will ensure their intensive reproduction, which meets the digital economy requirements. The need to transform the workforce quality, its dependence on the individual's competence and information orientation, the need to acquire new skills in order to meet new requirements, the need to bridge the digital gap, are confirmed by a number of domestic and foreign authors [13, pp. 166–167; 14, p. 108; 15, p. 110; 16; 17, pp. 382–383; 19, p. 3; 20, p. 2].

Under these conditions, the attitude towards the motivation concept should also change. Because people are also motivated, not resources. Under the motivation should not be understood the impulse to simply increase physical labor intensity of a person. It should be viewed not as a motivation to work, but as a motivation to change it, a motivation to change the work process due to new technologies, management methods, etc. If you motivate a person to more productive work, his productivity will increase when he connects their efforts with more advanced technologies and more sophisticated tools. If you correctly combine work skills with more advanced technology, then it is then that increases activity efficiency. Currently, in agriculture, there is a gap in work skills with modern technical capabilities. Bridging this gap is the main task of motivating intensive labor resources reproduction in agriculture.

In addition, the motivation concept should not be considered in a narrow sense as a set of material and moral incentives and motives and should be applied only at the organization level. It is necessary to motivate the system itself, which strives for efficiency, so that it «from the top to the bottom» solves the problem: the agrarian policy at the country level should encourage technologies renewal at the same time as they improve workers skills and competence. When these two factors develop simultaneously and synchronously, then efficiency is achieved. You can motivate by creating credit conditions, subsidizing, purchasing modern equipment, training personnel in necessary skills for working with new equipment, etc. An important condition for this is simultaneity of these actions.

In the agriculture of regions of the Russian Federation at the moment, the following problems are observed related to labor resources reproduction: a decrease in population, a decline in interest in agricultural work, a decline in the level of training and a low level of return of professional personnel to villages.

As a way out of this situation, it is necessary to create a motivation to preserve the labor potential and aimed at reproducing a positive attitude towards agricultural work. For this, each district should have special attention to social infrastructure, health care, culture and education.

The motivation concept should not be perceived in a narrow sense. The motivation system should cover not just the totality of all techniques and methods, incentives used in individual economic entities. The motivation system has a broader meaning, covering different levels of management – from the governmental to the particular enterprise level.

3   Results

The motivation system is not just remuneration, it is a broader concept. Motivation, even when applied at the individual organization level, can be: moral and material, individual and team, «encouraging» and «punishable», expected and unexpected?

But motivation cannot be considered only at the organizational level. All levels begin to act in concert. This creates a general policy of «attention» to the agricultural sector. This gives a response from the population, changes its behavior and attitude to agricultural work. The cessation of attention to the agriculture development, motivation of organizations and the population on the part of the state leads to demotivation of rural residents working in this area.

Motivation should be considered as the cumulative effect of the authorities efforts of different levels. Labor resources should be considered systematically at different organizational and economic levels: at the level of industry, region, municipality and individual organizations (see Fig. 1).

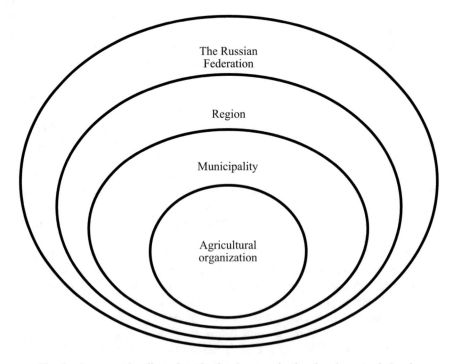

Fig. 1. A concentric effect of motivation by organizational and economic levels.

The first level is the state motivation at the level of livestock, crop, fish farming and forestry sectors. It can be represented by various forms of state support for agriculture, for example, in the form of government subsidies, and various targeted programs at the federal and regional levels, allocation of priority areas of activity in agriculture, priority sectors or agricultural products. Motivation is also possible through the education system.

The second level is the scale of a region (a separate region, or regions). Here, the motivation system should be aimed at developing rural areas sustainability, supporting young personnel, implementing targeted programs, etc. impact and control, taking into account each individual region specifics. This level should continue the general policy of the government.

The third level is a micro level or a separate organization level. It is here that various forms and systems of remuneration, various incentives and motives can be applied in all their diversity.

The described «three scale» idea re-opens the motivation system understanding for the labor resources reproduction. At different levels, it must be solved by different motivational mechanisms. The cumulative effect has a concentric effect on individual agricultural organizations and operates through multiplying the effect of each level (enterprise – municipality – region – the Russian Federation).

The cumulative effect has a concentric effect on individual agricultural organizations and operates through multiplying the effect of each level (see Fig. 2).

It follows from the proposed model that the accumulation of certain incentives and influence occurs «from the top to the bottom» and it is enough to stop supporting this accumulation in any of the links, as their number tends to zero, and as a result the accumulation effect is lost.

The study revealed stable recurrent set of relations between efforts to maintain agricultural efficiency at a country, region, municipal district and organization levels, which can be expressed in quantitative form as a model that reflects this stable relationship:

$$X = \sum\nolimits_{i=1}^{n} x_i \rightarrow X = x_1 \times x_2 \times x_3 \qquad (1)$$

where x is the accumulated (cumulative) effect of motivation;

x1, x2, x3 - the results of the motivation actions, emerging from individual activities at different organizational and economic levels (country, region, organization).

It follows from the proposed model that the accumulation of certain incentives and influence occurs «from the top to the bottom» and it is enough to stop supporting this accumulation in any of the links, as their number tends to zero, and as a result the accumulation effect is lost. The cumulative effect in relation to the motivation for labor resources reproduction is achieved precisely as concerted actions of all levels of government and budgets. From this point of view, motivation is relatively synchronous, purposeful, accumulating effect, should be made at all levels of government.

Thus, the main condition for the cumulative effect of motivation is the consistency, complexity and synchrony of various actions at all levels. They will lead to the achievement of the desired result.

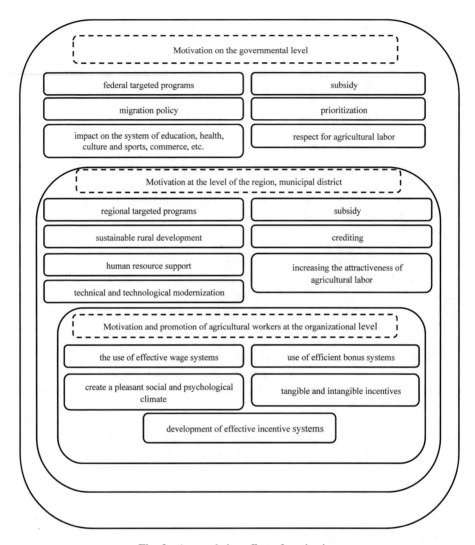

Fig. 2. A cumulative effect of motivation.

4   Conclusion

Each organizational and economic level uses its own terminology. At the national economy level as a whole, it is said about «labor resources» reproduction. This refers to their demographic, educational, gender, and other properties. At the level of a region or municipal district, it is more correct to use the terms «labor potential», «human resources» and «human capital».

Thus, to understand motivation in the narrow sense is wrong. So the idea of «three scales» was proposed. It is proposed to use the cumulative effect of motivation to improve the efficiency of labor resources reproduction of villages. The proposed idea will allow to understand the problem of the reproduction of agricultural labor resources more correctly and outline ways to solve it at the modern level.

The breakthrough in efficiency of agriculture outlined by the country's leadership, connected with economy digitization and the use of the latest achievements of genetics, robotics, development of innovations, ultimately depends on competences and motivation of people. In general, the proposed ways to improve the motivation system will improve the financial situation of agriculture as a result of achieving the overall economic effect.

References

1. Nesterenko, E.S.: The employment dilemma of labor resources in the conditions of digitalization of the economy. Bull. North Caucasus Fed. Univ. **6**(63), 104–109 (2017)
2. Odegov, Yu.G.: Labor transformation: 6th technological mode, digital economy and employment trends. Living Level Popul. Russ. Reg. **4**(206), 19–25 (2017)
3. Vasilenko, N.V.: The impact of the economy digitalization on employment and the labor market. In: Babkin, A.V. (ed.) Scientific-Practical Conference with International Participation. Industrial Policy in the Digital Economy: Problems and Prospects, pp. 34–37. Publishing House of Polytechnic University, Saint-Petersburg (2017)
4. Golovina, T.A.: Personnel aspects of the development of the digital economy in Russia. In: International Scientific Conference Dedicated to the 90th Anniversary of S.P. Kapitsa, Human Capital in the Format of the Digital Economy, pp. 316–323. Editorial and Publishing House RosNOU, Moscow (2018)
5. Forecast of the socio-economic development of the Russian Federation for the period up to 2024. http://economy.gov.ru/minec/activity/sections/macro/201801101. Accessed 15 Dec 2018
6. Decree: The forecast of the socio-economic development of Nizhny Novgorod Region in the medium term (for 2019 and for the planning period of 2020 and 2021), dated 24.10.18, No. 703. https://government-nnov.ru/?id=222376. Accessed 15 Dec 2018
7. The Ministry of Labor and Social Protection of the Russian Federation. https://rosmintrud.ru. Accessed 15 Dec 2018
8. The Ministry of Social Policy of Nizhny Novgorod Region. http://www.minsocium.ru. Accessed 15 Dec 2018
9. Presidential Decree of May 7, 2018 No. 204: The national goals and strategic objectives of the development of the Russian Federation for the period up to 2024. https://www.garant.ru/products/ipo/prime/doc/71837200. Accessed 15 Dec 2018
10. Law of Nizhny Novgorod Region of November 1, 2008 No. 149-Z: Measures of State Support for the Personnel Potential of the Agro-Industrial Complex of the Nizhny Novgorod Region. https://government-nnov.ru/doc. Accessed 15 Dec 2018
11. Resolution of the Government of the Russian Federation of July 14, 12, No. 717: The State Program for the Development of Agriculture and Regulation of Agricultural Products, Raw Materials and Food Markets for 2013–2020 (as amended up to March 1, 2018). https://ohotnadzor.government-nnov.ru/?id=111421. Accessed 15 Dec 2018

12. Decree of the Government of Nizhny Novgorod Region of 08.07.15, No. 430: The organization of work to implement on the territory of the Nizhny Novgorod Region measures to improve the living conditions of citizens living in rural areas, including young families and young professionals, using social benefits under the direction (subprogrammes) 'Sustainable Development of Rural Territories' of the State Program for the Development of Agriculture and Regulation of Agricultural Products, Raw Materials and Food Markets for 2013–2020 years (as amended on April 28, 2018). https://government-nnov.ru/?id=170502. Accessed 15 Dec 2018

13. Yasinsky, D.Yu.: Identification of the workforce quality in the information economy. Basic Res. **10**, 161–167 (2017)

14. Lovchikova, E.I., Solodovnik, A.I.: Digital economy and personnel potential of the agroindustrial complex: strategic interconnection and prospects. Bul. Agrarian Sci. **5**(68), 107–112 (2017)

15. Lyaskovskaya, E.A., Kozlov, V.V.: Personnel management in the digital economy. SUSU J. **12**(3), 108–116 (2018). Series «Economics and Management (comment)»

16. Shirnova, S.A.: New qualities of digital capital in the conditions of formation of the digital labor market. In: Kuznetsov, S.V., Ivanov, S.A. (eds.) All-Russian Scientific-Practical Conference 2018, Multifactor Challenges and Risks in the Context of the Implementation of the Strategy of Scientific-Technological and Economic Development of the Macro-Region North-West, pp. 458–463. Saint-Petersburg State University of Aerospace Instrumentation, Saint Petersburg (2018)

17. Sizova, I.L., Khusyainov, T.: Labor and employment in the digital economy: problems of the Russian labor market. SPSU Bul. Sociol. **10**(4), 376–396 (2017)

18. Nofal, M.B., Coremberg, A., Sartorio, L.: Data, measurement and initiatives for inclusive digitalization and future of work. Economics Discussion Papers, No. 2018-71, Kiel Institute for the World Economy (2018). http://www.economics-ejournal.org/economics/discussionp-apers/2018-71

19. Annunziata, M., Bourgeois, H.: The future of work: how G20 countries can leverage digital-industrial innovations into stronger high-quality jobs growth. Econ. Open-Access Open-Assessment E-J. **12**(2018-42), 1–23 (2018). http://dx.doi.org/10.5018/economics-ejournal.ja.2018-42

20. Chetty, K., Qigui, L., Gcora, N., Josie, J., Wenwei, L., Fang, C.: Bridging the digital divide: measuring digital literacy. Econ. Open-Access Open-Assessment E-J. **12**(2018-23), 1–20 (2018). https://doi.org/10.5018/economics-ejournal.ja.2018-23

Artificial Intelligence Research

Whether Be New "Winter" of Artificial Intelligence?

Leonid N. Yasnitsky[1,2]([⊠])

[1] Perm State University, 15, Bukirev Street, 614600 Perm, Russia
yasn@psu.ru
[2] Higher School of Economics, National Research University,
38, Studencheskaya Street, 614070 Perm, Russia

Abstract. The article analyzes the formation and development of artificial intelligence as scientific industry, identifies cycles of leaps and drops of its popularity. It's concluded that the decline in the popularity of artificial intelligence in the near future is inevitable.

Keywords: AI winter · Future · Artificial intelligence · Crisis ·
Decline in popularity · History · Development cycles

To look into the future, you need to explore the past.

There is an opinion [1] that the history of artificial intelligence began with the invention in the 13th century by Raymond Llully of a mechanical expert system capable of making horoscopes, making medical diagnoses, making crop forecasts, and providing legal advice. The intellectual system of R. Llully was very popular. To make a look on the miracle of technology and get useful tips and answers gathered people from afar.

However, no signification events in the history of the development of Artificial Intelligence (AI) were observed over the next seven centuries. This period in some articles has name "The artificial intelligence winter" (see Fig. 1).

The next surge in the popularity of artificial intelligence came in the middle of the XX century. It began with the invention of the Mathematical neuron by W. McCullock and W. Pitts and the creation by F. Rosenblatt of a neural network capable of recognizing Latin letters. Journalists and writers publicized this success. The USA government allocated large subsidies for the development of a new scientific direction. Special hopes were placed on the creation of a neural network recognition system "Friend-or-Foe", which is of strategic importance in connection with the approaching Caribbean crisis.

In addition to congressional representatives, businessmen and doctors were interested in the possibilities of neural networks too. The first were interested in the possibility of predicting stock prices and exchange rates, the second – the automatic interpretation of electrocardiogram data. Young scientists got down to business.

However, despite solid financial injections, the promises of young scientists were not to come true. They could not overcome the "XOR problem". The learning processes of neural networks did not converge.

© Springer Nature Switzerland AG 2020
T. Antipova (Ed.): ICIS 2019, LNNS 78, pp. 13–17, 2020.
https://doi.org/10.1007/978-3-030-22493-6_2

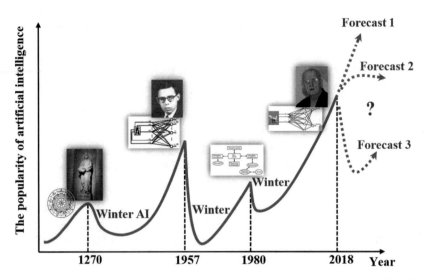

Fig. 1. Periods of surges and falls artificial intelligence. The variants of predictions of the future.

When it became clear that ambitious projects were stalled and taxpayers 'and businessmen' money was wasted, the public declared neural networks a "dead-end scientific direction" [2]. The popularity of artificial intelligence has plummeted.

This was the second "artificial intelligence winter", which this time did not last long. By the end of the 1970s, projects related to the creation of expert systems began to gain momentum. Again, "startups" could not resist. Again, young scientists have started to make promises to all. The third wave of enthusiasm for artificial intelligence ended in the early nineties, when many companies failed to meet high expectations and "burst" [3]. The third "Winter of artificial intelligence" has come.

We are witnessing another surge in the popularity of artificial intelligence today. Its prerequisites were the works of Soviet (A.I. Galushkin, A.S. Zak, B.V. Tyukhov, V.A. Vanyushin, etc.) and American (P. Verbos, D.E. Rummelhardt, etc.) scientists who almost simultaneously and independently invented algorithms for training multilayer neural networks. Due to this, "The XOR problem was solved [4]. It is thanks to these discoveries over the past 15–20 years, one after another, there are reports of the successful application of neural network technologies in industry, Economics, medicine, political science, sociology, criminology, psychology, pedagogy, etc.

Artificial intelligence has become popular again. As in past centuries are made, breathtaking predictions. On AI pay attention to government officials. "The one who will become a leader in the field of artificial intelligence will be the ruler of the world" – this statement by Russian President Vladimir Putin, made by him on September 1, 2017 (https://www.ntv.ru/novosti/1922597/), literally shook the whole world.

And now, as last time, large subsidies are allocated for projects to create intelligent systems. Many Russian foundations, such as RFBR, RGNF, NTI, etc. announce competitions of grants focused on the creation of intelligent systems. The largest Russian companies, Sberbank, 1C, etc., have already created AI laboratories. About the "success" of young scientists can already be found on the Internet. These are neural networks designed to detect terrorists from a person's photo or crooks trying to get a Bank loan. From a huge number of training examples neural networks have learned "brilliant knowledge", such as: "if a person in the photo has a beard, it means that he is a terrorist, "or" if a person smiles, then he is a crook."

Such mistake knowledge in the theory of mathematical statistics is usually called "false correlation dependencies" and take special measures to eliminate them, but ambitious young scientists, apparently, do not know about it.

To find out what will end today's surge in popularity of artificial intelligence, just look at Fig. 1 and remember the events of past centuries. It seems, as in previous times, we are waiting for disappointment and another "The artificial intelligence winter".

Today, along with great financing large-scale projects it is mass-produced scientific literature with a strong PR focus. Thus, in early 2018, the publishing house "Peeter" was released the book "Deep learning" [3]. It is necessary to pay tribute, the book is written by fine easy language. It is felt that its author is a competent specialist, but surprising fact – on 420 pages, the word "revolution" has been used more than 42 times in this book.

Let's try to understand what is the modern revolution of artificial intelligence:

1. Is it using of neurons with sigmoid activation functions?
   No, it is not.
   Widrow and Hoff first used sigmoid activation functions in 1960 [5]
2. Is it in using of large number of hidden layers (deep neural networks)?
   Nope.
   The training algorithms of multilayered neural networks was opened by Galushkin, Verbos, Rummelhardt etc. in the 70-ies of XX century [6–8].
3. Maybe, revolution is the invention of the "convolution", which allowing to take into account the topology of the images?
   Facing the wrong way again.
   Cognitron and neocognitron were invented by Fukushima [9] in the late 1980s.
4. Is it in the application of self-learning input layers?
   No, it is not.
   Hybrid neural networks with self-learning input layers, and the output – perceptron, described, for example, in the book Osovsky [10].
5. Is it in the application of regularization techniques?
   No, it is not.
   The use of weighting factors in objective functions in solving optimization problems can be found, for example, in the book [11].
6. Revolution is the fact that currently accumulated large amounts of information suitable for neural network processing, is it?
   But, it is not a revolution. It is evolution.
7. In an attempt to turn Neuroinformatics from science into a craft.
   Yes. Such attempts take place.
   A lot of libraries and tools have appeared on the Internet, allowing to create and apply various neural network architectures for solving many practical problems by "pressing buttons".

Let me say! What is the transformation of Neuroinformatics from science to craft, if as a science it has not yet formed? Where are the theorems and methods based on them, with the help of which for each specific problem it is possible to form an optimal neural network that provides a solution to the problem with a given accuracy?

To date, there are no such theorems. It is on this occasion that experts joke, calling Neuroinformatics art rather than science, and even, to some extent – religion, because it is never possible to predict in advance whether the project will be successful or not lucky. The result depends not only on the knowledge of the theoretical foundations, but also on the experience (including negative) and on the intuition of the researcher. Young scientists should not forget that neural networks are created "in the image and likeness" of the brain and inherited from its prototype many properties, in particular – the ability to deceive their creators. For example, they often reveal false correlations, as happened with the above-mentioned projects for the detection of swindlers and terrorists.

Conclusion

We have shown that AI in its development is subject to the law of cyclicity, as well as any natural phenomenon. Winter gives way to spring and summer. Summer gives way to autumn and winter.

The reason for the "spring" of AI are new ideas of brilliant scientists and a long hard work on their implementation. The reason for the onset of "autumn" and "winter" of AI is the failures of young inexperienced people who took up the implementation of ambitious AI projects. Excessive PR and ADV of AI, and reckless funding AI projects from business and States contribute to the onset of AI "winter".

Currently, AI is at the peak of its popularity. But signs of "autumn" already are observed.

Will there be a new artificial intelligence "winter"?

Gratitudes

The article was prepared with the support of the grant of the Russian Foundation for basic research 19-010-00307.

References

1. Yasnitsky, L.N.: Introduction to Artificial Intelligence. Academy, Moscow (2005)
2. Andrew, A.M.: Artificial Intelligence. Abacus Press, Turnbull Wells (1983)
3. Nikolenko, S., Kadurin, A., Arkhangelsk, E.: Deep Learning. Peter, St. Petersburg (2018)
4. Yasnitsky, L.N.: On the priority of Soviet science in the field of Neuroinformatics. Neurocomput. Dev. Appl. **21**(1), 6–8 (2019)
5. Widrow, B., Hoff, M.E.: Adaptive switching circuits. 1960 IRE WESTCON Conferenction Record, New York (1960)
6. Galushkin, A.I., Zak, A.S., Tyukhov, B.P.: Comparison of criteria for optimization of adaptive pattern recognition systems. Cybernetics **5**, 122–130 (1970)
7. Werbos, P.: Beyond regression: new tools for prediction and analysis in the behavioral sciences. Ph.D. thesis. Department of Applied Mathematics, Harvard University, Cambridge, Mass (1974)

8. Rummelhart, D.E., Hilton, G.E., Williams, R.J.: Learning internal representations by error propagation (1986). McClelland et al.
9. Fukushima, K.: Neocognitron: a self-organizing neural network for a mechanism of pattern recognition unaffected by shift in position. Biol. Cybern. **36**(4), 193–202 (1980)
10. Osovsky, S.: Neural Networks for Information Processing. Finance and Statistics, Moscow (2002)
11. Yasnitskii, L.N.: Superposition of base solutions in methods of Trefftz type. Mech. Solids **24**(2), 90–96 (1989)

Intelligent System for Prediction Box Office of the Film

Leonid N. Yasnitsky[1,2(✉)], Igor A. Mitrofanov[1],
and Maksim V. Immis[1]

[1] Perm State University, 15, Bukirev Street, 614600 Perm, Russia
{Leonid.Yasnitskyyasn, Igor.Mitrofanovyasn,
Maksim.Immisyasn}@psu.ru
[2] Higher School of Economics, National Research University,
38, Studencheskaya Street, 614070 Perm, Russia

Abstract. This article describes a mathematical model designed to predict the box office of movies. The model is based on a neural network trained on data about films obtained from open sources. Computer experiments were performed by the method of "freezing": with the help of a neural network calculations were performed with a gradual change in the value of one of the input parameters of the model, while the remaining input parameters were not changed. It is established that the size of the film budget has the greatest impact on the amount of box office among all other input parameters. However, its impact is not always positive. It is established that the United States, as a country that takes part in the production of the film, is able to have the greatest impact on box office compared to other countries. According to the research the duration of the film to varying degrees can affect the amount of box office. Moreover, the power of influence depends on such a factor as the genre of the film. As a rule, the power of this influence increases with the increasing semantic load of the film. The practical value of the study is that the created mathematical model can be used to optimize costs when planning the production of new films.

Keywords: Film business · Box office · Profitability · Genre ·
Neural networks · Scenario forecasting

1 Introduction

The film industry is a big part of modern culture. Many companies seek to capitalize on the success of the film. One of the reasons why forecasting box office receipts is an important part of film production is the need for high cash costs. Therefore, it is important to know whether the project will be profitable. Forecasting the box office of movies allows you to find out what factors affect the amount of profit, and to establish how much this influence. Thus, it is possible to optimize the budget of the film in order to increase potential revenues.

The creation of an intelligent system of forecasting the box office of cinema is currently relevant, as the technology is constantly improving, therefore, changing the approach to shooting movies, which affects the final product. Also, the movie is quite a

T. Antipova (Ed.): ICIS 2019, LNNS 78, pp. 18–25, 2020.
https://doi.org/10.1007/978-3-030-22493-6_3

risky investment project, because there are a large number of factors that affect the payback of the project. In addition, the film business is a difficult formalized area of human activity, so the traditional methods of mathematical modeling used to solve the problem of this problem are not so effective.

To solve this problem, we decided to use neural networks. Currently, neural networks have found application in many spheres of human activity. Methods of artificial intelligence are effectively used in such difficult formalized areas as medicine, biology, zoology, sociology, cultural studies, political science, business, medicine, criminology, etc. A special place is occupied by the problem of scenario forecasting [1–8], i.e. the study of the influence of the input parameters of the model on its output parameters. Scenario forecasting allows you to find out what will be the performance of the simulated object in the future, and how they can be affected to get closer to the desired result.

This article describes the continuation of early research in the field of neural network modeling in the film business [9–10].

2   Problem Statement and Formation of Training Data Set

To build a neural network, there are many input parameters that affect the amount of box office: the year of release of the film; the Country (USA, UK, France, Germany, Russia, Canada, Italy, Australia, Spain, Japan, China, India, Belgium, Hong Kong, Ireland, Denmark, Sweden, the Netherlands); the number of Directors, screenwriters, producers, operators, composers, artists, editors, actors with "Oscar" who participated in the production of the film; the film belongs to the genre (fiction, action, Thriller, adventure, melodrama, Comedy, family, music, drama, crime, fantasy, sports, detective, biography, history, military, horror, musical, Western, cartoon, film Noir, documentary, anime, short film, concert, children's, news); the film budget, expressed in us dollars; the duration of the film, expressed in minutes.

The output parameter of the model was the value of the box office of a particular movie, expressed in dollars.

We have compiled a training sample based on real data from 1915 to 2017 and collected data from various Internet sources that store information about movies. The training set contained 4329 examples. The testing set consisted of 764 examples - 15% of the total number of examples. The set included data on films of different years, countries, genres with different budget size and the size of box office.

3   Formation of a Set of Input Parameters

After determining the set of input parameters, we have determined which of them significantly affect the output parameter. To do this, for each of the input parameters we calculated the Pearson correlation coefficient. The results are shown in Fig. 1.

As can be seen from the graph, the greatest impact on the amount of box office has the size of the budget, because the movie is an investment project that requires investment. Further, it is clear that movies released in the United States, relatively

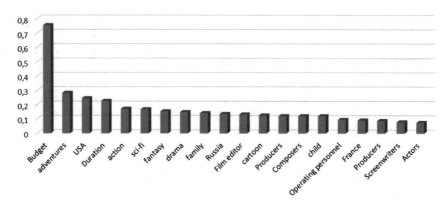

Fig. 1. Dependence of the output parameter on the input parameters

strongly affect the amount of box office. American cinema is a huge film industry, which includes many well-known reputable film companies such as 20th Century Fox, Paramount Pictures, Warner Bros. As a Result, the demand for films produced in the United States is high, which contributes to an increase in profits. Russian-made films have less influence on the size of box office, as the share of Russian films in the box office is relatively small, and also because some people are skeptical about the Russian cinema due to the smaller diversity of the cast and the lower level of development of video processing technologies compared, for example, with the United States.

More influential at the box office are genres such as adventure, action and fantasy because the films of this genre cover the widest audience. Demand for them is higher than for films of other genres. In addition, films of these genres occupy a relatively large share of film distribution.

A significant impact on the amount of box office has the duration of the film, because, for example, in a film with a longer duration can fit a more developed story, which will affect the film as a whole, and, consequently, the perception of the viewer.

Parameters associated with the presence of "Oscar" in those or other members of the crew, affect the amount of box office almost equally. This influence is quite natural, because the winners of this award attract the attention of the press, so that more people know about them. In this case, the viewer is more likely to choose a particular film from those offered by the film rental at the moment, if Oscar laureates take part in its production.

However, countries that were not mentioned above cannot be excluded from the model, as each country has its own mentality and history, which is reflected, among other things, in the production of films. For example, the Italian cinema as a rule, does not differ in brightness of a plot, but looks quite easily. The Russian cinema demonstrates patriotism, pride in their homeland, the unity of the nation, the ease and immediacy of communication between people with each other.

It should be noted that, for example, the presence of such a genre as melodrama can significantly affect the course of the plot, and, consequently, the whole picture as a whole. Similarly, the absence of any of the genres can affect.

Thus, we used all these input parameters for further neural network design.

4 Training and Research of Neural Network Model

We have designed a neural network of perceptron type with the following characteristics:

- 56 neurons on the input layer. The activation function is a hyperbolic tangent;
- 13 neurons on a hidden layer. The activation function is a hyperbolic tangent;
- 1 neuron on the output layer. The activation function is a hyperbolic tangent.

Then we trained neural network by the method of resilient propagation.

After training the neural network, we tested it for adequacy. To do this, we used a testing set consisting of 764 elements that are not included in the training set. For Fig. 2 to illustrate, the difference between the actual and projected cash collection values in a small portion of the test sample is presented.

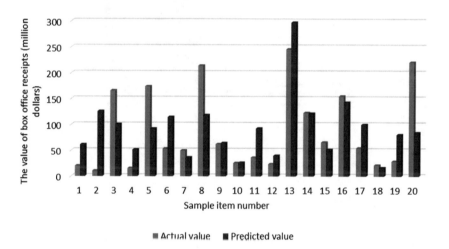

Fig. 2. Difference between actual and predicted values

After testing for adequacy, we found that the created neural network predicting box office fees of movies has training and testing errors of 3.46% and 9.17%, respectively. To calculate the errors, we used the following formula:

$$E = \frac{\sqrt{\frac{\sum_{n=1}^{N}(d_n - y_n)^2}{N}}}{\max(d_n) - \min(d_n)},$$

where N is the number of sample items, d_n is the actual cash collection of the n-th movie, y_n is the projected cash collection of the n-th movie.

After we checked the adequacy of the neural network model, we used it to predict the box office of movies and study the impact of various factors on the final result.

Five films were selected for forecasting:

1. "X-Men"
2. "Zootopia"
3. "Swiss Army Man"
4. "Titanic"
5. "The Revenant"

These films were not included in the training and testing set.

As seen in Fig. 3, the amount of cash fees does not depend much on the year of production, which is quite logical and confirms the results of the correlation.

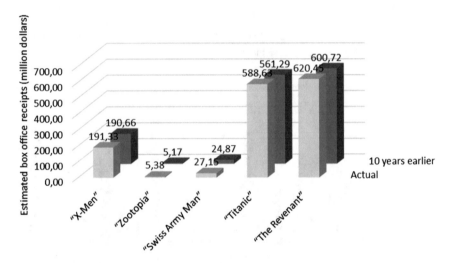

Fig. 3. The dependence of box office on the year of release of the film

We also studied the impact of budget changes. For comparison, the actual budget, the budget increased by 10% and the budget reduced by 10% were taken. The results are shown in Fig. 4.

As can be seen from the graph, the increase in the budget has a positive impact on the amount of cash fees. Budget reduction, on the contrary, contributes to the reduction of cash fees. Also from the figure it is noticeable that not all of the movies the budget changes affected equally. Presumably, this is due to the fact that the films "Titanic" and "The Revenant", on which the change in the budget affects more significantly, carry a stronger semantic load than the others. Consequently, the quality of the film changes significantly with the change of the budget.

Then we studied the dependence of the box office on the duration of the film. The results are shown in Fig. 5.

As can be seen from the graph, the increase in the duration of the film contributes to an increase in the size of the box office. The strongest increase in absolute value is observed in the first, third and fifth films. Apparently, this is due to the fact that the second and third films have a comedic character. As a rule, comedies have a relatively short duration (1.5–2 h).

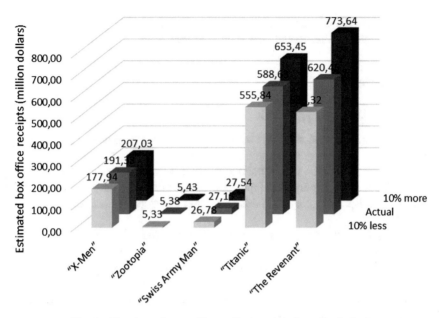

Fig. 4. The dependence of box office receipts from the budget

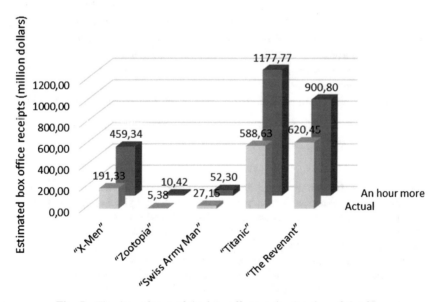

Fig. 5. The dependence of the box office on the duration of the film

We made the assumption that in the case of the movies "Titanic" and "The Revenant" the increase in duration would allow to reveal the inner world of the characters more deeply, which in turn would help the viewer to feel more strongly what is happening on the screen. Similarly, increasing the length of the movie "X-Men" could make this movie more exciting, for example, by detailing some elements of the script and showing scenes containing special effects.

5   Conclusion

As the results of the study, we have done the following:

1. Created neural network the mathematical model of the film business that allows you to fulfill the patterns of the simulated domain.
2. We have found that the size of the budget has the greatest impact on the amount of box office fees among all other input parameters. However, its impact is not always positive, so it is necessary to choose a budget that does not lead to a reduction in cash fees.
3. The United States, as the country involved in the production of the film, is able to have the greatest impact on box office compared to other countries.
4. We have found that the length of the film to varying degrees can affect the increase or decrease in box office. Moreover, the power of influence depends on such a factor as the genre of the film. As a rule, the power of this influence increases with the increasing semantic load of the film.
5. Developed neural network mathematical model of the movie business has pretty low accuracy of the forecast value of box office receipts of the film, so it cannot be used directly on the quantitative level, i.e. to calculate the magnitude of the profits from the production and distribution of specific films. However, as shown above, the model has learned the patterns of the simulated domain. Therefore, it can be used at a qualitative level, i.e. at a qualitative level. with the help of our model, you can identify General trends, for example, to find out to increase or decrease the profitability of the business will change one or another parameter of the film.

References

1. Yasnitsky, L.N., Dumler, A.A., Cherepanov, F.M.: Dynamic artificial neural networks as basis for medicine revolution. In: Antipova, T., Rocha, A. (eds.) Digital Science, DSIC 2018. Advances in Intelligent Systems and Computing, vol. 850, pp. 351–358. Springer, Cham (2019)
2. Alexeev, A.O., Alexeeva, I.E., Yasnitsky, L.N., Yasnitsky, V.L.: Self-adaptive intelligent system for mass evaluation of real estate market in cities. In: Antipova, T., Rocha, A. (eds.) Digital Science, DSIC 2018. Advances in Intelligent Systems and Computing, vol. 850, pp. 81–87. Springer, Cham (2019). https://doi.org/10.1007/978-3-030-02351-5_11
3. Yasnitsky, L.N., Dumler, A.A., Cherepanov, F.M.: The capabilities of artificial intelligence to simulate the emergence and development of diseases, optimize prevention and treatment thereof, and identify new medical knowledge. J. Pharm. Sci. Res. 10(9), 2192–2200 (2018)
4. Yasnitsky, L.N., Yasnitsky, V.L.: Technique of design of integrated economic and mathematical model of mass appraisal of real estate property by the example of Yekaterinburg housing market. J. Appl. Econ. Sci. XI(8(46)), 1519–1530 (2016)

5. Yasnitsky, L.N., Vauleva, S.V., Safonova, D.N., Cherepanov, F.M.: The use of artificial intelligence methods in the analysis of serial killers' personal characteristics. Criminol. J. Baikal Nat. Univ. Econ. Law **9**(3), 423–430 (2015)
6. Yasnitsky, L.N., Dumler, A.A., Poleshchuk, A.N., Bogdanov, C.V., Cherepanov, F.M.: Artificial neural networks for obtaining new medical knowledge: diagnostics and prediction of cardiovascular disease progression. Biol. Med. **7**(2), 8 p. (2015). BM-095-15
7. Yasnitsky, L.N., Dumler, A.A., Bogdanov, K.V., Poleschuk, A.N., Cherepanov, F.M., Makurina, T.V., Chugaynov, S.V.: Diagnosis and prognosis of cardiovascular diseases on the basis of neural networks. Biomed. Eng. **47**(3), 160–163 (2013). https://doi.org/10.1007/s10527-013-9359-0
8. Yasnitsky, L.N., Plotnikov, D.I.: Economic and mathematical neural network model for optimization of financial expenses in film business. Fundam. Study (11, part 2), 339–342 (2016). https://www.fundamental-research.ru/ru/article/view?id=40977
9. Yasnitsky, L.N., Beloborodova, N.O., Medvedeva, E.Yu.: The method for forecasting box-office grosses of movies with neural network. Digest Finan. **22**(3), 298–309 (2017). https://www.fin-izdat.com/upload/iblock/a66/df0317-298.pdf

Business and Finance in Digital Age

Digitalising Tax System in Sri Lanka: Evidence from Inland Revenue Department

Dayananda Ambalangodage[1], Chamara Kuruppu[2(✉)],
and Konstantin Timoshenko[3]

[1] University of Sri Jayewardenepura, Nugegoda 10250, Sri Lanka
[2] School of Business, University of South-Eastern Norway,
3603 Kongsberg, Norway
cku@usn.no
[3] Nord University, 8049 Bodø, Norway

Abstract. The adoption of state-of-the-art information technology and digi-talised tools has gained prominence in recent years. This now applies equally to private sector and public sector institutions, including central governments. This paper explores the implementation of a novel digitalised tool, namely Revenue Administration Management Information System (RAMIS), in Sri Lanka. Rogers' diffusion of innovation theory serves as a sensitizing theoretical lens through which we approach this innovation tool. The empirical evidence is gathered through multiple sources, encapsulating semi-structured interviews and document search. The results of the study are twofold. On the one hand, the study shows that a great deal of interest has been expressed in the digitalised tax system at the very top. The Asian Development Bank as well as administrators having international experience have taken the decisive lead in promoting the RAMIS. On the other hand, the implementation of the RAMIS has not yet moved beyond the pilot project and achieved the desired ends. On the contrary, a myriad of implementation barriers and challenges have been unveiled, amongst which special mention should be made of poor infrastructure; resistance to change, especially amongst elderly employees; as well as the current legal framework that hinders the full-fledged functioning of the system.

Keywords: Revenue Administration Management Information System ·
Inland Revenue Department · Diffusion of innovation

1 Introduction

Public sector institutions, encompassing central governments are often blamed for having inefficient management and wasting public resources. Reform efforts under the banner of New Public Management (NPM) have convinced politicians and adminis-trators worldwide to streamline the process of managing the government sector. Hood [10] has pinpointed seven dimensions of NPM reforms, namely, the disintegration of public institutions as per production, a shift towards private sector management practices, more focus on discipline and parsimony in resource use, a shift towards hands-on management, specific and calculable standards of performance, and focus on output control. Such reform initiatives are expected to increase efficiency and

© Springer Nature Switzerland AG 2020
T. Antipova (Ed.): ICIS 2019, LNNS 78, pp. 29–38, 2020.
https://doi.org/10.1007/978-3-030-22493-6_4

effectiveness, as well as to heighten accountability, to list just a few. Therefore, the government sector institutions, including central governments tend to increasingly rely upon business-like practices [8, 13]. Amongst the most notable initiatives have been the introduction of business-like accrual accounting [8, 13] and balanced scorecard [3]. Likewise, the privatization of government-owned business entities has been another strategy to follow [21, 22].

Following the NPM trend in the developed countries, international financial institutions, as varied as the World Bank, the International Monetary Fund (IMF) and the Asian Development Bank (ADB), have convinced developing nations and emerging economies to adopt business-like practices under the rubric of structural adjustment programmes [1, 11]. In addition to the privatization of government corporations since the adoption of economic liberalisation policy in the late 1970s [21, 22], Sri Lanka has attempted to launch business-like accounting and budgeting practices [1]. In particular, the country has embarked on a path toward participatory budgeting to strengthen the grassroots democracy since the early 2000s [12]. Perhaps more importantly for the sake of this paper, there have been assiduous efforts to adopt information technology and digitalisation on the island. Drawing on Rogers' diffusion of innovation theory, the present study seeks to explore the process of digitalisation at Inland Revenue Department (IRD) in Sri Lanka. In doing so, this study seeks to answer the question of what lessons can be learnt from this on-going process.

The remainder of this manuscript is structured as follows. The next section describes a theoretical framework. The third section elaborates on data collection methods. Following the data collection method section, the findings are analysed. The final section offers some concluding remarks.

2   Theoretical Framework

The government sector worldwide has striven to become more efficient and effective by introducing various innovations. An innovation can be an idea, practice or an object that is acknowledged as something novel to an individual or institution attempting to adopt it [15]. That said, digitalisation of the tax system is considered an innovation. De Vries et al. [6] have pinpointed the importance of exploring the diffusion of public sector innovations. As such, the diffusion of innovations occurs through a specific process [2, 6, 7]. Rogers [15] has presented a five-stage model to explore the diffusion of innovations.

In the first stage, coined as 'knowledge', policy makers and other important stakeholders become aware of the existence of innovations. Prior work shows that politicians and top administrators are often influential in the introduction of public sector innovations [7, 20]. Similarly, their awareness of existing practices' weaknesses and a comprehension of a need for innovative practice are key characteristics of this stage [2, 4, 5, 7]. Hence, attention should be focused on what alternative exists, and how and why it works [15].

'Persuasion' is the second stage in which stakeholders tend to generate either positive or negative attitudes towards the innovation [7, 15]. Besides consultants [5, 6], administrative professionals can play a significant role at this stage, which, however, depends on their theoretical understanding of innovation and analytical skills to justify the importance of innovation [6, 9]. In this stage, proponents of innovations tend to criticize the existing practice, whilst underscoring the benefits of innovative practice [2, 7]. The third stage is referred to as 'the decision stage', which may range from straightforward acceptance of innovation to immediate rejection and from later adoption to continued rejection [15]. The penultimate (i.e. fourth) stage envisages the implementation of innovation, which can rather be complex and problematic [2, 7, 15]. Therefore, it is likely to experiment the innovation at a specific section or unit for a predetermined period in the outset [15]. The last stage of the diffusion trajectory is 'confirmation' [2, 6, 7, 15]. Having realised the benefits of innovation, the adopters confirm the continuous use of innovation throughout the whole organization. Nevertheless, the failure to garner the intended benefits may result in the rejection and abandonment of innovation [15].

3 Data Collection

The IRD is an entity responsible for collecting and administering taxes levied by the Government of Sri Lanka. The use of information technology by the department has been confined in scope as it has largely relied upon a manual tax administration system. This manual system has been identified as the major drawback in hindering the revenue collection potential of the department.

The present case study seeks to answer the question of what lessons can be learnt from the process of digitalising the IRD in Sri Lanka. As argued in the literature, scholars tend to focus less attention on public sector innovation and appeal for more studies to explore public sector innovation [6, 9]. This descriptive case study is a response to such an appeal. In the process of collecting data, documents were initially gathered which also proved to be valuable in identifying and selecting relevant key officials for further interviews. Amongst collected documents have been the IRD's performance report, the country's central bank report and reports of the ADB. Having reviewed the above documents, we then contacted senior administrators in the IRD to get their free consent. In the process of seeking permission to interview officials, we promised not to disclose interviewees' identity with a view to avoiding any negative consequence to them. Ten semi-structured interviews were then conducted with administrators at the IRD. Each interview lasted 45 min on average.

Discussions with officials were first and foremost concerned with tracing how the idea of digitalisation came to light on the island and who expressed convincing arguments to adopt the RAMIS. Implementation concerns were also discussed. All interviews were digitally recorded and subsequently transcribed. We then thematically coded the findings. Finally, these themes were clustered and links established so as to discover the relationships between them. Based on this, narratives were developed and attempts made to construct meanings by interlinking them with the diffusion of innovation theory.

4 Background for a New System: Initiation, Implementation and Challenges

The IRD in Sri Lanka, as in any emerging economy, relies upon a manual tax management regime inherited from its colonial legacy. The ADB pinpointed several loopholes in a manual tax administration system [16]. Amongst its main limitations are the minimum number of taxpayers due to undisclosed income and pain taking task of assessing tax liability of individuals because of a large volume of documents in the island's tax management system. The top administration of the IRD and politicians have been keenly interested in increasing the tax revenue of the government, as the government is heavily indebted. Both politicians and administrators have become familiar with the success stories of tax management in other countries. The awareness of limitations in the existing practice and knowledge of alternative superior practices are crucial in the process of diffusing innovation [2, 7, 15]. As one administrator recalled:

> *Our seniors visiting the developed countries have become aware of how they have increased their tax revenue with the help of information communication technology (ICT).*

As elaborated by the ADB [16], the use of a computerized tax management system renders benefits to taxpayers and the governing authority. The USAID's Leadership in Public Finance Management shows that a computerised management information system for tax administration enables officials to receive reports without any delay that contributes to identifying risk and internal problems in advance, whilst increasing the efficiency and effectiveness in the tax management system. The ADB has expressed its willingness to provide funds and technological assistance in the process of digitalising the existing outdated tax system in the IRD. In the process of persuading to embrace any innovation, its advantages are highlighted, whilst showing the weakness of prevailing practice [2, 7, 15]. As noted by an administrator:

> *We have observed that the existing tax system does not assist in improving the tax revenue of the government and mitigating those who avoid tax payment through loopholes of the current tax system. It has been necessary to adopt a system that is capable of overcoming our prevailing situation. Similarly, representatives of the international financial institutions like the ADB have also underscored in several discussions the limitations of our manual tax management system and the benefits of ICT based tax system.*

The decision stage occurs when the decision-making authority engages in the activities leading to a choice of adoption or rejection of the innovation [15]. The highest leadership in the IRD has made up its mind to introduce a digitalised tax system – Revenue Administration Management Information System, whilst some administrators are sceptical of getting active involvement of all stakeholders in its adoption. Nevertheless, having observed a discussion concerning a potential new system, the Cabinet of Ministers officially endorsed the RAMIS Project in April 2014 [19]. This move paved the way for making further strives to initiate implementation strategies.

The IRD was intended to facilitate the tax administration and tax compliance for taxpayers through the RAMIS. Additionally, increasing the revenue collection and reaching taxpayers in more effective and efficient way was also anticipated. The benefits of the RAMIS have been considered from the perspectives of multiple stakeholders - taxpayers, IRD and its employees [18]. It is expected to be convenient for taxpayers to pay their annual tax obligation or handle tax matters and access tax information, whilst the IRD, in turn, may increase tax compliance, revenue collection and serve taxpayers' complaints more efficiently. Likewise, the staff of the IRD may avoid the complexity arising during tax audits and access taxpayers' information.

Some of the required infrastructure to migrate to an ICT based system was already available at the IRD. For example, a Data Centre and a Disaster Recovery Centre were established at the beginning of 2010 under the initiatives of the Ministry of Finance on the Fiscal Management Reforms Programme and the Fiscal Management Efficiency Project. Further, Local Area Network and Wide Area Network were also introduced in order to link all the regional offices to the head office and to link branch-to-branch [17]. However, computer-based applications were yet to be developed. For this purpose, a global tender process was initiated to select a suitable supplier. The main selection criterion was adequate experience on automation of tax systems. However, the tender did not result in the selection of a supplier. As the tender process failed, the Cabinet of Ministers approached the Ministry of Finance of Singapore seeking assistance to implement the RAMIS as a Government to Government (G2G) project. As echoed by a Commissioner at the IRD:

> There were several reasons to select Singapore. The main reason is the availability of one of the best functioning automated tax system in Singapore. Moreover, the geographical proximity between the two countries and the Singapore's involvement in implementing similar tax automation projects in Australia and in some other countries were also considered.

A Memorandum of Understanding was signed to implement the RAMIS project. Based on the Memorandum of Understanding, Infocomm Development Authority, the Government of Singapore's execution arm for collaboration with overseas Governments on ICT projects, submitted a proposal. The Infocomm Development Authority engages subcontractors to plan, analyse, design and implement the tax system with the participation of relevant tax authority. A Cabinet Appointed Negotiation Committee and a Project Committee were responsible for making a technical evaluation of the proposal submitted by the Infocomm Development Authority. Both committees revealed the technical acceptability and compatibility of the proposal as per the requirements of the IRD. Moreover, a further due diligence process was then conducted by the Commissioner General and a Core Team of the IRD by visiting the Inland Revenue Authority of Singapore. This delegation endorsed the involvement of the Infocomm Development Authority and its subcontractors to undertake the task of developing the RAMIS. The process of negotiation then focused its attention on the project's work scope, the terms of contract etc. Both partners had a clearer understanding on the specifications of the RAMIS and the timeframe of the project. In this way, decision makers have sought to avoid unnecessary delay in adopting the new system.

The commitment of senior management is reckoned essential in diffusing innovations [15]. A Change Management Committee was formed to communicate expected changes under the RAMIS project. The main task of the committee was to organise change management programmes for the IRD's employees. The Change Management

Committee consists of the IRD change agent, team change manager, the IRD Deputy General (Head of the Team) and Project Director. Changes are communicated through RAMIS News, RAMIS Mass briefing sessions, Monthly Commissioner's and Senior Commissioner's meetings and Monthly Steering Committee meetings which involve all the Deputy Commissioner Generals and Commissioner General. As observed by a commissioner:

> The team proposed several initiatives to improve the awareness of the project amongst employees by publishing magazines, distributing leaflets and displaying posters at the IRD. Similarly, seminars were arranged to generate positive attitude towards the RAMIS. In addition, the IRD has sent 1200 officers to Malaysia for a five-day training programme to understand how an automated system operates. Moreover, the committee organised meetings for branches at the IRD and practical training sessions for both IRD staff and users.

The IRD has conducted a series of in-house seminars for its staff members concerning the Change Management during weekends. Furthermore, the mass briefings for taxpayers and supporting organizations such as Chamber of Commerce, professional accounting institutes, Tax Agents and Tax Consultants were also scheduled and are still in progress to maintain a continuous communication. As noted by an administrator:

> The mass briefings with taxpayers and related institutions are very important with a view of ensuring the smooth functioning of the platform. This is a new initiative not only to the employees but also to the taxpayers. There could be system failures, fear of using IT systems. The mass briefings help the IRD as a point of departure to track weak points in the process of experimenting.

The implementation stage of any innovation is often threatened or obstructed by its opponents [2, 7, 15]. Sri Lanka was no exception to this trend. The new system was not appreciated by all the staff members at the IRD. A commissioner revealed how some employees, bearing negative attitudes towards the new system, behaved and how the IRD strove to overcome the resistance:

> I regret to state that some of our employees were resisting the project. They spread bad word of mouth and did not contribute to the implementation process. The elderly workforce or employees, who have no computer literacy, raised objections. The departments evaluate the performance of employees biennially. The employees, who failed to complete the given assignments, receive negative marks. In this way, our leadership expects employees to adjust their behaviour automatically as per the RAMIS.

The ADB as the international financial institution has extended financial assistance to launch the new system. The ADB granted funds to acquire hardware. Its grants enabled to set up local area network and wide area network that have linked all the regional offices to the head office and branches or sections within the head office. The system developer aims to streamline each process as per the perspectives of the IRD and taxpayer. Key changes through the RAMIS were e-registration opportunity for taxpayers to avoid the need of visiting the IRD, e-payment opportunity through the online banking payment system, a centralised processing centre at the Head of the IRD for centralised data entry and regional processing centres for all regional offices to enter data through external interfaces. Moreover, it permits multilingual communication, which is pivotal for a multi-ethnic country like Sri Lanka. An officer highlighted the linkage with external interface as follows:

Actually, the linkage with external interfaces is the most important feature under the RAMIS and the IRD have identified 23 such agencies to be interfaced. The two main government banks, which are Peoples bank and Bank of Ceylon, have already been connected. A memorandum of understanding has been signed with the Excise Department and the Department of Customs.

Numerous hurdles were observed throughout the implementation stage that could impede the full-fledge adoption of innovations [2, 15]. In fact, the RAMIS's implementation has been behind the intended timeline and there have been considerable deviations from the intended plan. Budget changes and delays due to the rigorous procedure to be followed in obtaining the approval of the government for funds have been claimed the main causes behind the delay. Likewise, the unavailability of compatible IT systems in other organizations (e.g. Department of Registration of Companies) has adversely affected the timely execution. Furthermore, unexpected technical issues occurred in the testing phase and considerable time consumed to address them have obstructed the implementation. Despite the delay, the pilot project has been implemented and appreciated by a majority of employees, taxpayers and some institutions. A staff member noted:

We could not implement this project as per the time plan. This considerable deviation occurred in the testing phase due to arising minor issues. However, we are now progressing towards the final stage. Web portal and E-filing have now been introduced.

It is pivotal to adopt context-specific strategies to change the negative attitudes of actors towards the innovative practice [2, 15]. As envisaged by the ADB [16], the older workforce will not see the benefits of computerization to the same extent like young administrators. Besides the technical issues, faults occurred in the process of selecting staff members for training programmes. Albeit young administrators at the IRD are eager to embrace a digitalised system, their seniors have been offered the opportunity to attend training programmes in Singapore as per the norms of the government sector. They were required to identify crucial issues that should be addressed, whilst adopting the new system in Sri Lanka. Some of them have not observed any problem or issue that can arise in the operational process. Moreover, some administrators, who attended the training sessions in Singapore, have reached retirement age before the implementation. Furthermore, no administrator has been trained to gain the overall knowledge of the entire system. Therefore, the adoption endeavour of the RAMIS has resulted in some unintended obstacles. As noted by an assistant commissioner:

The IRD officers were offered opportunity to learn the Singaporean tax system and to identify specifications for the RAMIS in Sri Lanka. However, the officers selected for training were at their retirement age and no intention to understand the process in detail. Some officers, who participated in the training programs, retired even before the implementation of the RAMIS. Hence, there were difficulties in founding a tax management information system, which is suitable for our local context.

The RAMIS should connect twenty-six institutions or departments with the IRD. Nevertheless, only two main government banks – The Peoples Bank and The Bank of

Ceylon have been incorporated into the RAMIS by now. The Central Bank's rules and regulations should be amended to link private banks' data with the IRD. Similarly, the existing systems in the respective organisations are not compatible with the RAMIS system. Moreover, there have been some practical and legal restrictions for a number of the governmental departments to share information. For instance, the island's Department of Registration of Companies, Land registry, Department of Registration of Persons, Department of Census & Statistics and Department of Motor Traffic still maintain manual records. The Department of Motor Traffic (DMT) is unable to share information as per the Motor Traffic Act. However, this issue had been discussed with the respective Minister and since the Inland Revenue Act supersedes all other Acts in the country, the IRD will obtain permission to interlink their database with the DMT in the near future. However, other issues are likely to further delay the complete adoption of the RAMIS. The full-fledged diffusion of innovation occurs when a wide range of stakeholders accept its benefits and use it in day-to-day operations [15, 20].

5 Concluding Remarks

The present paper has sought to trace the introduction and implementation of a new digitalised tool, namely Revenue Administration Management Information System (RAMIS), at the Inland Revenue Department in Sri Lanka. Rogers' diffusion of innovation theory has been applied as a theoretical frame to examine the rise and diffusion of this novel tool. The empirical evidence has been collected through, amongst others, semi-structured interviews and document search. The results of our study are twofold. On the one hand, the study shows that a great deal of interest has been expressed in the digitalised tax system at the very top. Under these circumstances, the ADB as well as administrators having international experience have taken a decisive lead in promoting the RAMIS across the nation. This has made them gradually turning into "pervasive communicators" who provide expert advice on how the novel system should function.

On the other hand, the implementation of the RAMIS has not yet moved beyond the pilot project and achieved the desired ends. On the contrary, a myriad of implementation barriers and challenges have been unveiled, amongst which special mention should be made of poor infrastructure; resistance to change, especially amongst elderly employees; as well as the current legal framework that hinders the full-fledged functioning of the system. All this indicates that the success of the reforms depends largely on proper implementation, meaning that reforms need to be implemented as intended and 'are not blocked or watered down' [14].

The diffusion of innovations is a process in which some stakeholders could feel that their stake is at jeopardy [2, 15]. As envisaged by the previous studies dealing with the adoption of Western accounting practices or Japanese cost management practices in Sri Lanka to measure performance in a privatised textile mill and a privatised telecommunications company, both native managers and factory employees have cooperated closely to override the Western practice in the former case [21]. However, trade unions, bureaucrats and political leadership have impeded the diffusion of performance measurement in the latter case [22]. In both cases, political patronage and clientelism have played the decisive role in the process of undermining innovative performance measurement practices [21, 22]. Similarly, despite the commitment of Treasury's administration to adopt accrual accounting since 2003, efforts to diffuse accrual accounting in the Sri Lankan central government have not crowned with success amid the lack of political leadership [1]. In contrast, both political and administrative leaders have expressed their common interest in the introduction of the RAMIS, thereby making its opponents rather weak.

It is noteworthy that the diffusion of the RAMIS is taking place before it becomes taken-for-granted in the Sri Lankan public sector if any. Future studies must examine the challenges associated with the implementation process that individual offices have encountered and/or are about to experience throughout this journey. Similarly, the existing practice of offering foreign training as per seniority could contribute to demotivating the young administrators at the IRD. It is, therefore, vital to explore how the highest authority of the IRD endeavours to retain the young administrators' commitment for the island wide adoption of the RAMIS.

Hopefully, the present paper has an applicable nature and shows the process of transformation of management in the public sector through an institutional and mental lens, which is very important for understanding the process of promoting of reforms. Moreover, the article is deemed pivotal for researchers who deal with the issue of public finance, tax administration and other related institutional areas. Last, but not least, it may allow to make useful practical conclusions about the goals of implementing reforms and the problems of their implementation in practice.

References

1. Adhikari, P., Kuruppu, C., Matilal, S.: Dissemination and institutionalization of public sector accounting reforms in less developed countries: a comparative study of the Nepalese and Sri Lankan Central Governments. Acc. Forum **37**(3), 213–230 (2013)
2. Adhikari, P., Kuruppu, C., Wynne, A., Ambalangodage, D.: Diffusion of the cash basis international public sector accounting standards in less developed countries – the case of the Nepali Central Government. Res. Acc. Emerg. Econo. **18**(15), 85–108 (2015)
3. Askim, J.: Performance management and organizational intelligence: adapting the balanced scorecard in Larvik Municipality. Int. Pub. Manage. J. **7**(3), 415–438 (2004)
4. Christensen, M.: The third hand: private sector consultants in public sector accounting change. Eur. Acc. Rev. **14**(3), 447–474 (2005)
5. Christensen, M., Parker, L.: Using ideas to advance professions: public sector accrual accounting. Financ. Account. Manag. **26**(3), 246–266 (2010)
6. De Vries, H., Bekkers, V., Tummers, L.: Innovation in the public sector: a systematic review and future research agenda. Pub. Adm. **94**(1), 146–166 (2016)
7. Ezzamel, M., Hyndman, N., Johnsen, A., Lapsley, I.: Reforming central government accounting: an evaluation of an accounting innovation. Crit. Perspect. Acc. **25**(4/5), 409–422 (2014)
8. Guthrie, J., Olson, O., Humphrey, C.: Debating developments in new public financial management: the limits of global theorising and some new ways forward. Financ. Account. Manag. **15**(3/4), 209–228 (1999)

 9. Hansen, M.B.: Antecedents of organizational innovation: the diffusion of new public management into Danish Local Government. Pub. Adm. **89**(2), 285–306 (2011)
10. Hood, C.: The new public management in the 1980s: the variation on a theme. Acc. Organ. Soc. **20**(2/3), 287–305 (1995)
11. Hopper, T., Lassou, P., Soobaroyen, T.: Globalisation, accounting and developing countries. Crit. Perspect. Acc. **43**, 125–148 (2017)
12. Kuruppu, C., Adhikari, P., Gunarathne, V., Ambalangodage, D., Perera, P., Karunarathne, C.: Participatory budgeting in a Sri Lankan urban council: a practice of power and domination. Crit. Perspect. Acc. **41**, 1–17 (2016)
13. Lapsley, I., Mussari, R., Paulsson, G.: On the adoption of accrual accounting in the public sector: a self-evident and problematic reform. Eur. Acc. Rev. **18**(4), 719–723 (2009)
14. Polidano, C.: Why civil service reforms fail. Pub. Manag. Rev. **3**(3), 345–361 (2001)
15. Rogers, E.: Diffusion of Innovations, 5th edn. The Free Press, New York (2003)
16. http://www.adb.org/sites/default/files/publication/150133/tool-kit-tax-administrationmanagement-information-system.pdf. Accessed 25 Apr 2016
17. http://www.cbsl.gov.lk/pics_n_docs/10_pub/_docs/efr/annual_report/AR2014/English/10_Chapter_06.pdf. Accessed 20 May 2016
18. http://www.ird.gov.lk/en/publications/Annual%20Performance%20Report_Documents/PR_2013. Accessed 19 Mar 2019
19. http://www.ird.gov.lk/en/publications/Annual%20Performance%20Report_Documents/PR_2014. Accessed 19 Mar 2019
20. Walker, R.M.: Innovation type and diffusion: an empirical analysis of local government. Pub. Adm. **84**(2), 311–335 (2006)
21. Wickramasinghe, D., Hopper, T.: A cultural political economy of management accounting controls: a case study of a textile mill in a traditional Sinhalese village. Crit. Perspect. Acc. **16**, 473–503 (2005)
22. Wickramasinghe, D., Hopper, T., Rathnasiri, C.: Japanese cost management meets Sri Lankan politics: disappearance and reappearance of bureaucratic management controls in a privatised utility. Acc., Auditing Account. J. **17**(1), 85–120 (2004)

Risk-Based Project Management Audit

Julia Klimova$^{(\boxtimes)}$

Kuban State University, Krasnodar 350040, Russia
ladycat23@mail.ru

Abstract. The project management is considered as a tool for the implementation and development of strategic transformation. Therefore, it finds application in various fields of activity such as manufacturing, marketing and innovation activities, the social sphere. The existence of developed organizational measures and procedures, including internal audit, allows in a preventive manner to focus management attention on risky areas of project management and determine activities to optimize the performance of company to achieve its goals and strategy. In these conditions, a risk-based project management audit becomes a priority for increasing the efficiency of the business. In present work we examine the key areas and control procedures of project management audit based on the analysis of risks inherent in the project.

Keywords: Project management · Risk-based approach · Internal audit

1 Introduction

The influence of the uncertainty on business necessitates the timely identification and assessment of risks in order to prevent and reduce the possible negative consequences for the project management. In this connection, the application the control procedures of internal audit becomes the necessary process, which helps management to focus attention on risky areas of project management and optimize activities to achieve the project's goals.

Studies of recent years [18–20] have noted the increasing role of risk management as an integral part of management functions in the company. Timely identification of risks and the development of a set of measures for preventive exposure make it possible to create a reliable basis for decision-making and planning and improve the management of the company to achieve the objectives of business processes [12, 14–16, 21].

Despite a significant number of works [1, 4, 5, 17] devoted to the application of risk-based approach in the activities of companies, the issue of practical implementation of control procedures in the company's business processes remains relevant. The development of elements of the internal control system as a whole and internal audit in particular is becoming more urgent for improving the risk-management process [2, 4, 5, 8–11, 13, 17, 22].

In present work we suggest the main procedures of project management audit as a practical guide to prevent failures in order to improve project performance. Firstly we examine the scope of project management and its risky areas in order to address

© Springer Nature Switzerland AG 2020
T. Antipova (Ed.): ICIS 2019, LNNS 78, pp. 39–49, 2020.
https://doi.org/10.1007/978-3-030-22493-6_5

relevant control instruments. In second section we suggest the program of project management audit based on the analysis of risks inherent in the project.

2 The Scope of Project Management

To implement effective control and audit of the project, control system should be established. Its composition should include monitoring of the actual performance of work, analysis of the status of work by comparing the available results and bench-marks, a set of corrective actions until the project is stopped.

Project management audit is considered as a method and set of procedures based on accepted management principles that are used to plan, evaluate and control work assignments in order to obtain the desired end result in a timely manner, within the allocated funds and in accordance with the project requirements.

By definition, the project has the following characteristics [3, 6]:

- the presence of complex and numerous tasks
- unique sequence of events
- finiteness (the dates of commencement and completion of work are set)
- limited resources and budget
- involvement of a large number of people, usually from several functional units organizations
- established order of tasks
- orientation to achieve the ultimate goal
- result (receiving the final product or providing services).

The project management system can be structured by functional areas. In the most general form, they include: subject project area, deadlines, cost and financing, labor, risks project, information and communication, quality, project support resources and services [7].

The management of the project domain implies the initiation, planning and control of the achievement of the main goal of the project.

Time management of the project includes the processes of planning, monitoring, making changes, as well as further analysis of the timing and reserves.

Restrictions on the use of financial resources are laid when managing the cost of the project. Cost management is the development of methods and procedures by which financial indicators will be planned and monitored. At the planning stage, the basic principles of cost accounting are laid out; during the implementation, the project budget is monitored, cost analysis and regulation is carried out. At the stage of completion, an important link is the comparison of actual financial indicators, with those that were originally laid.

Project quality management involves processes that ensure that it meets customer requirements, existing technologies, and the design decisions that the organization has asked the project.

Project risk management is an activity aimed at increasing the sustainability of a project by reducing the impact of risks. Reducing the impact of project risks is carried out by identifying them, ranking by significance, and drawing up a program to reduce them.

Management of communications in the project is carried out in the form of collection, processing and timely provision of information necessary for project participants to perform work on time. To improve communication management, the manager needs to have feedback from the project participants.

The management of supplies and contracts in a project includes processes whose goal is to provide the project with resources from outside. This functional area includes: conducting marketing of the market for products and services, planning deliveries and contracts to meet the needs of the project, organizing and preparing contracts, administering them in the implementation process, and closing contracts.

Project personnel management is probably one of the main functional areas of the project. The project manager provides the project with the necessary human resources, motivates and controls his team. The main goal of managing this functional area is the most effective management of the human resources involved in the project.

3 The Key Areas and Procedures of Risk-Based Project Management Audit

Each project is unique, and therefore the project management audit should be started after a detailed analysis and study of its high-risky functional areas, which are determined at the stage of preliminary analysis by answering following questions:

1. What is business strategy? What business goals does the project meet?
2. What are the goals and expected results of the project? What are the project benefits for business?
3. Is there a correlation between KPIs and business tasks reflected in the project plan?
4. Who are the leaders interested in this project? What is the support expressed by the managers of this project?
5. Has the market and competitive environment been studied at the pre-project activity stage?
6. Has the target model and project implementation plan been developed (business case; the organizational structure and distribution of administrative and functional responsibility)? What are the data sources for the project's business case?
7. Has an agreement on the adoption of a project budget been received by financiers?
8. Is the project schedule set? Does the project documentation fix the person responsible for each stage in the project schedule? Is the project deadline monitored in the information system?
9. Does the project documentation define the list of goods/services purchased?
10. What are the procedures for quality control and compliance with the requirements of the technical requirements of the purchased products for the project (securing responsible, availability of verification criteria, etc.)?
11. Is the composition of the project team fixed in the project passport, are competencies and responsibilities distributed, is personal responsibility established? Are KPI for employees interconnected with project KPI?
12. Are risk management measures developed and approved? Are responsibilities for risk prevention, assessment, monitoring and reporting defined and delimited?

13. How is the accounting and monitoring of project results/indicators ensured?
14. Is the ability to mark income/opex/capex related to the project implemented in IT-systems?
15. How is the exchange of information and materials on the project organized?
16. Are there different levels of access to project information in accordance with the administrative level and the functionality being executed, including in information systems?
17. What is the reporting on the project? Is reporting formed manually or unloaded from information systems?
18. Is the responsibility for the accuracy and quality of information provided? Are the reports signed by the supervisor?

According to received information about the project the auditor assesses the effectiveness of the implementation and management of the project at the stages of initiation, planning, performance, monitoring and control, completion, including:

– expediency of project implementation/compliance with the Company's strategy;
– project office management;
– preparation of project documentation;
– project risk management;
– resource management (financial, human, time);
– level of automation of the implementation process;
– methodological support of the project;
– formation of management reporting within the project;
– implementation of stages/tasks/obligations under the project;
– procurement procedures and interaction with counterparties;
– achievement of goals/objectives/KPI of the project;
– documenting the project results.

The key areas and procedures of project management audit according to risk-based approach are systematized in Table 1.

As we can see, the suggested audit program is based on the main areas of project management and proposes wide range of control procedures to cover the main project risks. Risk-based project management audit can act as an early warning system to ensure control procedures at all stages of project, to detect the main risks and to control failure on a timely basis. The goal of implementing project management audit is to enhance the overall visibility of the organization to risk and project performance through the effective use of technology.

The company can establish automated collection of audit evidence and indicators from an entity's IT systems, processes, transactions, and controls that allow real-time analysis of project with a significant reduction in time and resources. This information enhances auditor capabilities and helps to ensure compliance with policies, procedures, and regulations.

To develop a project management audit, it is necessary to create an automated information system to aggregate all procedural and financial reports, which allows to detect deviations from the targets of projects. The results of project management audit should be submitted to management in a timely manner in a consistent official report,

Table 1. The key areas and procedures of risk-based project management audit.

Project stages	Risk	Scan area	Audit procedures
Initiation	Failure to achieve the strategic goals of the company	Level of decision making and authorization	Verification of the presence of the Steering Committee/Starting Order Protocol with a decision on the feasibility of project implementation/project financing
			Analysis of the goals/KPI of the project of the Company's Development Strategy and the goals/KPI of the project portfolio
	Failure of project management objectives	Corporate project management system	Evaluation of the effectiveness of the organization of the project management system
Planning	Lack of expected returns from project/failure to achieve project goals	Project documents	Verification of the completeness of the design documentation in accordance with internal regulatory documents
			Verification of compliance of project targets with economic justification/costing/budget
			Checking the validity of project targets (income/opex/capex)
			Evaluation of the correctness of the algorithm for calculating KPI project/planned indicators and criteria for their achievement
		Control environment	Analysis of the completeness/sufficiency of the regulation: key procedures, tasks, functions, membership, owners of business processes, principles of interaction, separation of powers of responsible project participants
		Project risk management	Checking the existence of the risk register/risk accounting in the business case and the plan of measures aimed at eliminating and/or minimizing project risks
			Assessment of the completeness of the implementation of response measures for the reporting period
		Project team management	Verification of project KPI decomposition to the level of participants

(continued)

Table 1. (*continued*)

Project stages	Risk	Scan area	Audit procedures
Performance	Failure to achieve project goals	Deadlines of the stages/tasks	Verification of compliance with project deadlines and availability of documentary evidence
			Evaluation of the effectiveness of key control procedures of business processes that ensure project implementation
		Automation level	Assessment of the level of automation of business processes that ensure the implementation of the project
	Lost income, fines	Interaction with contractors	Verification of compliance of the procurement procedures with the project requirements
			Verification of the compliance of the terms of the concluded contracts with the results of the procurement procedures (including the substantive conditions: the subject of the contract, price, volumes, composition of work/supply, terms of fulfillment of obligations, technical solution)
			Inspection of penalty for non-fulfillment of contractual obligations by counterparties in the amount stipulated by the contract
			Checking the timeliness of the payment deadlines established by the contractual agreements
	Unreasonable investments		Verification of compliance of results/deadlines/indicators/SLA with contractual obligations
			Verification of the completeness of the fulfillment of contractual obligations for design and exploration work
			Verification of the completeness/quality of the implementation of the volume of construction and installation works
	Inaccuracy in reporting		Analysis of the correctness of the formation of fixed assets
	Ineffective expenses/loss of assets		Analysis of the status of registration permits for objects

(*continued*)

Table 1. (*continued*)

Project stages	Risk	Scan area	Audit procedures
	Additional analytical procedures for the project with the income contract		
	Lost income		
	Revenue contract obligations	Evaluation of the timeliness of the conclusion of the income contract	
		Lost income, penalties	Verification of compliance of the results/deadlines/indicators/project SLA with the requirements of the income contract
		Lost income	Checking the reasonableness of the reduction of payments under the income agreement
	Lost income, fines	Interaction with contractors	Verification of compliance with the terms of reference/deadlines/indicators/SLA established by the contractual agreements, the terms of the income contract
Inefficient costs	Verification of compensation of penalties imposed by the customer, by the contractor of the income contract, by the contractor		
Monitoring and control	Inaccuracy in reporting	Methodical support of the project	Analysis of the methodological security of accounting of income and expenses in the accounting and management accounting for the project
	Inaccuracy in reporting/failure to achieve project goals	Project budget management	Verification of the correctness/reasonableness of the allocation of income/opex/capex on the project
	Lost income/failure to achieve project goals		Evaluation of the achievement of the plan for revenue, agreed in the business case and its impact on the operating performance indicators (OIBDA/EBT), project performance indicators (NPV)
	Unreasonable costs/failure to achieve project goals		Analysis of plan/fact expenditure (OPEX) for the project, agreed in the business case

(*continued*)

Table 1. (*continued*)

Project stages	Risk	Scan area	Audit procedures
	Inaccuracy in reporting	Management reporting	Assessment of the level of automation of the collection and consolidation of project reporting data
			Analysis of the procedure for reporting on the project (indicators/frequency/responsible/deadlines/results of consideration)
	Failure to achieve project goals	Project management efficiency	Checking the timeliness/completeness of the financial and non-financial goals/project KPIs
Completion	Inaccuracy in reporting	Registration of the project results	Assessment of the completeness of the control procedures at the completion of the project in accordance with internal documents
			Verification of documenting the transfer of project results to the customer

including observations and understanding of the risks, control measures and consequences associated with the findings.

In order to prevent significant costs, it is advisable to create automated information system on the basis of existing network resources and information databases in the company. Modern information technologies make it possible to create digital software that aggregates data from information systems with both financial and technical parameters according to given algorithms and regulations. Using digital technologies will allow to detect inconsistency with the indicator and to receive the detailed information in real time. It is possible that basic data analysis can be performed using a range of tools, including spreadsheets and database query and reporting systems.

4   Conclusion

As we can see, an audit is necessary for periodic evaluation and analysis of the overall state of the project, identifying reserves for improving project performance. An audit of the project is carried out to obtain up-to-date and as accurate information as possible on the most important areas of the project in order to take timely and correct management decisions.

The project management audit helps to:

- reveal the strengths and weaknesses in the principles of project management;
- assess the degree of elaboration of management documentation;
- identify areas requiring special attention and priority improvement;
- get recommendations on how to improve project management processes.

The suggested control procedures of project management audit were developed by the author according to common business practice, internal regulatory documents and audit procedures. The suggested audit program includes the description of the key areas and procedures of project management audit according to risk-based approach and may be used as a practical guide to improve project performance in companies. The use of this approach jointly with the digital automated technologies provides an integrated access to real-time risk assessment in order to enhance risk and control oversight capability through monitoring. The move to automated testing and audit procedures also changes the traditionally cyclical nature of the project management audit process. Comprehensive testing of transactions and controls effectiveness, on an ongoing automated basis, enables audit to move to a more risk-based approach. The results of auditing techniques provide visibility into whether risk is increasing in specific areas and warrants additional audit focus.

References

1. Belousov, S.A.: Risk-oriented internal audit as an element of key risk management of the company. Risk Manage. **2**, 14–18 (2011)
2. de Aquino, C.E., Lopes da Silva, W., Sigolo, N., Vasarhelyi, M.A.: Six Steps to an Effective Continuous Audit Process. https://www.researchgate.net/publication/266059571_Six_Steps_to_an_Effective_Continuous_Audit_Process

3. Degaltseva, Y.A., Lysenko, E.G., Atamas, M.V.: Development of project management in Russia. Scientific community of students of the XXI century. In: Economic Sciences: Proceedings of the XLIV International Student Scientific-Practical Conference No. 7 (44). https://sibac.info/archive/economy/7(44).pdf

4. Firova, I.P., Bikezina, T.V.: Modern problems of integrated risk management in order to reduce the financial risks of economic subjects. Sci. Bus.: Ways Dev. **11**, 35–38 (2016)

5. Grischenko, O.V., Efimenko, A.: The role and place of internal audit in corporate governance system. Bull. Taganrog Inst. Manage. Econ. **1**, 46–53 (2009)

6. Weiss, J.W., Robert, K.: 5 stages of project management. A practical guide to planning and implementation. http://web.krao.kg/7_menejment/0_pdf/1.pdf

7. Kolpakova, M.A.: Improving the project management system in the organization: functional project management areas and requirements for qualification of the project manager. In: Scientific Forum: Economics and Management Collection of Articles on the Materials of the XI International Scientific and Practical Conference, pp. 45–50 (2017)

8. Kovalenko, A.I.: Features of financial risk management of international corporations. Econ. Sustain. Dev. **4**(28), 36–44 (2016)

9. Kuter, M.I., Sokolov, V.Y.: Russia. In a Global History of Accounting, Financial Reporting and Public Policy: Eurasia, the Middle EST and Africa, pp. 75–106. Emerald Group Publishing Limited, Sydney (2012)

10. Luchakova, E.V., Tuvaeva, A.M., Sergienko, L.V.: Risk management in accounting. Actualscience **12**, 277–279 (2016)

11. Nigrini, M.J.: Continuous auditing. Ernst & Young Center for Auditing Research and Advanced Technology and Advanced Technology University of Kansas. http://aaahq.org/audit/midyear/01midyear/papers/nigrini_continuous_audit.pdf

12. Mihret, D.G, Khan, A.A.: The role of internal auditing in risk management. In: Seventh Asia Pacific Interdisciplinary Research in Accounting Conference. Kobe (2013)

13. Kokemuller, N.: The Advantages of Continuous Auditing. http://small-business.chron.com/advantages-continuous-auditing-39568.html

14. Paschenko, T.V., Tarasova, K.J.: Methodical approaches to assessing financial investments for the purpose of the financial reporting and expertise of the balance sheet of assets. Prob. Mod. Econ. **4**(64), 82–86 (2017)

15. Pislegina, N.V.: Problems of estimation of financial investments in accounting and financial statements. Proc. Altai State Univ. **2**, 104–106 (2002)

16. Pravkina, E.I.: The role of risk management in the process of business planning. Bull. Univ. (State Univ. Manage.) **7–8**, 247–250 (2016)

17. Selezneva, E.S.: Information support of risk-oriented internal audit. Bull. Saratov State Soc. Econ. Univ. **1**(50), 107–110 (2014)

18. Spira, L.F., Page, M.: Risk management: the reinvention of internal control and the changing role of internal audit. https://poseidon01.ssrn.com/delivery.php?ID=285088095002025121 09506406807710309103302003907204508902808711202310111507711808307504906 3-09701511202301602706610301408909501808305500007910308508909400612110909 5-06901400912709206509000800901610810910602008800007309308807611910612610 8117118119026&EXT=pdf

19. The changing role of internal audit. https://www2.deloitte.com/content/dam/Deloitte/in/Documents/audit/in-audit-internal-audit-brochure-noexp.pdf

20. The Role of Internal Audit in Enterprise-wide Risk Management. https://www.ucop.edu/enterprise-risk-management/_files/role_intaudit.pdf
21. Vasile, E., Croitoru, I., Mitran, D.: Risk management in the financial and accounting activity. Intern. Auditing Risk Manag. **1**(25), 13–24 (2012)
22. Stippich, W.W.: Continuous Auditing = Continuous Improvement. http://www.corporate-complianceinsights.com/continuous-auditing-continuous-improvement

A Systems Approach to Comprehend Public Sector (Government) Accounting

Konstantin Timoshenko[1](✉), Chamara Kuruppu[2], Imtiaz Badshah[3], and Dayananda Ambalangodage[4]

[1] Nord University, 8049 Bodø, Norway
konstantin.y.timochenko@nord.no
[2] University of South-Eastern Norway, 3603 Kongsberg, Norway
[3] Østfold University College, 1757 Halden, Norway
[4] University of Sri Jayewardenepura, Nugegoda 10250, Sri Lanka

Abstract. The present paper seeks to elaborate a framework for investigating public sector (government) accounting in terms of its relationship with the traditional budget. Indeed, a global shift towards a more accountable public sector under the banner of New Public Management in general and accrual accounting in particular is somehow expected to encroach on this relationship, gaining a strong impetus to research it. By employing a conceptual framework for what accountants do or are anticipated to do, we first examine accounting as a system of interrelated elements. We then turn our attention to exploring the nature of the relationship between the budget and accounting in the light of the machine metaphor. Thereafter, some empirical evidence on the topic is succinctly reviewed as reflected in the literature. We end this paper by posing some questions to be addressed in future studies.

Keywords: Public sector (government) accounting · Budgeting · Systems approach

1 Introduction

Accounting is a global phenomenon that is found in all kinds of entities around the globe regardless of ownership. However, accounting still lacks a single comprehensive paradigm, meaning that it is a complex multiple paradigm science [5]. As a result, a plethora of images of accounting as varied as a *system, symbol of legitimacy, colonizing force, ideology, myth, power, institution, language* and *law*, and that *assisting change* under conditions of uncertainty are among those appearing in contemporary accounting literature[1]. To state with certainty that one of the aforementioned approaches or images of accounting is the best or all-inclusive would undoubtedly be oversimplifying the discussion, as there are a myriad of prevalent accounting issues that need to be taken into consideration.

---

[1] See e.g. [18, 25] for more discussion on this topic.

© Springer Nature Switzerland AG 2020
T. Antipova (Ed.): ICIS 2019, LNNS 78, pp. 50–59, 2020.
https://doi.org/10.1007/978-3-030-22493-6_6

Perhaps, one of the most widely spread and taken-for-granted images of accounting in the literature (especially in accounting textbooks) is that of a *technology*, implying the existence of specific calculative practices and procedures for handling and monitoring economic and financial activities. This fits squarely with treating accounting as a measuring instrument, whose scientific task is to calibrate, polish, and clarify that instrument so that it generates true measures of reality [20, 34]. Likewise, accounting may be looked upon as "a mirror or picture which neutrally and objectively records the "facts" about what has happened in an organization over a particular period of time" [33], or as a "neutral arbiter of organizational truth" [34].

On the basis of this approach, accounting is *a system* of various elements, inextricably intertwined to each other in one way or another and intended to fulfill certain objectives. To define the term, a system is simply an assemblage or combination of elements which when in interaction, constitutes a unitary whole. Changes in one element of the system are anticipated to lead to those alterations in other elements. It is in this manner that we strive to examine the relationship between the government budget and accounting in this paper. As resources in the public sector are – or should be – allocated through the budgetary process, a tight link may be assumed between budgeting and accounting [40]. As Chan has clearly accentuated, one auspicious way of characterizing government accounting is to formulate its models in terms of its divergence from the traditional budget [11]. That emphasized, a fundamental use of government accounting is for budgeting and the concomitant budgetary control, a use that is often referred in the literature to as "budgetary accounting" [22]. Hence, changes in government accounting may be treated as confronted to this traditional budgetary accounting. This is what we coin '*a systems approach to accounting*' for the purpose of this inquiry.

The present paper seeks to elaborate a framework for investigating public sector (governmental) accounting in terms of its relationship with the traditional budget. Indeed, a global shift towards a more accountable public sector touted under the banner of New Public Management generally and accrual accounting particularly is somehow expected to encroach on this relationship, gaining a strong impetus to research it. By employing a conceptual framework for what accountants do or are anticipated to do, we first examine accounting as a system of interrelated elements. We then turn our attention to exploring the nature of the relationship between the budget and accounting in the light of the machine metaphor. Thereafter, some empirical evidence on the topic is succinctly reviewed as reflected in the literature. We end this paper by posing some questions to be addressed.

2   Accounting as a System of Elements

Viewed as a system, accounting can be defined as a framework, consisting of various theoretical dimensions, closely linked to one another in a hierarchical order [5]. Specifically, these theoretical components are the objectives of financial reporting, the fundamental postulates and principles of accounting, the particular accounting methods and techniques, as well as the accounting reports[2] [10]. It is these theoretical components which make accounting capable of translating the world into financial values which seek no further referent [28].

Taking objectives of financial reporting as a point of departure, they serve as the basis for propositions on accounting. As a matter of fact, there is a clear-cut consensus among scholars on the overall objective of financial reports, that is, to furnish financial information about the reporting unit in a systematic way [20]. This information should be *useful*, and based primarily on the needs of users of the accounting reports. Such viewing corresponds closely to what is known in the accounting literature as the decision-usefulness perspectives [1]. To be more precise, the conventional [mainstream] approach to accounting considers the phenomenon of interest as a pivotal element of formal organizations, the intended function of which is to curb uncertainty in the relationship between *the accountee* (i.e. the user of accounting information) and *the accountor* (i.e. the supplier of accounting information) with a view to heightening control and decision-making by providing a faithful representation of economic phenomena [18, 24–27, 41].

The *control* objective means that the accounts should furnish information about the management of resource use. Through this process, accounting enables actors or responsible bodies to juxtapose where they are in relation to where they should be, and what they have achieved in comparison with what they should have achieved [28]. This is what is also known in the extant literature as the accountability or stewardship objective [24, 25]. The *decision-making* objective implies that the accounts should provide a solid basis for decision-making [2, 3]. The accounts are in this sense vital as they "… make actions imaginable and consequences interpretable by … [defining] the meaning of history, the options available, and the possibility for actions. Accounts are used both to control events and to provide reassurance that events are controllable" [23]. Taken together, these two objectives are reckoned as the rhetoric or mantra of accounting and create thereby a solid foundation for the development of normative accounting theory and accounting standards [1, 24, 30, 32].

Next, the fundamental accounting postulates are regarded as assumptions about environments in which accounting is enmeshed. The postulates of accounting, amongst other things, define the accountant's area of interest and confine the number of objects and events that are to be incorporated in financial reports; determine the span of the accounting period; and introduce a unit of exchange and of measurement needed to account for the transactions. All the fundamentals conform closely to the objectives of financial reports from which they derive. In their turn, the accounting principles are defined as general decision rules, based on both the objectives of financial reporting and the fundamental accounting postulates. The principles of accounting are key in the sense that they keep us informed about *when* and *how* transactions and events are registered, measured, and communicated to users (see Table 1 for a brief overview of

---

[2] Viewing accounting as a system is not confined to one discussed in this paper. For example, Bergevärn *et al.* [6] suggest to divide accounting in action into two systems – a norm system and an action system. The latter can further be elaborated and classified into two sub-systems, namely a practice sub-system and a use sub-system. This distinction is, however, beyond the scope of the present research.

both the fundamental accounting postulates and principles of accounting). Depending upon when the transactions and economic events are recognized for financial reporting purposes, a wide range of accounting systems is unveiled around the world. Simply speaking, this variety goes from pure cash accounting to full accrual accounting, with a series of the "intermediate" versions in-between, labeled either modified cash or modified accrual.

Table 1. The fundamental accounting postulates and principles.

The Fundamental Accounting Postulates	The (Separate) Entity Postulate	Each organization is an accounting unit separate and distinct from its owners and other organizations/entities
	The Going-Concern (Continuity) Postulate	The entity will continue its operations long enough to realize its projects, commitments, and on-going activities
	The Unit-of-Measure (Monetary) Postulate	Accounting is a measurement and communication process of the activities of the organization that is measurable in monetary terms
	The Accounting (Time) Period Postulate	An organization should disclose its financial reports periodically
The Accounting Principles	The (Historical) Cost Principle	The acquisition (historical) cost is an appropriate valuation basis for recognition of the acquisition of all goods and services, etc.
	The Accrual Principle	Transactions and events are recognized when they occur irrespective of when cash is received or paid
	The Matching Principle	Expenses/expenditures should be recognized in the same period as the associated revenues
	The Consistency Principle	The similar economic events should be recorded and reported in a consistent manner from one period to another
	The Conservatism (Prudence) Principle	The accountant displays a generally pessimistic attitude when choosing accounting techniques for the purpose of financial reporting

Furthermore, the accounting methods and techniques are viewed as specific rules derived from the accounting principles that serve the basis for handling economic transactions and events encountered by the accounting entity. They are comprised of a myriad of accounting books and registers calibrated to record entries in them. When books are closed, accounting reports are prepared and then disseminated to a constellation of users. Such reports envisage the outcome of accounting, and are intended to aid the users of accounting information with a view of augmenting control and decision-making [24, 25]. Once again, the concept of "usefulness" is paramount to the perspective in focus: "... accounting data is seen as resting in the needs of its users; accounting exists only because there is a user who needs useful information..." [14]; "a user service is at the core of its concern" [21]. Quite clearly, the principle of user primacy holds, implying that the interests of users of financial reports take priority over the interests of preparers of financial reports. This principle is normative, grounded in Rawls' (1971) theory of the "veil of ignorance" [17], according to which a decision-maker is required to construct a just (fair) system from the position of a disinterested individual.

To summarize, treating accounting as a set of elements helps us comprehend what a government accounting and financial reporting system does. For example, **who** supplies **what** financial information to **whom**? **when**? **how**? and **to what ends**? [11, 12]. And what is clearly implied from the discussion above is that changes in one component of a system are to be followed by those alterations in other components. Wherever the change comes from, various practices of economic calculation and measurement would generate contrasting outcomes. For instance, a successful shift from cash accounting to accrual accounting would hardly be possible without the corresponding changes in the objectives of financial reporting and the content of final accounts themselves. More specifically, whereas cash accounting rests predominantly on the control objective, there is a robust preoccupation for using accounting information based on accruals for the purpose of decision-making. This also has direct consequences for the relationship between the budget and accounting, which is the subject-matter of the subsequent section.

3   Accounting in Relation to the Budget

From the systems perspective, accounting is intended to operate like a machine, that is, as a routine, in an efficient, reliable, and predictable way [19, 29, 30]. That is to say that the elements of the system have to do their preprogrammed jobs and that the machine has to be designed to run in the most efficient way. When describing this system (with respect to government accounting), it is vital to define its elements, i.e. procedures and roles, with specified interrelationships (see Fig. 1).

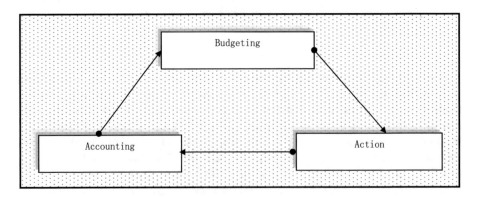

Fig. 1. Interrelationships between Budgeting, Action, and Accounting [24]

Taking budgeting as a point of departure, it may, in fact, be defined in many different ways. Welsch [42] views it as a useful tool in decision-making, planning, and control. In a more or less similar vein, Flamholtz [16] looks upon it as the process of deciding about the goals of an organization as well as the means to attain those goals. Apart from the machine metaphor, Wildavsky [43] relies upon metaphors of gaming and combat, pointing to the political character of budgeting. Yet, Czarniawska-Joerges and Jacobsson [13] propose to gaze at budgeting as a way of communicating rather than a means to control. Above all, images of budgeting as a ritual of reason, language of consensus, or a tool for accountability and transparency, to mention just a few, are also widespread [13, 30] hereby challenging the instrumentality of budgeting.

Broadly conceived, through budgeting one determines "… a plan expressed in quantitative, usually monetary, terms covering a specified period of time" [4]. It provides accounting for past expenditures and revenues, controls current spending and revenues, as well as estimating those of the future [8]. Equally speaking, budgeting precedes accounting chronologically, and provides accounting with a benchmark to measure and communicate actual plan accomplishments, that is, the outcome of actions. In so doing, accounting can aid actors to rationalize their previous actions by providing a justification for undertaking similar actions in the future [25], so it can assist in comprehending the world and in forming visions about the future in terms of budgets [7, 25].

Chan [11] proposes staring at government accounting in terms of its relationship with the traditional budget. Among the main features of this traditional budget are an aggregation of line items or inputs, one period (usually one year), as well as cash receipts and cash outlays, even though appropriations may be expressed in terms of obligations authorized to be incurred. Consequently, a fundamental use of government accounting is for budgeting and the concomitant budgetary control, a use that is often referred to in literature as "*budgetary accounting*" [22]. As such, it merely measures and controls the use of budgetary resources as promulgated by law and records receipts by source. Budgeted and actual amounts are juxtaposed line-by-line and, if a discrepancy is unveiled, a corrective action then follows. Hence, *control* is said to be a key function of accounting in this traditional mode. As such control requires a legal basis, politicians and lawyers are reckoned as major professions supporting and operating this type of accounting system [11].

Alternatively, the accounting system may be calibrated to carry out other functions transcending budget execution. One of them is to provide data for "true and fair" presentation of the government's financial position and operations to a wide constellation of *external users*. Under the influence of this external orientation, the basis of accounting used in external reporting does not need to be identical to the one used in the budget. And, as a plethora of studies [31, 35, 38] documented, especially with the rise and proliferation of NPM reforms around the globe, the business sector accounting solutions have been extended into the government sector at an accelerated pace, making accountants and auditors – especially those outside of government – play the pivotal role [11].

4 Some Empirical Evidence: A Succinct Overview

While Morgan's model prescribes a robust integration between the elements shown in Fig. 1 with equal attention devoted to both the budget and accounting, this is in effect subjected to immense criticism in the literature. To illustrate, Wildavsky [43] is first and foremost preoccupied with budgeting as the chief instrument for public allocation and control, whereas "accounting lives a far less conspicuous life" [30]. Surprisingly, no specific reference is made by Wildavsky to accentuate the instrumental role of accounting in maintaining and updating budget bases. Moreover, Høgheim *et al.* [19] speak of "The Two Worlds of Management Control" emphasizing thereby a loose coupling of decisions (as read from budgets) from actions (as read from accounts). As they put it themselves:

> "Between these two worlds [the world of decisions and the world of action] there is a coupling, but it is systematically switched on and off which has the effect that the signals it sometimes produces are very ambiguous" [19].

Next, Olsen's empirical study [30] strongly supports the findings above, manifesting that accounts in Norwegian municipalities are loosely integrated with budgets. In his study of Russian public sector accounting reform, Timoshenko [36, 37, 39] juxtaposes changes at the macro-level with those alterations at one state-sponsored university of tertiary education. As far as the macro-level is concerned, it is inconceivable that Russian central government accounting can be comprehended without any reference to the budget. In line with the current legislation, it is still officially labeled *budgetary accounting* to emphasize the vital role of the budget in Russian economy and society. Remarkably, even the ongoing efforts to reinvent public accounts in favor of accrual accounting are carried out under the aegis of streamlining the Russian budget process. Albeit still inextricably intertwined, it is not self-evident today that budgeting remains as dominant in relation to accounting as it was a decade or two ago. Indeed, there are definite signs of enlarging the use of Russian central government accounting beyond budgeting and concomitant budgetary control, a use that was prevalent and indeed the only one throughout the 1990s. Furthermore, budgeting itself is also being revived, seeking to adopt the methods and techniques widely advocated in the West.

Although new rules and regulations touted by the Russian state signalize a clear-cut shift towards information usefulness for economic and financial decision-making at the macro-level, the benefits of this reform effort are seemingly not so appreciable if at all at the micro-level. Indeed, remaining essentially budget-oriented in its nature, the chosen university is more preoccupied with the budget and the use of appropriations rather than the action-oriented accrual-based accounting figures. The evidence presented by Timoshenko exhibits that cash-based data such as payments from students and other funding possibilities prevails in the day-to-day financial planning and control at the university, suggesting that budgetary accounting wins the battle over accrual-based financial information locally, and that the accounting reform itself is rather for the state than for the university.

All the above fits squarely with a common observation that organizations gather and display huge amounts of information unused and hardly relevant for decisions, that required statements are not scrutinized, and that information is asked for after actions have taken place [9, 13, 15]. Additionally, a great deal of inconsistency and ambiguity is deemed to exist, not just between accounting and budgeting, but between the budget and actions, as well as between actions and accounting in public sector organizations [24].

5 Summary

The present paper has sought to elaborate a framework for investigating public sector (government) accounting in terms of its relationship with the traditional budget. Stemming from the discussion above, exploring the linkage between the budget and accounting is believed to provide a promising theoretical apparatus through which to examine changes in government accounting within a specific context. How are the budget and accounting related to each other in this specific context? Is the latter heavily dominated by the budget, thereby making accounting an appendix of the budget? It may well be the case, that it exerts a great deal of ascendancy over the budget, granting accounting more autonomy and independent life. Equally speaking, is the accounting information used for the purpose of control only? Perhaps, it encompasses other possibly more important functions. Does digitization of the data have any impact on the relationship between the budget and accounting, and if so, how? This is what future empirical studies of public sector accounting within various settings can in principle deal with.

References

1. American Accounting Association (AAA): Statement of Accounting Theory and Theory Acceptance (1977)
2. American Institute of Certified Public Accountants (AICPA): Objectives of Financial Statements (1973)
3. American Institute of Certified Public Accountants (AICPA): Objectives of Financial Statements, vol. 2, Selected Papers, New York (1974)
4. Anthony, R., Young, D.: Management Control in Nonprofit Organizations. Irwin, Homewood (1988)
5. Belkaoui, A.: Accounting Theory, 5th edn. Cengage Learning EMEA, London (2004)
6. Bergevärn, L., Mellemvik, F., Olson, O.: Institutionalization of accounting – a comparative study between Sweden and Norway. Scand. J. Manag. 11(1), 25–41 (1995)
7. Burchell, S., Clubb, C., Hopwood, A.G.: Accounting in its social context: towards a history of value added in the United Kingdom. Acc. Organ. Soc. 10(4), 381–413 (1985)
8. Caiden, N., Wildavsky, A.: Planning and Budgeting in Poor Countries. Transaction, New York (1974)
9. Carpenter, V.L., Feroz, E.H.: Institutional theory and accounting rule choice: an analysis of four US State governments' decisions to adopt generally accepted accounting principles. Acc. Organ. Soc. 26, 565–596 (2001)
10. Chan, J.L.: Budget accounting in China: continuity and change. In: Chan, J.L., Jones, R.H., Lüder, K. (eds.) Research in Governmental and Nonprofit Accounting, vol. 9, pp. 147–174. JAI Press, Greenwich (1996)

11. Chan, J.L.: Global government accounting principles. In: Bräunig, D., Eichhorn, P. (eds.) Public Management, Accounting Standards and Evaluation Methods. Baden-Baden, Nomos Verlagsgesellschaft (2002)
12. Chan, J.L.: Government accounting: an assessment of theory, purposes and standards. Pub. Money Manag. **23**(1), 13–20 (2003)
13. Czarniawska-Joerges, B., Jacobsson, B.: Budget in a cold climate. Acc. Organ. Soc. **14**, 29–39 (1989)
14. Davis, S., Menon, K., Morgan, G.: The images that have shaped accounting. Acc. Organ. Soc. **7**(4), 307–318 (1982)
15. Feldman, M., March, J.: Information in organizations as signal and symbol. Adm. Sci. Q. **26**, 171–186 (1981)
16. Flamholtz, E.: Accounting, budgeting and control systems in their organizational context: theoretical and empirical perspectives. Acc. Organ. Soc. **8**(2/3), 153–169 (1983)
17. Gaa, J.C.: User primacy in corporate financial reporting: a social contract approach. Acc. Rev. **61**, 435–454 (1986)
18. Gårseth-Nesbakk, L., Timoshenko, K.: The functions of accounting revisited – new meanings and directions, Chap. 9. In: Bourmistrov, A., Olson, O. (eds.) Accounting, Management Control and Institutional Development, pp. 141–156. Cappelen Damm Akademisk, Oslo (2014)
19. Høgheim, S., Monsen, N., Olsen, R., Olson, O.: Two worlds of management control. Financ. Accountability Manag. **5**(2), 163–178 (1989)
20. Jönsson, S.: Accounting Regulation and Elite Structure. Willey, Chichester (1988)
21. Laughlin, R.C., Puxty, A.G.: The decision-usefulness criterion: wrong cart wrong horse. Br. Acc. Rev. **13**(1), 43–87 (1981)
22. Lüder, K., Jones, R.: The diffusion of accrual accounting and budgeting in European governments – a cross-country analysis. In: Lüder, K., Jones, R. (eds.) Reforming Governmental Accounting and Budgeting in Europe, pp. 13–58. Fachverlag Moderne Wirtschaft, Frankfurt am Main (2003)
23. March, J., Olsen, J.: Democratic Governance. Free Press, New York (1995)
24. Mellemvik, F., Bourmistrov, A., Mauland, H., Stemland, J.: Sluttrapport for Prosjektet: Regnskap i Offentlig (Kommunal) Sektor. Høgskolen i Bodø, Bodø (2000)
25. Mellemvik, F., Monsen, N., Olson, O.: Functions of accounting – a discussion. Scand. J. Manag. **4**(3/4), 101–119 (1988)
26. Mellemvik, F.: Accounting: the hidden collage? Accounting in the dialogues between a city and its financial institutions. Scand. J. Manag. **13**(2), 191–207 (1997)
27. Miller, P., O'Leary, T.: Governing the calculable person. In: Hopwood, A., Miller, P. (eds.) Accounting as Social and Institutional Practice, pp. 98–115. Cambridge University Press, Cambridge (1994)
28. Miller, P.: Accounting as social and institutional practice: an introduction. In: Hopwood, A., Miller, P. (eds.) Accounting as Social and Institutional Practice, pp. 1–39. Cambridge University Press, Cambridge (1994)
29. Morgan, G.: Images of Organizations. Sage, Beverly Hills (1986)
30. Olsen, R.: Ex-Post accounting in incremental budgeting: a study of Norwegian municipalities. Scand. J. Manag. **13**(1), 65–75 (1997)
31. Olson, O., Guthrie, J., Humphrey, C.: Global Warning: Debating International Developments in New Public Financial Management. Cappelen Akademisk Forlag, Oslo (1998)
32. Olson, O.: Accounting in a context of financial crisis – three research questions, Chap. 12. In: Bourmistrov, A., Olson, O. (eds.) Accounting, Management Control and Institutional Development, , pp. 193–197. Cappelen Damm Akademisk, Oslo (2014)

33. Roberts, J., Scapens, R.: Accounting systems and systems of accountability – understanding accounting systems in their organizational contexts. Acc. Organ. Soc. **10**(4), 443–456 (1985)
34. Roberts, J.: The possibilities of accountability. Acc. Organ. Soc. **16**(4), 355–368 (1991)
35. Timoshenko, K., Adhikari, P.: A two-country comparison of public sector accounting reforms: same ideas, different paths? J. Pub. Budg. Acc. Financ. Manag. **22**(4), 449–486 (2010)
36. Timoshenko, K., Adhikari, P.: Exploring Russian central government accounting in its context. J. Acc. Organ. Change **5**(4), 490–513 (2009)
37. Timoshenko, K., Adhikari, P.: Implementing public sector accounting reform in Russia: evidence from one university. In: Tsamenyi, M., Uddin, S. (eds.) Research in Accounting in Emerging Economies, vol. 9, pp. 169–192. Emerald Group Publishing House, Bingley (2009)
38. Timoshenko, K.: Accountability and accounting: insights from Russian public sector reform. In: Farazmand, A. (ed.) Global Encyclopedia of Public Administration, Public Policy, and Governance. Springer, Cham (2018)
39. Timoshenko, K.: Russian public sector reform: the impact on university accounting. J. Bus. Econ. Manag. **9**(2), 133–144 (2008)
40. Vela, J.M., Fuertes, I.: Local government accounting in Europe: a comparative approach. In: Caperchione, E., Mussari, R. (eds.) Comparative Issues in Local Government Accounting. Kluwer Academic Publisher, Boston (2000)
41. Watts, R.L., Zimmerman, J.L.: Positive Accounting Theory. Prentice Hall, London (1986)
42. Welsh, G.: Budgeting: Profit Planning and Control. Prentice-Hall, Upper Saddle River (1976)
43. Wildavsky, A.: Budgeting: A Comparative Theory of Budgetary Process. Little Brown, New York (1975)

Due Diligence Planning as Technology for Business Risk Assessment

Nadezhda Antonova(✉) [iD]

Department of Accounting, Audit and Automated Data Processing,
Kuban State University, Stavropolskaya Street 149, 350040 Krasnodar, Russia
antonova_n@list.ru

Abstract. Nowadays, audit companies are introducing new professional services. There is due diligence among such audit services. Due diligence is a modern business technology, that helps to identify business-risks and make decision on how to improve business processes. The study of the problem is of interest because it helps to identify new problems and trends in the market of audit services. The study objective is to develop the risk-oriented approach methodology for planning due diligence. Based on the results of the study, special matrix for performing due diligence is presented. All stages of due diligence are described. The algorithm of risk-oriented planning of due diligence is disclosed. Due to the absence of special standards for planning agreed-upon procedures, including due diligence, the scheme and the due diligence algorithm are proposed that provide an understanding of the process of conducting an expertise of due diligence object, as well as the features of the pre-due diligence planning.

Keywords: Due diligence · Risk-oriented approach · Business risk assessment

1 Introduction

The activity of economic entities entails business risks. In order to ensure effective operation and sustainable development, the companies need information on the impact of business risks on internal structure, business processes and external economic interrelations, as well as information on financial implications for companies that may arise due to the occurrence of business risks. Auditing and consulting companies offer a modern service that allows identifying and assessing of these risks – due diligence. The procedure helps to specify various risks inherent to activities of the company, implementation of an investment project or a merger and acquisition transaction. Due diligence stands out as an expert examination, the purpose of which is to study in detail financial and non-financial data on the project (company, investment project or M&A transactions), identify possible threats and make recommendations on how to prevent them, develop proposals for change in the organizational structure and optimization of business processes, cost optimization and reduction and to come up with some conclusions on feasibility of implementing the investment project or conducting M&A transaction.

Theory and application of due diligence had already been described in works of Denison [1], Gole [2], Cumming [3], McGee [4], Ramsinghani [5], Reed [6], Sacek [7], Shain [8], Sherman [9]. Over the past ten years, some Russian scientists focused their

T. Antipova (Ed.): ICIS 2019, LNNS 78, pp. 60–71, 2020.
https://doi.org/10.1007/978-3-030-22493-6_7

papers to exploration of the subject under consideration, among them Gerasimova [10], Guzov [11–13], Zakhmatov [14], Kerimov [15], Piskunov [16], Savenkova [11, 12], Tuikina [14], Sharkov [17], Kuter [18, 19], Kondratenko [20]. Evaluation of relevant scientific sources revealed that international multidisciplinary bibliographic and abstract databases contain a lot of materials on due diligence. For example, as at October 10, 2018, Science direct electronic library contained about 19,300 articles, Wiley online library contained about 25,400 articles. At the same time, the Russian Science Citation Index contains about 200 works of Russian scientists and practitioners. This is almost 100 times less than the overall number of foreign publications. In this regard, conducting research on due diligence seems appropriate as it provides data for better understanding of the existing problems in the field of due diligence, helps to expand knowledge on the subject under consideration, creates conditions for the implementation of new ideas on identifying business risks and improving business processes of companies.

It should be noted that standardization of due diligence as a form of auditing and consulting services is still insufficient. Specific provisions, instructions and guidelines developed by audit methodologists for delivery of due diligence service and final due diligence reports containing data on critical areas of the business are unavailable for a wide number of users. Therefore, there are difficulties in understanding of the process of procedure implementation, as well as in understanding of due diligence planning. In this regard, the purpose of this study is to develop an algorithm for due diligence planning by risk-oriented approach application.

2   Materials and Methods

Auditing and consulting companies apply specific methods to perform various procedures in compliance with the requirements of the related standards. It should be noted that due diligence is a relatively young field of professional activity. At the same time, the procedure for due diligence planning is not clearly established by rules and standards, and scientists have different opinions regarding the classification of due diligence in the system of audit services.

Kerimov [15] describes essential nature of due diligence and comes to the conclusion that the service under consideration "…is complex and mixed: on the one hand, it partially refers to audit services, as it provides for application of the audit procedures… and, on the other hand, it refers to other services related to auditing activities, since it includes implementation of a range of services that are not directly related to auditing activities."

Guzov and Savenkova [11, 12] analyzed in detail the specific criteria and identified due diligence as an agreed-upon procedure, taking into account the following key features: confidentiality (due diligence report is not available to the public), users of the report (limited circle of persons defined by the contract), confidence (due diligence is not an audit, it does not ensure confidence on the assumption of credibility; the recipient of due diligence report must draw its own conclusions in order to make a decision on capital investment). In this study, due diligence is considered as an agreed-upon procedure.

Preparing this article the author analyzed results of the research of Guzov [13]. It is worth mentioning that the present research is based on risk-oriented audit, since risk-oriented approach is poorly covered by the Russian economic literature. As part of his study, Guzov [13] defines and describes in detail risk-oriented audit stages, and then demonstrates the process of implementation of the procedure. According to Guzov, planning of risk-oriented audit increases the efficiency of audit procedures and optimizes the total cost of the audit. Kuter and Antonova [19] conducted a study in which they performed a comparative analysis of audit and due diligence. The authors came to the conclusion that there are significant similarities (methods, informational basis, indicators, subjects), but at the same time there are significant differences. Therefore, due diligence should not be equated with audit. These services are completely different procedures. Based on the results of the research of Guzov [11–13], Kuter [18, 19], a Table 1 is formed.

Table 1. The concept of the due diligence procedure as modern audit and consulting services.

Element of concept	Description
1. Definition	Audit-related service that must be agreed with a potential investor to be performed as comprehensive investigation to identify and assess risks
2. Purpose	Identification, analysis and assessment of current and potential risks of the proposed capital transaction (or investment project), the proposal of measures aimed at minimizing, eliminating and preventing the risks of investing
3. Functions	Systemic, informational, complex-analytical, research, innovative, evaluative, preventive
4. Aspects	Accounting, auditing, analytical
5. Types	Accounting due diligence, financial due diligence, tax due diligence, legal due diligence, managerial due diligence, marketing due diligence, IT due diligence, environmental due diligence
6. Principles	Honesty, objectivity, professional competence and due care, confidentiality, professional behavior, future orientation
7. Key approach	Risk-oriented approach
8. Procedure provider	Expert working group - employees of audit and consulting companies
9. Subjects	Proposed capital transaction (or investment project)
10. Objects	EBITDA, Net Debt, Working Capital
11. Methods	General logical: analysis, synthesis, induction, deduction, analogy, modeling, probably statistical methods; specific: survey, observation, request, confirmation, recalculation, comparison, tracking, evaluation

Table 1 *describes the key elements of the concept of the due diligence procedure as modern audit and consulting service.*

A large amount of information on audit planning is revealed in international auditing standards, specific manuals of auditing companies, as well as in scientific articles. According to the classical view, audit consists of the following stages: planning, implementation of the planned procedures, preparing of audit results and drawing up of audit report. As noted earlier, due diligence should be classified as an agreed-upon procedure rather than audit. The implementation of agreed-upon procedures should be guided by ISRS 4400. But ISRS 4400 does not describe algorithm for due diligence planning. There is only one proposal in ISRS 4400: it is necessary to plan the procedure to ensure that the objectives will be achieved. Therefore, the specific actions of experts in the course of planning and implementation of agreed-upon procedures to which due diligence belongs are not defined and described in the standard. Obviously, it complicates the task. A question arises: what guidelines should be followed when planning and implementing due-diligence? Research of Yu. N. Guzov on risk-oriented audit was taken as a basis for development of the algorithm of risk-oriented due-diligence planning.

There are differences of opinion among scientists concerning conducting of due diligence. Ramsinghani [5] made a very interesting observation: at present time, due diligence is rarely performed by a specific algorithm. The procedure is usually carried out casually. At the same time, planning determines quality of professional services and, consequently, the success of the investment project. The purpose of due diligence is to help the customer to form a reasoned opinion on project risks by identifying, analyzing and evaluating them, and to avoid costly and sometimes detrimental consequences. It is also interesting to note the opinion of Sherman [9], who believes that due diligence requires a more creative and strategic approach, as well as a deeper study of the recipient of investments. The study should cover mission, values, culture and other aspects of the company, rather than conducting of a formal review of key contracts and other corporate documents. Denison and Ko [1] noted that conducting of due diligence should not include traditional approaches.

Taking the above into consideration, in the present study, the modern risk-oriented approach applied in the audit activity planning has been adopted as a basis for algorithm of due diligence. Application of a risk-oriented approach ensures that the experts focus on the critical most-at-risk areas of due diligent project. It allows establishing, analyzing, evaluating business risks and making recommendations and suggestions for the customers. When developing the algorithm of risk-oriented due diligence planning, the analogy method has been applied. It provides for establishing of the sequence of stages in the procedure as it is in audit. Method of scientific cognition and interpretation has been applied for an explanation of each stage of the developed algorithm for due diligence planning.

3   Results and Discussion

A specific planning tool – due diligence PLAN-matrix – has been developed as a part of the study (see Table 2). The interpretation of the process of conducting due diligence in matrix form combines both plan and strategy for implementation of the procedure. Due diligence plan is revealed through the description of the sequence of operations and actions conducted by experts. Therefore, using the matrix, one can easily determine what stages an expert group should go through during the test, and what tasks it should perform. The strategy determines the direction that should be followed in order to achieve the objectives of due diligence through the selection of procedures that allow identifying and assessing business risks of a potential investment project in order to develop proposals for optimization and reengineering of business processes (taking into account established potential risks). Just like for the auditing procedure, the strategy should be determined at the planning stage, as there are certain limitations, such as labor resources, timeframes, volumes of information to be verified, availability of the requested information and others.

Table 2. Due-diligence PLAN-matrix.

Stage	Title of stage	Step 1	Step 2	Step 3
1	P - Pre-analysis	P1. Initial meeting with the customer to familiarize experts with the objectives and specifics of due diligence project	P2. Preliminary analysis of available information in order to determine possibility of conducting due diligence	P3. Agreement of terms and conditions of due diligence
2	L - List	L1. Preparation of a checklist for collecting and analyzing of data in corresponding modules	L2. Development of a master plan, strategy determination for due diligence	L3. Preparation of separate plans for separate modules of due-diligence, selection of risk identification procedures
3	A - Assessment	A1. Development of due diligence risk map	A2. Identification of business risks	A3. Business risk assessment
4	N - Notification	N1. Preparation of separate reports for due diligence modules	N2. Preparation of consolidated due diligence report and sending it to the customer	N3. Conducting a final meeting to explain to the customer the results of due diligence

Table 2 *shows the algorithm for due diligence. It gives an opportunity to understand the whole process of conducting due diligence from beginning to the end. The scheme of due diligence given above allows distinguishing two action blocks, such as planning (stages 1 and 2) and implementation (stages 3 and 4).*

Let us consider in detail each stage of the developed algorithm of risk-oriented due-diligence planning.

Stage P1. An auditing and consulting company, a group of experts who will be responsible for due diligence project, conducts negotiations with the customer. During the meeting, main objectives of due diligence project should be determined, as well as actions in the course of further negotiations. It is important to establish a framework for mutual cooperation in order to provide experts with information necessary for conducting due diligence, especially if the information is commercially sensitive.

Stage P2. At this stage the expert working group, depending on the tasks set by the customer, carries out a preliminary analysis of the available non-financial and financial information on due diligence project, gets a comprehensive framework for understanding of the possibility of the procedure, determines form of the final report and the circle of users to whom this report is addressed, estimates and discusses costs of conducting due diligence.

Stage P3. Experts conduct additional negotiations with the customer and representatives of the due diligence project. During the negotiations, the experts discuss further questions arising from the results of the preliminary analysis, determine the order of participation of the parties in the course of due diligence implementation, clarify conditions of the previously concluded due diligence agreement in order to ensure that all parties fully understand all the terms and to prepare the final contract. The result of all negotiations is the agreement on the all essential conditions of the forthcoming investigation of the transaction and the conclusion of a contract.

Table 3. Sample checklist for "accounting due diligence".

Key questions and documents	Information obtained		Expert comments on results of investigation
	Yes	No	
1. Have audits been carried out? Audit reports for the last three years	✓		
2. Are there quarterly activity reports? Quarterly reports for the agreed check-out period	✓		
3. Are there existing financial budgets and operating budgets or business plans? Plans for the past three years	✓		
4. Were there any changes in accounting methods in the past three years? Explanation of the changes		✓	
5. Were there any significant write-offs? Write-off information	✓		
6. Are there credit/debit repayment schedules for the last three financial years? Plans and graphs	✓		

An example of a checklist for "Accounting Due Diligence" is presented in Table 3. *Due diligence checklist developed on the basis of the specific characteristics of the project is sent to the parties in order to obtain financial and non-financial information which is necessary for conducting detailed procedures and establishing business risks.*

Stage L1. Experts prepare a specific checklist that contains questions, list of necessary documents and materials for investigation of each field of interest of due diligence. The set of questions in the checklist is individual for each procedure.

Stage L2. At this stage the following key features of the project should be determined: expert group that will carry out due diligence project (see Fig. 1), time budget and duration of works, directions of due diligence on which some specific procedures should be taken, limitations and difficulties that experts may encounter when conducting due diligence.

Type of due-diligence \ Expert	Auditor	Financial analyst	Appraiser	Tax Advisor	Lawyer	Marketer	IT expert
Accounting due-diligence	■	▨	■	▨			▨
Financial due-diligence	▨	■	▨			▨	
Tax due-diligence	▨			■			
Legal due-diligence					■		
Managerial due-diligence	■		▨	▨	▨	▨	
Marketing due-diligence		▨				■	
IT due-diligence							■
Environmental due-diligence				■			

Legend: ■ responsible expert   ▨ secondary expert   ☐ another expert attracted if necessary

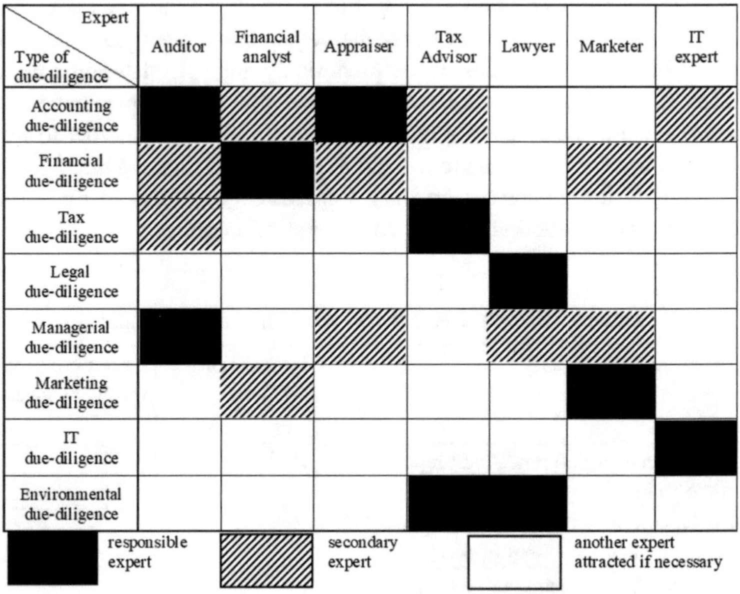

Fig. 1. Allocation of responsibilities among the members of expert team.

While pondering about the experts who should conduct due diligence, Gole [2] is of the opinion that the team should consist of specialists, each of whom is responsible for research of the project in terms of his area of activity or professional duties.

Gole [2] also notes that all specialists conducting due diligence look at the target company through the prism of their individual professional experience. To conduct effective due diligence, it is necessary to perform cross-functional analysis. That means all checking procedures should be carried out in conjunction to ensure an integrated approach, which was taken into account when developing the algorithm of risk-oriented due diligence planning (see Table 2).

In this article we propose to allocate responsibilities among the members of expert group. It helps to understand roles of experts during due diligence (see Fig. 1). One can see what areas of due diligence each expert investigates, what he checks first of all, in what fields he may intersect with other experts during the investigation, what areas are secondary for him, and in investigation of what areas he may be engaged if necessary. After determining the composition of labor recourses and the amount of time, assessment of the quality and quantity of financial and non-financial information received in the check-list, a consolidated due diligence plan is developed.

Stage L3. In accordance with the consolidated plan, responsible experts prepare detailed due diligence plan for specified areas. Such plans allow you to specify a range of applicable procedures and focus on identification of established risks. After establishing circumstances that affect due diligence conducting, it should be determined whether the strategy of investigation process needs to be adjusted. Besides, the areas of responsibility should be adjusted as well, and the workload should be redistributed.

Stage A1. After obtaining data specified in the checklist (see Table 3), experts conduct data analysis. An object risk map is prepared with regard to the targeted values (see Fig. 2).

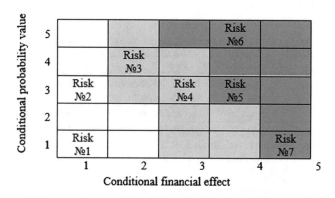

Fig. 2. Sample risk map for due diligence project.

Table 4 *shows that risks on the map are evaluated by experts basing on the possibility of their occurrence and degree of financial effect on the due diligent project. Values of conditional possibility and values of financial effect of business risks are established taking into account the specifics of the due diligence project.*

Table 4. Example of interpretation of conditional values for risk assessment of due diligence project.

Conditional probability value	Expert assessment of risk probability, %	Conditional financial effect	Expert assessment of financial effect of risk, mln cu
1	10	1	10
2	30	2	20
3	50	3	30
4	70	4	40
5	90	5	50

For example, experts will indicate in the report that the risk №4 associated with equipment breakdown and IT-system disruption will likely occur, since its evaluated probability is 50%. But even if the risk №4 occurs, its financial effect in losses related to the due diligent project will not exceed 30 million cu. Therefore, application of the risk map and explanations of experts given in the final report will help the customer to indicate critical areas of the due diligence project.

Stage A2. Experts carry out special procedures for each module of due-diligence plan (Table 5). As a result of the implementation of the procedures, areas requiring special attention are specified and business risks are identified. This is the way of realization of a risk-oriented due diligence planning. After identification of critical areas experts compare the business risks with the typical ones and determine their types. Basing on the results of the identification, experts offer risk response measures and describe due diligence recommendations on the reengineering of the business processes of the subject under investigation.

Table 5. Modules of risk-oriented due diligence planning.

Module name	Due diligence procedure
Accounting due diligence	Analysis of the state of the accounting system and the internal control system, analysis of audit reports, etc.
Financial due diligence	Analysis of the reasonableness of equity to debt capital, review of interest due calculation, verification of compliance with covenants, etc.
Tax due diligence	Analysis of the state of the tax accounting system, review of calculations of tax payments, assessment of the results of tax audits, etc.
Legal due diligence	Analysis of constituent documents, related contracts and agreements, verification of availability and validity of licenses and permits, analysis of information on current court proceedings, etc.
Managerial due diligence	Assessment of the implementation of human capital investment programs, efficiency assessment of the company management, efficiency assessment of different categories of employees, analysis of the collective agreement, etc.
Marketing due diligence	Product analysis, customer analysis, supplier analysis, competitiveness assessment, etc.
IT - due diligence	Performance analysis of IT system and communication channels, verification of system maintenance, proficiency examination of operating personnel, etc.
Environmental due diligence	Verification of compliance with environmental requirements, verification of environmental tax assessment and payment, etc.

Table 5 *shows modules of due-diligence plan. Team of experts carries out special procedures for each module based on risk-oriented approach. As a result expert identify critical areas and potential business risks and recommend risk preventive measures.*

Stage A3. After identification, the experts analyze and evaluate the impact of the identified business risks. The analysis establishes risk factors, or causes of risks, and the expected consequences. At the assessment stage, experts should determine the following: financial implications of business risks, how dangerous they are for the due diligence project, how they can affect the existing structure of the business processes.

Stage N1. After carrying out the planned procedures aimed at identifying business risks and business risk assessment for each module of due-diligence program, the experts prepare separate reports. When preparing individual reports, experts can formulate proposals for reengineering of business processes, bearing in mind the existing business risks. The implementation of measures to reengineer the business processes of an investment project will minimize business risks by making the processes more transparent and understandable and optimizing the costs.

Stage N2. Consolidated business risk report on the due diligence project should be based on these individual reports. An important supplement to the report is a risk map (Fig. 2), which shows the identified risks considering the impact on the business processes of due diligence project. The consolidated report may also include proposals for optimization and reengineering of business processes.

Stage N3. Based on the results of the procedures, a final working meeting with the customer is organized. At the meeting, the results of the assessment of the impact of the identified business risks and conclusions developed by experts based on the analysis of financial and non-financial information should be discussed. The customer receives all the necessary explanations and recommendations, as well as answers to all his questions, that are required for taking effective management decisions regarding the due diligence project.

4   Conclusion

Guzov developed a risk assessment algorithm for due diligence objects (EBITDA, Net Debt, Working Capital) in the study [12]. This algorithm is specific, it allows you to understand the main actions of group of experts when analyzing the impact of risks on the due diligence objects. Zakhmatov and Tuikina [14] did not offer a specific algorithm for due diligence planning, but they described in detail certain features of the group of experts that should be considered in practical interaction with investors and the management of the company under study.

In this paper the author has developed an algorithm for planning due diligence procedure by adopting a risk-based approach, which complements the algorithm of Guzov [12] on the risk assessment of due diligence objects and expands the presentation of Zakhmatova and Tuykina [14] on the sequence of actions of the group of experts during the due diligence.

Matrix consisting of four stages was taken as a main planning tool for due diligence. Each stage of the procedure was described on the basis of the matrix, which, as a result, forms the algorithm of risk-oriented due diligence planning. The tables worked out in the context of the study allow us to understand the key features of the preparatory work conducted by the experts and to get a general idea of the planning and implementation of due diligence procedure. Subjects of further research on the issue under consideration can be as follows:

– development of documentation for the purposes of due diligence standardization;
– development of technical solutions for implementation of due diligence procedure to expand business opportunities and create new development prospects;
– development of IT-system for due diligence implementation.

References

1. Denison, D., Ko, I.: Cultural due diligence in mergers and acquisitions. Adv. Mergers Acquisitions **15**, 53–72 (2016)
2. Gole, W., Hilger, P.: Due Diligence: An M&A Value Creation Approach, 1st edn. Wiley, New Jersey (2009)
3. Cumming, D., Zambelli, S.: Due diligence and investee performance. Eur. Financ. Manage. **23**(2), 211–253 (2017)
4. McGee, J.A., Byington, J.R.: Due diligence issues in China. J. Corp. Acc. Finan. **20**(2), 25–30 (2009)
5. Ramsinghani, M.: The Business of Venture Capital: Insights from Leading Practitioners on the Art of Raising a Fund, Deal Structuring, Value Creation, and Exit Strategies, 2nd edn. Wiley, Hoboken (2014)
6. Reed, A.: M&A due diligence in the new age of corporate governance. Ivey Bus. J. **71**(3), 1–10 (2007)
7. Sacek, A.: Critical factors of pre-acquisition due diligence in cross-border acquisitions. Contemp. Stud. Econ. Financ. Anal. **98**, 111–134 (2016)
8. Shain, R.: Hedge Fund Due Diligence: Professional Tools to Investigate Hedge Fund Managers. Wiley, Hoboken (2008)
9. Sherman, A.J., Welch, J.E.: Due Diligence in a new era of accountability. J. Corp. Acc. Finan. **20**(6), 65–70 (2009)
10. Gerasimova, A.V.: Methods of analysis and risk assessment on financial due diligence. Int. Res. J. **1**, 28–36 (2016)
11. Guzov, Yu.N., Savenkova, N.D.: Due diligence: theory and practice. Audit J. **7**, 56–69 (2015)
12. Guzov, Yu.N., Savenkova, N.D.: Due diligence: identification and assessment of risks associated with the acquisition of an investment object. Audit Financ. Anal. **2**, 201–206 (2013)
13. Guzov, Yu.N., Strelnikova, O.V.: Planning practice in a risk-focused audit. Audit J. **1**, 29–41 (2015)
14. Zakhmatov, D.Iu., Tuikina, Yu.R.: Role of the appraiser in preparing purchase the business and assets through comprehensive audits (due diligence). Property Relat. Russ. Fed. **9**, 46–51 (2015)
15. Kerimov, V.V.: Due diligence as a special audit procedure and intellectual property as an object of intellectual property. Econ. Manage. Probl. Solutions **3**, 47–51 (2014)
16. Piskunov, A.I.: Conducting due diligence in mergers and acquisitions. Topical Issues Econ. Sci. **11**(1), 130–133 (2010)
17. Sharkov, D.A.: Due diligence: a fundamental procedure for structuring mergers and acquisitions. Probl. Mod. Econ. **1**(53), 123–125 (2015)

18. Kuter, M.I., Sokolov, V.Y.: Russia. In: A Global History of Accounting, Financial Reporting and Public Policy: Eurasia, the Middle EST and Africa. pp. 75–106. Emerald Group Publishing Limited, Sydney, Australia (2012)
19. Kuter, M.I., Antonova, N.A.: Benchmarking study of due diligence and audit. Audit J. **4**, 3–16 (2016)
20. Kondratenko, D.V., Muzhylivskyi, V.V.: Due diligence as a comprehensive system for protecting against business risks and building a strategy. Bull. Econ. Transp. Ind. **52**, 37–45 (2015)

Liquidation Financial Reporting of Russian Companies in Terms of Digital Economy

Ruslan Tkhagapso$^{(\boxtimes)}$ ⓘ, Mikhail Kuter ⓘ,
and Anastasiya Trukhina ⓘ

Kuban State University,
Stavropolskaya St., 149, Krasnodar 350040, Russian Federation
rusjath@mail.ru

Abstract. The paper presents a critical analysis of the formation and disclosure of liquidation reports of Russian organizations in the context of an active transition to a digital economy. The problems, developing with a significant decrease in the effectiveness of accounting data in terms of the effectiveness of decision-making process conducted by financial statement users in the modern information society, predetermined the fundamental research in the field of the theory and practice of accounting and auditing.

Under these circumstances, the issue regarding the formation of high-quality liquidation reports to satisfy the information needs of creditors and external users of a bankrupt organization within the framework of the economy is no less significant. The scientific article discusses the main directions of modernization of financial (liquidation) reporting in the context of the information society. The paper provides the analysis of the current requirements for reporting indicators, as well as recommendations on the formation and disclosure of individual indicators of the financial state of a bankrupt organization.

Keywords: Bankruptcy · Digital economy · Information society ·
Digital reporting · Liquidation balance sheet

1 Introduction

Global changes in the value of information in the social life have been the main trend of Russia's social and economic development and have identified the transition to the information society and the formation of the digital economy in the country.

At the present day the financial reporting is nearly the only source of information available, on the basis of which certain conclusions and decisions can be made by both the management and external users. Poor quality and falsity of financial information can result in serious implications for its users – investors, owners, creditors, government authorities, etc.

The problems of formation and disclosure of accounting information in financial statements have concerned a lot of researchers. The problems of balance sheet formation in accordance with the goals pursued were first examined in the works of Savary [1]. Sher [2] wrote about the dominant role of the balance sheet. Scher regarded the balance as the basis of accounting. "The initial balance and the final balance, – he wrote, – make up the

T. Antipova (Ed.): ICIS 2019, LNNS 78, pp. 72–81, 2020.
https://doi.org/10.1007/978-3-030-22493-6_8

alpha and omega of all accounting" [2]. Accounting begins with a balance and ends with the liquidation of the organization final balance. Johann Scher called imperfect and inappropriate according to economic and legal characteristics any accounting, which is not built from the opening balance sheet. Johann Scher regarded the assessment as one of the main characteristics of the reliability of the balance compiling.

Betge [3], Oberbrinkman [4] can be noted among the modern researchers of financial statements. Kuter and Gurskaya [5, 6] examine the problems of improving financial reporting from the perspective of its historical development. Controversial issues of the formation and disclosure of corporate reporting indicators in Russian practice are examined in the works of Efimova, Rozhnova and Antipova [7, 8].

2 Research Results

Nowadays, many experts note that the content of the information disclosed by financial statements, which is formed by traditional accounting systems limits the decision-making ability of both internal and external users, and at the same time makes it significantly more difficult to assess investment prospects and market cost of commercial organizations.

With the increasing complexity of the economic relations in the modern society, there has been a need for actuarial calculations which can justify the long-term financial relationship between the participants.

For a long time, there was a legal void regarding the Russian accounting system and the formation of indicators of the financial statements of organizations to be liquidated, and in 2013 only with the introduction of the Federal Law "On Accounting" [9], and especially Art. 17, the situation has changed. It should be noted that in Art. 17 minor attention is paid to the necessity and frequency of reporting in the context of business liquidation. In this regard, the issues of formation and disclosure of indicators of liquidation financial statement are extremely relevant to this day.

From the conceptual point of view, the process of liquidation of an organization implies the rejection of one of the presumptions – the *going concern* basis of accounting. The impossibility of the practical application of this principle in accounting and compiling reports leads to certain changes in the methodology and organization of accounting.

In accordance with the principle of a *going concern* during the liquidation of an organization, special rules (which distinguish liquidation balance sheets from others) of property valuation of an economic entity based on *net realizable value* of each asset, come into force. According to the theory of static balance of Heinrich Niklisch, the balance must give an exact calculation of the value of the assets of the enterprise, and for this purpose the balance sheet must give the account of the possible price of realization of the assets on the date of the balance sheet formation [5, p. 243].

Liquidation balance is one of static balances and deals with the problems of property nature: the detailed reflection of the property of an organization, stating the owners of the property and liabilities of the organization. The basis of the static concept of the balance is the periodic evaluation of the property and the verification of the adequacy of the funds received from the fictitious sale of the organization's assets to cover the creditors' claims. Thus, the main goal of a static balance is reduced to the protection of creditors' rights, the preparation of objective information on the state of property and debts, which is also the main goal of the bankruptcy procedure.

According to Le-Kutre, the balance shows the assets and liabilities of the enterprise in a state of instant rest and therefore is static in its nature. Manfred Berliner, one of the contemporaries with Johann Scher, proposed to consider each balance as a liquidation balance, where the right side of the balance is the fractional distribution of an organization's property. After M. Berliner, the asset was considered as a set of values to be distributed according to the plan presented in the liabilities side of the balance-sheet. R. Chambers defined the assets side of the balance sheet as a set of objects, each of which could be removed from the enterprise. Therefore, goodwill, deferred expenses, losses cannot be included in the asset.

Accounting for liquidation of an organization involves several interrelated steps:

(1) reflection of the business operations conducted by the organization from the moment of the decision to liquidate it till the moment when the liquidation balance sheet is compiled;
(2) preparation of the interim liquidation balance sheet;
(3) reflection of operations regarding the satisfaction of creditors' claims;
(4) preparation of the final liquidation balance sheet.

The result of each stage of the liquidation of an economic entity must be reflected in the interim and final liquidation balance sheets, the need for which is stipulated in the Russian civil law [10].

It is possible to note some distinctive features of the formation of liquidation balance sheet in comparison with current financial statements.

Firstly, they regard the liquidation balance as an inventory, i.e. compiled from the inventory data. At the same time, a statement on the need to have ownership of the property represented in the balance sheet is made, as the property should have a real ability to be sold.

Secondly, based on the limited life span of an organization, the authors propose to change the order of use of regulatory and control bookkeeping account.

Thirdly, they emphasize the existence of differences in the way the liquidation assets are estimated, since the liquidation process determines the value of the property that enables users of financial statements to calculate the coverage of claims with the maximum precision.

Fourthly, the authors highlight the need to apply a new grouping of 'Asset' and 'Sources' articles, which will be corresponding to the actual degree of property liquidity and the established procedure for satisfying creditors' claims, as a distinctive feature. In the Asset of liquidation balances, the things of value should be presented separately, those intended to cover existing requirements, and those accounting objects that ensure the activity of the liquidation commission (liquidation costs) or are subject to cancellation in the prescribed manner due to the impossibility of their implementation; and the list of submitted and not submitted claims of the creditors of an organization of bankruptcy should be presented in Sources.

Meanwhile, the current financial reporting forms do not provide exhaustive information on the liquidation procedure, proved debt to creditors and satisfaction of them at the expense of the organization's assets, as they are aimed at a normally operating enterprise.

The data presented in the current reporting forms do not solve the problems of providing information to the managers of a bankrupt organization during the liquidation procedure, do not reflect the interests of the organization's creditors, since they do not provide exhaustive information about the possibility of satisfaction of debts at the expense of the organization's assets. This fact stipulates the necessity in improvement of the use of information potential of the current reporting based on the creation of its derivatives, which would serve various management objectives.

The development of computer technology, means of communication, global networks and, subsequently, the digital economy made it possible to digitally process input accounting information and use software applications for accounting and financial reporting. Active digitalization of the economy offers exciting possibilities of using software for the implementation of complex techniques used in the formation of financial statements.

The interim liquidation balance sheet is one of the main forms of accounting reports of the liquidation commission for the whole period of the liquidation process in terms of completeness of information, content and functions performed.

The process of formation of bankruptcy assets in the interim liquidation balance and its subsequent analysis aimed at assessing the property complex of an economic entity and determining the proportion of property that can provide real money to cover the debts of a bankrupt organization must go through several stages.

The first stage of determining the asset of the interim liquidation balance sheet is the adjustment of the asset balance of the bankrupt organization by excluding property which does not belong to it by right of ownership, lost valuation due to the opening of liquidation process, etc. Formation of assets of the interim liquidation balance sheet is conducted with taking into account the degree of liquidity of each component.

Today, in the Russian system of accounting and reporting, the formation of liquidation balance sheet indicators remains a burning issue.

The authors believe that under the crisis circumstances of operation, *the asset balance sheet items arrangement in accordance with the degree of liquidity decrease is more consistent and informative for the creditors of the company being liquidated.*

When compiling an interim liquidation balance sheet, one should take into account that the bankruptcy estate is the property that is not only available at the time of the beginning of the procedure, but also revealed during the whole liquidation procedure. It is common knowledge that sometimes the accounting documents do not reflect all the property of a debtor due to intentional or unintentional violations in accounting. Therefore, the next step in the formation of the interim asset liquidation balance sheet is an inventory of property. In these situations, the active use of digital technologies can improve the quality and reduce the time spent on operations.

When inventorying the assets of a bankrupt organization, special attention should be paid to receivables in order to identify the share of uncollectible receivables, which should later be excluded from the interim liquidation balance sheet, since it cannot be considered as a source of covering payables. In this regard, receivables which statute of limitations has expired are written off to the profit and loss statement of the bankrupt organization.

It is possible to estimate the degree of probability of reimbursement of funds by the debtor of the organization by analyzing its financial condition and examining the contracts on the basis of which the debt arose. While studying the contracts of the economic entity, special attention should be paid to the subject of the contracts, the payment deadlines for a product, works, services provided, and the terms of the contracts. The assessment of receivables should be made in terms of the time passed from the moment when the contract was signed and how long the payments for this contract have been delayed for.

When analyzing the financial condition and the ability of debtors of a bankrupt organization to discharge their debts, it is necessary to identify the receivables for which real money is practically impossible to obtain due to various circumstances. For example, due to the termination of business activities of the debtor.

Based on the above, the authors offer a variant of the interim liquidation balance sheet of a bankrupt organization, which is presented in Table 1.

Table 1. Interim liquidation balance sheet

Asset		Sources	
Name of balance sheet items	Sum	Name of balance sheet items	Sum
I. Bankruptcy (liquidation) assets		**III. Liabilities**	
Cash on hand and account with banks		Current (Extraordinary) claims	
Financial investments		Claims of first-ranking creditors	
		Claims of second-ranking creditors	
Accounts receivable		Claims of creditors secured by a charge over the assets	
Inventories		Requirements in terms of financial sanctions	
Long-term assets		Other creditors claims	
		Undeclared claims	
II. Exceptions from bankruptcy assets		**IV. Equity**	
Legal expenses		Equity of organization	
Other expenses liquidation		Retained earnings (undistributed loss) before bankruptcy	
Excluded assets		Retained earnings (undistributed loss), after recognition of the organization as bankrupt	
Total assets	Σ	**Total equity and liabilities**	Σ

The "Asset" of the balance sheet represents bankruptcy estate, due to which creditors' claims will be satisfied, as well as the exceptions from the bankruptcy estate, represented mainly by the costs of liquidation process and illiquid intangible assets.

The "Sources" in the liquidation balance sheet are represented by the claims of creditors, which are arranged according to the priorities of the claims. The matter of discussionis the value of the property, on which it should be reflected in the interim liquidation balance sheet. The debtor's accounting data on the value of the property does not provide a realistic assessment of the amount of funds that will be received in the course of its realisation, and consequently, the amount of satisfaction of creditors' claims. In this case, the authors believe, it is more acceptable to evaluate the property as its realizable value in the interim liquidation balance sheet.

The process of forming the indicators of the "Sources" of the interim liquidation balance sheet of the bankrupt organization and its subsequent analysis should be aimed at *forming a list of creditors' claims and assessing the company's equity*. One of the main tasks of the bankruptcy procedure is to identify all the claims of the creditors by compiling an inventory of the liabilities of the organization and thus determining the entire set of claims that may be submitted to the bankrupt organization.

The vertical interrelationships of the "Asset" items of the interim liquidation balance sheet have a direct influence on the arrangement of the "Sources" items of the balance sheet. In this case, when forming items of liabilities, the authors consider it more appropriate *to proceed from the order of discharging creditors' claims by the means of available assets*. Due to the primary nature of the claims for monetary obligations and obligatory payments in the proposed version of the liquidation balance sheet, they are reflected first and foremost, and therefore, the residual nature of the claims of owners rationalize their reflection in the fourth section of the balance sheet.

After compiling of the inventory of the liabilities of the bankrupt organization, determining the volume of claims, cancellation of unrealistic liabilities of the bankrupt organization, grouping of the liabilities becomes possible.

"Sources" of the interim liquidation balance sheet include obligations arising prior to the composition in bankruptcy, as well as liabilities accrued but not paid for during the bankruptcy proceedings. Debt outstanding takes priority over the other liabilities and represents the current debt on payments to the budget and extra-budgetary funds, current salary debt, current debt to third parties. A distinctive feature of the "Sources" of the interim liquidation balance sheet is the absence of the division of liabilities into long-term and short-term ones, since with the opening of the bankruptcy proceedings, the deadline for fulfilling all obligations is considered to have occurred.

After the adjustment of the "Asset" and "Sources" of the balance sheet, aimed at identifying the claims of creditors and the bankruptcy estate of the debtor, which will cover the claims of creditors, an interim liquidation balance sheet is compiled.

According to the authors, *the capital of a bankrupt organization represents the residual interest of the owners*, which must be discharged in the last turn after the full repayment of the claims of all previous groups of creditors. Thus, according to the Russian civil law [10], the property of a legal entity remaining after satisfaction of the claims of creditors is transferred to its founders. In this connection, it would be more correct to reflect in the interim liquidation balance sheet the ownership capital in the amount actually paid and with allowances made for the cancelled shares repurchased by the organization.

The accumulated losses in the interim liquidation balance sheet is a balancing item, and it is necessary to separate the loss formed before the bankruptcy procedure was introduced from the loss incurred during the bankruptcy proceedings.

In order to increase the informative content of the interim liquidation balance sheet for users, an explanatory note is recommended to be attached with the following notes:

- the information regarding the property assets of the organization, including such indicators as the identification of the asset, the book value of the asset, and the liquidation value of the asset;
- the records of creditors' claims.

Interim liquidation balance sheet acts as a link between the accounting in the organization before bankruptcy and accounting in the course of bankruptcy proceedings. Since the liabilities are grouped according to the sequence of their repayment, and the property is reflected at the realizable value, the balance makes it possible to realistically assess the probability of repaying the claims of the creditors. To analyze the means of organization and the reality of the timing and degree of debt repayment, it suffices to compare the funds on the balance sheet asset with the liabilities. Preparing documents for an interim liquidation balance sheet, one should conduct an in-depth analysis of the financial condition of the organization at the liquidation stage in order to be able to take all measures to fully pay off debts and prevent litigation proceedings with lenders, which can significantly increase the liquidation period.

The main users of the information mentioned in the interim liquidation balance sheet are the organization's creditors, who are interested in the real market value of the bankrupt organization's property, as well as the amount of profits and losses that the company received during the period of bankruptcy proceedings, especially during the inventory period and its property valuation. This interest is caused, in our opinion, not only by the assessment of the possibility of creditors obtaining funds in the order of priority, but also by monitoring the activities of the bankruptcy trustee.

After compiling an interim liquidation balance sheet, the bankruptcy trustee sells the organization's assets included in the bankruptcy property, and, based on the funds received, makes settlements with creditors in the established order. After the settlements, the liquidation commission compiles the final liquidation balance sheet of the bankrupt organization.

The purpose of compiling the final liquidation balance sheet is *to show the losses incurred by the owners and creditors of the organization, as well as the results of changes in the property structure and the sources of its formation since the preparation of the interim liquidation balance sheet* [11]. The liquidation balance sheet provides information on the results of the bankruptcy proceedings, on the results of actions to collect and sell property and satisfy the creditors' claims, on the amount of unsatisfied claims. Liquidation balance sheet is the balance sheet of the organization, compiled after the completion of settlements with creditors, which reflects the data on the results of the liquidation procedures.

The main requirement for the preparation of accounting reporting of a liquidated debtor is its comparability with other periods, i.e. the data at the beginning of the period should correspond to the data of the interim liquidation balance sheet in order to provide a clear picture of the effectiveness of the bankruptcy proceedings. Since the final liquidation balance sheet is compiled after satisfying the creditors' claims, the total of the liquidation balance asset at the end of the reporting period should be equal to zero.

Before compiling the final liquidation balance sheet, all property must be sold or disposed of, and the receivables collected. There should not be any property in the asset balance after making all payments. Even if the property was offered for sale, but was not sold yet and the rights of the participants are not claimed, the authority of municipal structure should be notified of the property, accept it on their balance and bear the cost of its maintenance.

The compilation of financial statements of a bankrupt organization during the bankruptcy proceedings must meet the requirements and interests of those interested in or participating in the liquidation of the enterprise. However, the final liquidation balance sheet of a bankrupt organization reflects the final result of the bankruptcy proceedings only, i.e. the remaining amount of unsatisfied creditors' claims and losses of the organization, while creditors and the arbitration court are interested in information on the course of the bankruptcy proceedings, i.e. the amount of money received from the sale of the bankruptcy assets of the organization in liquidation.

The value of the property of an organization at the time of the preparation of the interim liquidation balance sheet and the value of the property sold may later be different. It is not reflected on the liquidation balance sheet and so does the information on the number of satisfied creditors' claims, as well as the funds spent by the bankrupt organization on the bankruptcy procedure itself.

Cash included in the bankruptcy assets of the debtor in liquidation is formed only through the sale of these assets, as well as the income received from financial and economic activities conducted by the bankrupt organization until the opening of the bankruptcy procedure. Therefore, in reporting the debtor in liquidation, attention should be paid to the reflection of cash flows, which include:

(1) revenue from the sale of products (goods, works, services);
(2) profit from the sale of fixed assets and other property of the organization;
(3) dividends, interest on financial investments received by the organization during the period of bankruptcy proceedings;
(4) other income, which includes: the amount of money returned by accountable persons, borrowers; amounts of penalties received for violation of the terms of business contracts, etc.

The proposed form of the final liquidation balance sheet is presented in the Table 2.

In accordance with the accepted nomenclature of asset and liability items, liquidation balances contain general data on the value of economically homogeneous objects. For disclosure of material items in Russian and international practice, notes to the balance sheet in the form of a separate form (appendix, transcript) or as part of an explanatory note (directors' report, financial review) are used.

Table 2. Final liquidation balance sheet

Asset		Sources	
Name of balance sheet items	Sum	Name of balance sheet items	Sum
I. Bankruptcy (liquidation) assets		**II. Liabilities**	
		Outstanding claims	
		III. Equity	
		Equity of organization	
		Undistributed loss	
Total assets	Σ	**Total equity and liabilities**	Σ

3   Conclusion

For effective decision-making in the present conditions, the process of forming reliable economic information becomes a top priority. The largest share in the total aggregate of economic information is information provided by accounting, which plays a dominant role in the management system as a whole. Accounting is the link connecting the financial and economic activities of both the economic entity and its officials, who make management decisions.

The research showed that the impact of the development of the digital economy on the process of forming and using financial statements cannot be overestimated. Moreover, the digital economy undoubtedly has a positive effect on the quality of the information compiled in the financial statements.

By means of the use of computer technology, the information contained in the financial statements can be presented not only on paper but in electronic form as well. Reporting forms that are posted electronically on the organization's websites represent a set of data on the organization's performance indicators in digital form, which means that they are a necessary element of the digital economy. The process of expanding the digital economy provides great opportunities for improving the quality of reporting and its consumer properties, which are necessary for all users of reporting organizations in the modern conditions.

Modern users of financial statements strive not only to understand how indicators of financial statements were obtained, but to analyze the activities of an organization on the basis of indicators of its statements as well. Financial statements should reflect the realities of the economic life of a bankrupt organization as accurately as possible and be a reliable source of information for all bankruptcy creditors.

The compilation of a digital liquidation report of a bankrupt organization under the conditions of bankruptcy proceedings must meet the requirements and interests of persons interested in or participating in the liquidation of an enterprise.

References

1. Savary, J.: Le parfait négociant ou instruction générale pour ce qui regarde le commerce … et l'application des ordonnances chez Louis Billaire…; avec le privilège du ROY. (Reproduction en fac similé de la 1ère édition par Klassiker der Nationalökonomie, Allemagne) (1993)
2. Schar, J.F. Buchhaltung und Bilanz. Auf wirtschaftlicher, rechtlicher und mathematischer Grundlage für Juristen, Ingenieure, Kaufleute und Studierende der Betriebswirtschaftslehre. Berlin (1921)
3. Baetge, J.: Bilanzen. überarbeitete Auflage. Düsseldorf: IDW-Verlag GBH (2000)
4. Oberbrinkmann, F.: Modern understanding of the balance sheet: Trans. with him. Ed. I'M IN. Finance and Statistics, Sokolov, Moscow 416 p. (2003)
5. Kuter, M.I.: Introduction to Accounting: A Textbook, p. 512. Krasnodar, Prosveshchenie-Yug (2012)
6. Kuter, M., Gurskaya, M., Andreenkova, A., Bagdasaryan, R.: The early practices of financial statements formation in medieval Italy. Account. Historians J. **44**(2), 17–25 (2017). https://doi.org/10.2308/aahj-10543
7. Efimova, O., Rozhnova, O.: The corporate reporting development in the digital economy. In: Antipova, T., Rocha, A. (eds.) Digital Science. DSIC18 2018. Advances in Intelligent Systems and Computing, vol. 850, pp. 71–80. Springer, Cham (2019)
8. Antipova, T.: Digital view on the financial statements' consolidation in Russian public sector. In: Antipova, T., Rocha Á. (eds.) Information Technology Science. MOSITS 2017. Advances in Intelligent Systems and Computing, vol. 724, pp 125–136. Springer, Cham (2018)
9. About accounting: Federal law of Russian Federation dated 6 December 2011 No402-FZ with changes and amendments
10. Civil Code of the Russian Federation (Part One): Federal law of Russian Federation date 30 November 1994 No51-FZ with changes and amendments
11. Kuter, M.I., Tkhagapso, R.A.: Accounting in Insolvency: A Textbook, p. 204. Kuban State University, Krasnodar (2005)

Macroeconomic Model of Banking Digitization Process

Irina Toropova[1] ⓘ, Anna Mingaleva[1](✉) ⓘ, and Pavel Knyazev[2] ⓘ

[1] GSEM, Ural Federal University Named After the First President of Russia
B. N. Yeltsin, Yekaterinburg 620002, Russia
mingaleva.ann@yandex.ru
[2] FSBEI of HE "Ural State Economic University",
Yekaterinburg 620219, Russia

Abstract. The widespread use of digitization in the field of credit and banking activities is based on the fact that it accelerates and reduces the cost of providing relevant services and also increases the security of operations for both banks and their customers. Among the most obvious advantages of banking digitization customers highlight the possibility of remote account management. However, this advantage is associated with serious problems. First of all, it is various embezzlement of funds from bank accounts also with the help of digital technologies, conducting fraudulent operations.

The aim of this research work is to analyze implications of credit and banking sector digitization through macroeconomic modeling.

A significant increase in the number of crimes committed with the help of digital devices and a multiple increase in the amount of damage from them were revealed in the article based on an empirical study. The cumulative negative impact of crimes committed with the help of digital technologies and digital devices was revealed not only on individual banks and the banking system as a whole, but also on the possibilities of the country's social and economic development.

Keywords: Digitization of credit and banking sector · Macroeconomic model · Digital devices · Computer crimes · Credential theft

1 Introduction

In the past few years, serious changes have happened in the financial and banking sector under the influence of ICT.

Firstly, it is the replacement of cash by non-cash, which become electronic records on magnetic and optical media.

Secondly, mass computerization and automation of management processes and accounting of operations within the bank, digitization of many operations, including customer identification. For example, a speech recognition system for clients' voice identification was embedded in the mobile application of the British bank HSBS in 2017.

Thirdly, settlements and payments of banks with other banks, corporate clients and individuals are carried out on the basis of ICT. Thus, HSBS has 90% of bank

T. Antipova (Ed.): ICIS 2019, LNNS 78, pp. 82–90, 2020.
https://doi.org/10.1007/978-3-030-22493-6_9

interactions with customers on digital channels. In February 2017, the UAE opened the first fully digital bank in the world - Liv. This is a bank that has no physical branches at all. Juniper estimates that in 2021 already 3 billion people on Earth will use banking services on smartphones and computers. Juniper's Digital Transformation in Banking Readiness Index identified the "leading banks for digital transformation in the sector: Banco Santander, Bank of America, Barclays, BBVA, BNP Paribas, Citi, HSBC, JP Morgan Chase, RBS, Société Générale, UniCredit and Wells Fargo" [1].

The introduction of ICT in the sphere of banking activity was justified not only by the fact that it speeds up and reduces the cost of providing relevant services: "Mobile Money Applications are thriving due to the ease and convenience it brings to people, where it offers to transfer money between people's bank accounts/cards with a few taps on a smartphone either in the form of Mobile Banking or Mobile Payment Services" [2]. The introduction of ICT in the sphere of banking activity was justified also by the fact that it increases the security of operations for both banks and their customers. However, the last thesis is increasingly being questioned, both in practical terms and through theoretical modeling of processes. "However, a key challenge with gaining user adoption of mobile banking and payments is the customer's lack of confidence in the security of the services, and that makes much sense because whenever people grant service access to their debit/credit cards or bank accounts that automatically opens the door for identity thefts, fraudulent transactions, and stolen money" [2].

In order to verify this thesis, we will further conduct a study of the banking system security against various kinds of embezzlement, fraud and other crimes that are committed specifically in the digital environment with using digital devices and which are already being researched in special literature [3–5]. Then, using macroeconomic modeling, we will analyze the impact of increasing costs of banks on rising their digital security in terms of their impact on the banking system as a whole and on the possibilities for the socio-economic development of society.

2 Digital Security of Banks in the ICT Environment

We will begin the study of the banking system security from various kinds of embezzlement, fraud and other crimes that are committed in the digital environment and with the help of digital devices with the analysis of statistics. To this end, several of the most prominent examples of such crimes are presented in the article. Figure 1 shows the individual major bank robberies associated with the theft of cash and electronic money (compiled by the authors).

Moreover, if in the "pre-digital" era, the largest thefts occurred once in several years, then in the "digital" era they occur annually and several significant cases per year in different parts of the world, and the amount of damage they cause exceeds the total damage from cash theft.

However, the negative consequences of the digitization of banking operations are not only in resonant thefts of money and personal data [6].

A huge number of crimes are committed in such a field of banking operations' digitization as the use of digital devices for remote account management [7]. In the case when the banks' clients are individuals, the instruments of such operations are plastic

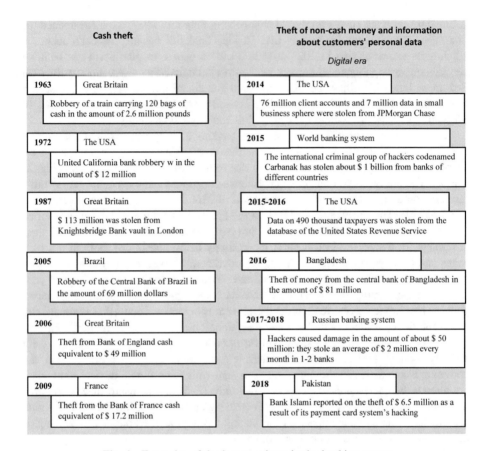

Fig. 1. Examples of the largest crimes in the banking sector

cards, mobile phones, computers and other technical means of communication through the Internet (mobile banking) [8–10].

All these devices are also a tool for theft, although the cumulative damage to individuals is less than that of companies. According to statistics, in Russia in 2017 the average amount of theft for legal entities was 1.6 million rubles, for individuals - 75 thousand rubles.

Among the methods of electronic money theft from people and enterprises Internet fraud has recently been the most widely used. Also, money theft from the bank accounts of individuals occurs through the hacking of remote banking systems, i.e. unauthorized person's access to protected information with the help of basic digital user identification tools (private key, private algorithm, digital certificate, user accounts (passwords), subscribers, or other protected items that allow to verify the identity of the information exchange participant). Many of today's studies are aimed at protecting the anonymous authentication scheme for roaming service in global mobile networks [11–13]. This is confirmed by official reviews and overview [14, 15].

Phishing is also widely used in Russian conditions. It is one of social engineering types based on the user's non-acquaintance of the network security basics: in particular, many Russian citizens do not know such a simple fact that official services do not send letters asking for their credentials, password, etc. [16]. The purpose of phishing is to gain access to confidential user data - logins and passwords. This is implemented through bulk mailing of emails on behalf of popular trademarks, as well as personal messages within various services, for example, on behalf of banks or within social networks. The letter often contains a direct link to a website that is indistinguishable from the original one or to a website with a redirect, i.e. automatic redirection of the user from one website to another. After a user gets to a fake page, fraudsters with the help of various psychological techniques try to prompt the user to enter its username and password on the fake page, which user set to access to a specific website, which allows fraudsters to access personal accounts and bank accounts.

The most common type of phishing websites is creation copy of banking sites. Schemes of crimes in such cases are as follows. A bank customer trying to access the online bank website usually gets on the page that tells about technical works on the website and the customer is asked to enter a phone number instead of SMS confirmation. After that, user begins to receive SMS-messages on the transfer of funds from his bank card. Then fraudsters call from allegedly the technical support of the bank and report that erroneous operations have occurred and asked to dictate the login, password and SMS code to cancel the operation. So the client provides necessary information and fraudsters get access to his bank account.

The bank client does various operations while being on website copy (for example, replenishing the subscriber number account, paying for various services, transferring money to another account, etc.). After entering username and password which client uses to access, they become known to fraudsters, who are able to debit funds from a bank card.

Such actions cause significant harm to the bank account holders, as well as to the banks themselves and to society as a whole. Let us prove this with the help of a macroeconomic model.

3 Model

We will conduct a macroeconomic analysis of the situation and analyze the consequences. A graphic image of this situation is shown in Fig. 2 (compiled by the authors).

The shape of the supply curve of banking products and services differs from the classical shape of the supply curve in that it is convex, but not concave, because the higher the price of banking products and services, the more banks are willing to provide their products and services. At the same time, even a small price increase can significantly expand both their list and volumes. The demand curve for banking products and services corresponds to the classical form, i.e. the lower the price, the greater the demand for various banking products and services, both in terms of nomenclature and volume.

In Fig. 2, the reference axes show the volume of banking products and services provided (axis X) and the total costs of the banking sector (axis Y). At the same time,

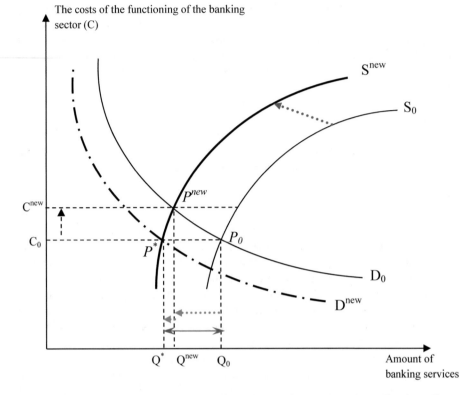

Fig. 2. Changes in the volume and price of banking products and services offer depending on the magnitude of the costs of their provision

the costs of maintaining banking security are included in the total costs. Their growth directly affects the growth of total costs due to the need to struggle digital crime.

The situation of equilibrium in the country's credit and financial market is reflected by the point P_0 until the appearance of banks' additional costs for maintaining digital security. P_0 is the intersection of the demand curve for banking services from commercial organizations and the public (D_0 graph) and the supply curve of various financial and credit products and services from banks (graph S_0). The magnitude of costs for the functioning of the banking sector corresponds to the point C_0, and the volume of provided banking products and services corresponds to the point Q_0.

Costs of banks for increasing banking security from "digital" crimes consist of several parts:

(1) Costs of creating and constantly improving the protection system against hacker attacks on clients' accounts and information theft about banks' customers (personal data of individuals, transaction information, credit histories, etc.).
(2) Costs of improving protection systems against the introduction of malicious software into banks' internal systems, that allow hackers to have access to the internal information of banks during long-run period;

(3) Direct losses of banks from funds' and information embezzlements;
(4) Expenses of banks for the compensation payment to affected customers;
(5) Costs of risk insurance, etc.

Thus, banks are forced to invest additional funds in security systems with the increase in hacker activity or the emergence of the danger of its gain. This leads to a cumulative increase of maintaining banking system costs (moving from point C_0 to point C^{new}). As a result, the graph of the supply curve of banking products and services is shifted to the left and up. A new equilibrium point (price of banking products and services) P^{new} now corresponds to the real volume of provided banking products and services Q^{new}, that is less than Q_0 ($Q^{new} < Q_0$).

However, not all bank customers (both business structures and individuals) are ready to receive banking products and services at new higher rates and refuse from part of services. In addition, banks cannot quickly change prices for services due to a number of legal restrictions (for example, lending rates, interest on deposits, etc.). Therefore, part of the services continues to be provided at a price corresponding to the previous level of bank costs (C_0). In Fig. 2, this corresponds to the price point for banking products and services P* (located on the same line with P_0). However, banks are ready to provide their products and services in smaller quantities at such price than at price P^{new}. As a result, the total amount of provided banking products and services is reduced even more (to the value of $Q*$), that as a result turns out to be significantly less than the real need of the economy for credit and financial resources and tools to maintain normal functioning ($Q* < Q^{new} < Q_0$).

Thus, a decrease in security in the credit and financial market due to an increase in digital crimes against the banking sector can significantly worsen the macroeconomic situation in the country, and the impunity of crimes in this area can attract additional criminals to it.

4 Results and Discussion

Transference of many banking operations into the digital realm requires greater attention to the security of these operations. This, in turn, implies an increase in costs of creating and maintaining computer systems of information and data protection, as well as customer accounts and internal bank management systems. Increasing banks' expenses for maintaining the digital security of their systems, customers' data and their accounts protection significantly increases the aggregate expenses of credit and financial and banking systems for functioning. This generally leads to rise in costs of banking products and services. A very high rise in prices may lead to a reduction of banking products and services consumption and a slowdown of economic growth in the country.

An analysis of losses caused by crimes in the banking sector, committed with the use of digital devices and digital technologies, showed their multiple excess over similar crimes with cash. It also increases bank expenses significantly.

Banks' compensation payments to millions of customers whose personal information became available to fraudsters due to hacking of banking information systems also increase the losses of banks, which also significantly increase the total expenses of banks for their maintenance. "Credential theft is a serious driver of cybercrime today" [6, 17] and credentials must be protected very strongly.

The banking security systems currently existing in Russia make it possible to prevent multiple unauthorized operations using remote banking services. On average, the total annual volume of such prevented unauthorized operations amounts more than 5 billion rubles. However, the cumulative amount of successful thefts reaches 2 billion rubles according to official data of the Central Bank of the Russian Federation. These amounts are much more, according to experts, but banks just do not advertise real losses.

The conclusions obtained on the basis of the constructed model about the significant negative impact of digitization on the security of not only individual banks and the banking system, but also the country's economy as a whole require the development of some institutional measures. In particular, it is necessary to make the punishment for computer crimes in the banking sector more severe, especially toughening the criminal legislation in the field of embezzlement from bank accounts. The maximum penalty for embezzlement of funds from a bank account on a large scale is 10 years of imprisonment according to the current criminal code of the Russian Federation, that is not enough to prevent such crimes according to lawyers.

A conclusion made on the basis of macroeconomic modeling that digitization having a significant impact on the cost of banking products and services can adversely affect the effectiveness of monetary processes, is an important result obtained in the process of research and being controversial. And if the damage from hacker theft is beyond anybody's doubt, the deterioration of banking services quality in the economy under (and because of) digitization due to the rising cost of products and services is more indirect and more controversial, that requires further research in this area.

5   Conclusion

In addition to the positive aspects, the digitization of the banking sector has a number of negative ones. In the most general form, they can be summarized by the following theses.

Translating many banking transactions into the digital realm requires increased security for these operations. This, in turn, implies a rise of the creating costs and continuous improvement of such security systems.

The thesis that the introduction of ICT in the banking sector accelerates and reduces costs of providing banking services and also increases the security of operations, both for banks and for their clients, was verified on the basis of building a macroeconomic model. The study showed the ambiguity of this thesis.

Thus, the results of the study allow us to make a following conclusion with respect to the Russian banking system. It is necessary to create a single center for ensuring the security of Russian banks and the establishment of uniform standards for such security for all credit institutions by the financial regulator in order to enhance the security of digital banking operations.

The identified problem suggests further research on digital security in the credit and banking sector, its development prospects in the conditions of digitization and increasing competition from new financial institutions that are full-fledged IT companies, and also issues of criminal penalties for computer crimes in the credit and banking sector.

References

1. Digital Banking Users to Reach nearly 3 billion by 2021, representing 1 in 2 global adult population. https://www.juniperresearch.com/press/press-releases/digital-banking-users-to-reach-nearly-3-billion-by. Last accessed 11 Feb 2019
2. Darvish, H., Husain, M.: Security analysis of mobile money applications on android. In: Proceedings - 2018 IEEE International Conference on Big Data, Big Data 2018, pp. 3072–3078. Seattle, United States (2019)
3. IBM Sponsored Study Finds Mobile App Developers Not Investing in Security IBM Security. https://www-03.ibm.com/press/us/en/pressrelease/46360.wss. Last accessed 1 Mar 2019
4. Khan, J., Abbas, H., Al-Muhtadi, J.: Survey on mobile user's data privacy threats and defense mechanisms. Procedia Comput. Sci. **56**(1), 376–383 (2015)
5. Data Breach Investigations Report 2017. www.verizondigitalmedia.com/blog/2017/07/2017-verizon-data-breach-investigations-report. Last accessed 1 Mar 2019
6. Top 10 Mobile Risks OWASP Mobile Security Paper. https://www.apriorit.com/dev-blog/435-owasp-mobile-top-10-2017. Last accessed 1 Mar 2019
7. Reaves, B., Scaife, N., Bates, A., Traynor, P., Butler, K.R.: Mo (bile) money, mo (bile) problems: analysis of branchless banking applications. In: 24th USENIX Security Symposium (USENIX Security 15), pp. 17–32. Washington, D.C. (2015)
8. Wazid, M., Zeadally, S., Das, A.K.: Mobile banking: evolution and threats: malware threats and security solutions. IEEE Consum. Electron. Mag. **8**(2), 56–60 (2019)
9. Kiljan, S., Simoens, K., De Cock, D., Van Eekelen, M., Vranken, H.: A survey of authentication and communications security in online banking. ACM Comput. Surv. **49**(4), 61 (2016)
10. Ghosh, S., Majumder, A., Goswami, J., Kumar, A., Mohanty, S.P., Bhattacharyya, B.K.: Swing-pay: one card meets all user payment and identity needs: a digital card module using NFC and biometric authentication for peer-to-peer payment. IEEE Consum. Electron. Mag. **6**(1), 82–93 (2017)
11. Mun, H., Han, K., Lee, Y.S., Yeun, C.Y., Choi, H.H.: Enhanced secure anonymous authentication scheme for roaming service in global mobility networks. Math. Comput. Model. **55**(1–2), 214–222 (2012)
12. Xie, Q., Hu, B., Tan, X., Bao, M., Yu, X.: Robust anonymous two-factor authentication scheme for roaming service in global mobility network. Wireless Pers. Commun. **74**(2), 601–614 (2014)
13. Zhao, D., Peng, H., Li, L., Yang, Y.: A secure and effective anonymous authentication scheme for roaming service in global mobility networks. Wireless Pers. Commun. **78**(1), 247–269 (2014)
14. FinCoNet: Online and mobile payments: supervisory challenges to mitigate security risks (2016). www.finconet.org/FinCoNet_Report_Online_Mobile_Payments.pdf. Last accessed 4 Mar 2019

15. FinCoNet: Online and mobile payments: an overview of supervisory practices to mitigate security risks (2018). www.finconet.org/FinCoNet_SC3_Report_Online_Mobile_Payments_Supervisory_Practices_Security_Risks.pdf. Last accessed 4 Mar 2019
16. Sypachev, A.Y.: Basic methods of theft using the internet. Concept 10, art. no. 15349 (2015)
17. Esparza, J.M.: Understanding the credential theft lifecycle. Comput. Fraud Secur. **2**, 6–9 (2019)

The Early Practice of Analytical Balances Formation in F. Datini's Companies in Avignon

Marina Gurskaya$^{(\boxtimes)}$ ⓘ, Mikhail Kuter ⓘ,
and Dmitriy Aleinikov$^{(\boxtimes)}$ ⓘ

Kuban State University, Stavropolskaya st., 149, Krasnodar 350040, Russia
marinagurskaya444@gmail.com

Abstract. The paper considers the procedure for calculation and distribution of financial result, based on the use of a static model, which is based on inventory and valuation of assets, cash, accounts receivable and payable, as well as invested and reinvested capital. This experience took place in Francesco Datini's companies in Avignon from 1366 to 1410. The information base used in the paper is oriented on the period of 1408–1409. This study was conducted using modern digital technologies. The use of the method of logical-analytical reconstruction allows the use of digitized archival sources in the formation of modern models of medieval accounting systems.

Keywords: Datini's archives · Companies in Avignon ·
Static model of accounting · Inventory and valuation · Financial result

1 Introduction

The reorientation to digital science and information technology supports a new approach to solving problems related to the study of the historical aspects of the birth of economic science, which primarily relates to the study of early accounting books. The digitization of primary sources stored in the archives of Italy, the development of modern methods of introduction of archival documents and the information contained in it allow us to reconstruct a visual representation of the origin and development of the given area of economic science. Distinguished modern researchers of the birth of double-entry bookkeeping Basu and Waymire (2017, 1–2) give the opinions of various researchers on the role of double-entry bookkeeping in the development of capitalism. As a rule, in such cases, the opinion of V. Sombart comes to the fore.

Sombart (1902) asserted that DEB played a central role in capitalism's emergence (Nussbaum 1933, 158–161) as did Max Weber (Carruthers and Espeland 1991). Schumpeter (1950, 123) argues that capitalism turns the monetary unit into a tool for rational calculations of costs and profits, the towering monument of which is the double-entry bookkeeping and Mises (1949, 231) suggests that DEB "makes success and failure, profit and loss ascertainable". In sharp contrast, Yamey (1949, 113) argued that the early use of DEB only "served limited objectives and… the striking of balances was performed primarily for narrow bookkeeping purposes," and attributed any use of

© Springer Nature Switzerland AG 2020
T. Antipova (Ed.): ICIS 2019, LNNS 78, pp. 91–102, 2020.
https://doi.org/10.1007/978-3-030-22493-6_10

DEB to the self-interest of accountants: "double-entry bookkeeping came to be accepted by teachers and accountants as the standard or desirable system of bookkeeping, and that their influence and employment helped to introduce its use even where simpler systems would have been adequate." Basu and Waymire come to the compromise conclusion: "The truth likely lies somewhere between these two extremes".

The given paper attempts to show one of the most widespread variants of the calculation and distribution of financial result, which was applied in the 14th century, along with double-entry bookkeeping. The paper (Kuter and Gurskaya 2018) states (according to preserved archival materials) that such a technique was used for the first time in Alberti's company (1304–1306) and was a prevailing one at that time. As for the use of double-entry bookkeeping, over the period from the end of the 13th century and the whole 14th century there are only four examples of its use G. Farolfi's branch in Salon (1299–1300); dell Bene's company (1318); the municipality of the commune city of Genoa (1340); F. Datini's company (first of all in Pisa, starting with 1383).

2   Review of Prior Literature

Many authors wrote about Francesco Datini's archives, both in the 19th and 20th centuries. The first references to Francesco di Marco's enterprises and companies are found in the works by Fabio Besta (1909, 317–320) and his disciples Corsani et al. (1922, 83–85) and Ceccherelli (1913, 1914a, b, 1939). Their studies were in some ways intertwined because, for instance, Ceccherelli referred to the work carried out by Corsani: "G. Corsani in his study on the Datini's books on the same page 142, which is devoted to the company in Avignon, formed by Francesco di Marco, Toro di Berto, Boninsegna di Matteo, Tieri di Benci and Andrea di Bartolommeo (1367–1410) …". The publications of the authors who did not focus on accounting issues but described Datini's business history or the history of the archives should be also referred to. They are Bellini Melis (1954), Carradori (1896), Guasti (1880), Nicastro (1914) and Bensa (1923, 1925, 1928). Of course, the most profound research was carried out in the second half of the 20th century by F. Melis, R.de Roover, T. Zerbi, A. Martinelli.

Although A. Martinelli did not directly work with Datini's archives, he convincingly showed the role of Datini's companies in changing-over to the "Venetian" account form, highlighting the priority of the companies in the 14th century. Initially he described the General state of Affairs in the Middle ages, in the archives where the oldest ledgers were found, kept in double-entry bookkeeping, especially in the archives of Florence, Siena and Prato in Tuscany, there are ledgers containing accounts with mixed sections, i.e. sections that mixed debits and credits on each other. The archives of Florence probably contain the largest number of ledgers stored in this way; hundreds of them cover the entire fourteenth century. By the beginning of the XV century began to appear the "Venetian method", where he had a lateral cross-section. The books of Alberti Del Giudice, as well as the books of Gallerani, Bardi, del bene, Frescobaldi, and Gianfilazi, containing records Dating back to the early fourteenth century, were all of your accounts in mixed sections (Martinelli 1974, p. 190). And further, with reference to Sapori (1932, 1934) only in 1382, the company Francesco Datini began to use a new form, possibly imported from Venice. Brothers Antonio and Bonifacio Peruzzi in 1430 kept their books "Alia veneziana" or the Venetian method, while in the XIV century members of the same family used accounts with mixed sections.

Martinelli, referring to Melis noted that, the case of Francesco Datini of Prato is remarkable. He founded a branch in Avignon, where from 1366 to 1401 he kept his books on the old model, that is, used accounts with mixed sections; a new "Venetian method" he introduced in his ledgers only in 1401. This new form, which used accounts with side sections, has already been introduced into the accounting system of the branches of Genoa, Florence, Pisa, Barcelona, Mallorca and Valencia, which he created and managed with other partners from 1383 to 1393. It should be added that in those cities where the heritage of ancient Roman customs and traditions was more obvious, such as Siena, Perugia and Rome itself, this new method of accounting was introduced much later, at the end of the XV century. In the archives of these cities, especially in Siena, there are numerous ledgers, which are maintained by the old method, using accounts with mixed sections that go beyond this period (Martinelli 1974, pp. 190–191).

As we can see, A. Martinelli was convinced that Datini's company in Avignon in 1401 stopped the practice of applying mingled accounts, widely used from the opening date, and reoriented itself to applying the accounts of the Venice form.

However, there is every reason to refute this statement and assert that many Datini's companies in Avignon continued to identify financial result beyond double-entry bookkeeping. The considered example of 1408–1409 confirms what was said above.

3   Research Method

The principal research method adopted in this study is archival. It uses material found in the State Archive of Prato. This research team has been working with the material in this archive for the past decade and many of the records have been recorded and linked together using logical-analytical reconstruction. This is an approach that we developed for the purpose of enabling entries in the account books to be traced visually between accounts and books and from page to page. It is described in the Appendix. By adopting this approach, we are able to see the entire accounting system electronically, making entries and their sources clear in a way that is not possible if all that you have is the original set of account books. This enables us to consider each transaction in detail, trace its classification, and so explain the bookkeeping and accounting methods adopted without possibility of misinterpretation. This approach represents a new paradigm in how to analyse and interpret accounting practice for periods when there was no concept of either a standard method or a unified approach to either financial recording or financial reporting.

4 Statement of Basic Materials

In Francesco Datini's archives in Prato many analytical notebooks (Quaderni di ragionamento) designed to calculate financial result are stored. The quality of the stored archival materials does not always comply with the possibilities of their research. One of Quaderno di ragionamento of the most appropriate quality (Prato, AS. D. 178/17) is dated 1408–1409. This Quaderno di ragionamento was of heightened interest for the authors, since it preceded the notebook, according to which in Avignon the first synthetic (condensed) balance (1411) was formed, which included only the totals of enlarged groups of indicators.

We should clarify at once that starting from 1366 in Datini's companies in Avignon financial result was calculated by means of inventory and valuation methods, and from 1398, analytical calculations were carried out using a combined method: some of the indicators included in the Quaderni di ragionamento, were identified through inventory and property valuation, the cash balance was drawn up in the special register called Entrata e Uscita. The information concerning the status of settlements with debtors and creditors, as well as invested and reinvested capital was selected from the Ledger. Separate data were transferred from the Memorial.

Figure 1 shows the photocopy of the first page, opening the first section "The sum of all goods and bales in transit" in the analytical Quaderni di ragionamento for 1409. We should note that the lower part of each double-page spread is damaged, as a result not all the information is readable. The photocopy of page AS. D. 178-17, c. 12r (Fig. 2) and folio c. 12v–13r (Fig. 3) can lend evidence to this.

Fig. 1. Photocopy of page Prato, AS D. 178/17, c. 1r

The first section, the inventory result, occupies 23 pages, includes 93 subsections, consisting of 592 entries. The formation of the section is not associated with the keeping of accounts by double-entry method. This statement is based on the absence of the cross-references to contra-entries in the records. The only exceptions are the last two entries on page Prato, AS. D. 178-17, c. 12r (Fig. 2), having cross-entries to pages

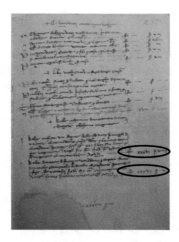

Fig. 2. Photocopy of page Prato, AS. D. 178/17, c. 12r, of the last page of the first section

36 (the sum of f. 36 s. 9, upper ellipse) and 37 (the sum of f. 36 s. 6, lower ellipse) of the Memorial, which, unfortunately, did not survive. We should note that for each page the totals are calculated, but there is no total of the first section on page Prato, AS. D. No. 178-17, c. 12r. The cost of the goods in the house and the shop and the goods that are in transit amounted to f. 2135 s. 2.

Page Prato, AS. D. 178/17, c. 12v (Fig. 3) contains the balance of cash: "We found cash in the big cash on Tuesday December 31 1409, there are gold and silver coins as it can be seen in Entrata e Uscita labeled "due" from c. 26 on, where the balance is made on this day. The cash was kept by Tieri di Benci himself. The total amount is f. 831 s. 2 d. 8 *correnti* of s. 24 each" (upper ellipse).

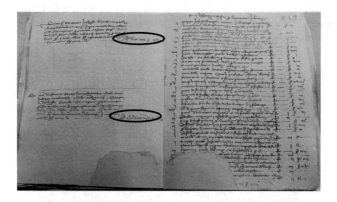

Fig. 3. Prato, AS. D. 178/17, c. 12v–13r.

The totals of the first two sections of the analytical balance formed according to data of 1409.

As it follows from the above entry, the sum of cash is confirmed by a cross-reference to page 26 in Entrata e Uscita. Here, there are a number of nuances. Of all the survived books of Entrata e Uscita belonging to this period, there is only one book available with the archive index Prato, AS. D. No. 130. It does not contain the men tioned "due" mark. In addition, the first section "Entrata" occupied pages 2r–16v (entries 1409 on pages 3v–8v), and the second section "Uscita" – pages 51r–60r (entries 1409 on pages 52r–57r). Consequently, the book Prato, AS. D. 130 did not contain the page with number "26" to which the reference is provided. And, most surprisingly, the accountant used Entrata e Uscita book to register the facts of Income and Outgoing of cash, but did not draw up any balances for specific dates.

Apparently, the reason is in the following: at this time there were several parallel companies, where partners were the same people in different combinations. Each company had their own accounting books and, in all likelihood, the folder Prato, AS. D. 178/17 and the book Prato, AS. D. No. 130, although they belonged to the same period of time, did not belong to the same company.

Page 12v is concluded with the total of two sections (the sum of f. 2966 s. 4 d. 8): "Total amount of all the merchandise we still have in our house and workshop and bales in transit and money in the big cash as it can be clearly seen in this Quaderno di Ragonamenti until here, that is from c. 1 until this page. The total amount is f. 2966 s. 4 d. 8 *correnti* of s. 24 each" (lower ellipse).

Here, we can make only one assumption – the accountant is still far from the idea of preparing a synthetic balance and he is not interested in the totals for each section. He is focused on one goal – to identify the total amount, as Pacioli will say later, "of everything that belongs to him in this world."

Pages Prato, AS. D. 178/17, c. 13r–15v of the analytical balance contain the information concerning the third section "Housewares and household equipment". The section occupies 6 pages, includes 14 subsections, combining 147 entries. Page AS D. 178/17, c. 15v (the photocopy is in Fig. 4), after the second subsection, crowning the third section, also does not contain any total. This confirms our assertion that the accountant was not interested in the totals of separate sections.

We managed to draw up the following total. It came to f. 262.

Fig. 4. Prato, AS. D. №178/17, c. 15v–16r. Totals of the first four sections of the analytical balance and beginning of the fifth section of "Debtors" section

The fourth section, "Food and Drinks," is also placed on Prato, AS. D. 178/17, c. 15v and occupies 13 positions. The total of the section is f. 55 s. 1 (marked with an ellipse). The accumulative total for the four sections is blurry and unreadable. According to our calculations it equals f. 3282 s. 5 d. 8.

Of course, special results were expected from the fifth section, "Those who owe us" (Settlements with debtors), since it is formed according to the data present in the double entry system. The first page of the section is shown in Fig. 4. The photocopy of the two final pages of the section, containing 34 entries, is shown in Fig. 5. The total of the section is f. 3899 s. 2 d. 9¼ (marked with an ellipse).

Fig. 5. Prato, AS. D. №178/17, c. 16v–17r. Final entries of "Debtors" section

It is known that the Ledger was lost. There was hope that particular entries were transferred from the Secret Book. However, not a single debtor was entered into the Secret Book.

As a result of our constructions, "Debtors" section of the analytical balance (Quaderni di ragionamento) was completely reproduced, the final indicators of which are introduced in Table 1.

At the top of page 17v (the photocopy is in Fig. 6) is the total of "Debtors" section: "Sum of all the merchandises and bales in travel and cash and furniture of the house and the workshop for food and beverage and whoever should give to the partnership as it can be seen in the notebook of the calculation from c. 1 to the last page, the total is f. 7181 s. 8 d. 5 ¼ of s. 24 current" (marked with an ellipse).

Table 1. Total indicators of "Debtors" section of Quaderni di ragionamento of F. Datini's company in Avignon according to data of 1409

No.	Section Name	Sum
1	Goods and bales in transit	2135.2.0
2	Cash	831.2.8
3	Housewares and shop supplies	262.0.0
4	Food and drinks	55.1.0
5	Those who must give us	3899.2.9¼
Total		7181.8.5¼

Now, we can proceed to "Creditors" section of the analytical balance. Before we do this, we should note that when studying the analytical balance, we deliberately omitted the description of page Prato, AS. D. 178/17, c. 17v.

"Creditors" section starts with page Prato, AS. D. 178/17, c. 18r (Fig. 6) and occupies two pages. The entries on the first page say:

+ whoever should have at the Big Black Book labeled A, started the 31st of December 1409:

- Messer Viscount d'Osses must have at the said Big Book at c. 7, f. 9 s. 4 current
- Piero Palmieri from Florence must have at the said Big Book at c. 8, f. 1
- Giufre Rostagno from Arles must have at the said Big Book at c. 8, f. 7 s. 18.

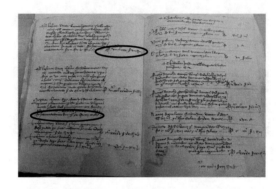

Fig. 6. Prato, AS. D. 178/17, p. 17v – 18r. Final entries in "Debtors" section

The references to the contra-entries refer to the lost Ledger and are not subject to verification.

Let us read further:

+ Whoever must have in the Secret Book of the partnership labeled 2:

- Francesco di Marco and Co. of Avignon of the old account until the 1st of November 1401 must have in the said book at c. 24, f. 2625 s. 15 for the merchandises and furniture assigned to that business period
- Francesco di Marco and Co. of Avignon of the new account from the 1st of November 1401 must have at the said book at c. 24, f. 3373 s. 19 d. 4¼ for the profit made from the 1st of November 1401 to the 31st of December 1408, which are 8 years
- Nanni di Girolamo who stays with us must have at the said book at c. 25 for 1 year of his wage, f. 18 current
- Donato di ser Giovanni himself must have at the said book at c. 25, f. 104 s. 1 d. 6 current for his wage
- Francesco di Marco and Co. o Avignon of the old account until the 1st of November 1401 must have at the said book at c. [...] f. 196 s. 22 d. 6 for debtors.

Unfortunately, the Secret Book of this period is also lost, but we found the traces of two indicators in the book Prato, AS. D. 161, c. 133v, which was kept from 1398 to 1408. The fact is that the sum of f. 2625 s. 15 for the furniture trade and the sum of f. 196 s. 22 d. 6 (Francesco di Marco and Co from Avignon according to the old account) came to us from the old company, which was closed on October 31, 1401. The first one is the profit transferred to the new company, the second one is the balance of settlements with the owners, which is also transferred to the new company, the beginning of its activities is November 1, 1401. These two sums are present in all the reports starting from December 31, 1402.

And, the last page of the analytical balance (Quaderni di ragionamento) – Prato, AS. D. 178/17, p. 18 V (the photocopy is on Fig. 7).

On the page there are only three positions transferred from the lost Secret book, united by the same name:

"Wages of labor masters and boys for 1 year and 1 month, from the 1st of January 1408 to the 31st of December 1409":

- Donato di Ser Giovanni from Arezzzo who stays in Milan for us must have the 31st of December 1409 for his wage from the 1st of January 1408 until the said day, which is one year for f. 40 each year
- Nanni di Girolamo who stays with us must have the said day for his wage from the 1st of January 1408 to the 31st of December 1409 the said day, which is one year for f. 15 each year.
- Giannino di Marchese who stays with us **must have** f. 7 s. 7 d. 6 for his clothes

Fig. 7. Prato, AS. D. 178/17, c. 18v. The last page the analytical balance formed according to the data of 1409

Sum of the page f. 62 s. 7 d. 6.

Here, we should remind that the New Year in the republic-cities of the medieval Italy began in March, and the interim reporting (or drawing up totals) was carried out for December 31. In our case, the indicators were calculated over the period from January 1, 1408 to December 31, 1409.

The final indicator, for the sake of which it was formed, finishes "Creditors" section: "Sum of whoever should have from us considering the merchandise, creditors and gain made in the 7 past years as it can be seen in this notebook of calculations in the previous page and in this one, the total sum is f. 6398 s. 15 d. 8¼" (помечено эллипсом).

"Traditional phrase" finishes the section: "The above-mentioned sum is put at c. 17 below the merchandises, the bales in travel and cash and food and a beverage and furniture of the house and the workshop, and debtors who must give. In the said page we do the balance from the 1st of January 1408 to the 31st of December 1409"

"Here ends the notebook of calculation for the said period, written by my own hand, of Tommaso di ser Giovanni who started writing at c. 1 until this page.

For the grace of God. Amen +".

The given total (f. 6398 s. 15 d. 8¼) is transferred to the second entry on page 17v. Thus, it is possible to make an intermediate generalization:

- Two separate sections are formed: "Debtors" or "Who *must give* us" and "Creditors" or "Who *must have* from us";
- the "central" page placed between "Debtors" and "Creditors" sections (Prato, AS. D. 178/17, c. 17v, the photocopy is in Fig. 6) is not included in any of these sections, but represents an independent and, one might say, the most important "Financial result account" section. Here, both totals (on "Debtors" and "Creditors" sections) are placed on one sheet and entered one below the other. The difference between them ("What we *must be given* – the sum of f. 7181 s. 8 d. 5¼" and "What they *must have* – the sum of f. 6398 s. 15 d. 8¼") is the financial result, as the appropriate entry says: "The profit made the last year as it can be seen in this Notebook of calculation made from the 1st of January 1408 to the 31st of December 1409, f. 782 s. 15 d. 9" (the second ellipse).

The last two entries in the financial result account is the distribution of profit. The entries say the following:

- Tieri di Benci himself is entitled to the ¼ of the sai profit f. 195 s. 16 d. 2 ¼;
- Francesco di Marco himself and Tommaso di ser Giovanni himself are entitled to the profit made, f. 587 s. – d. 6 ¾.

5   Conclusion

The paper contains the description of one of the most widespread models of the formation and distribution of financial result applied in the 14th century. The basis of its application is the inventory and valuation of assets, active and passive debts. From the stored archival documents, it follows that the first practice of its use took place in Alberti's company (1304–1306).

The company with Francesco Datini's participation in Avignon was established in 1363. Initially, accounting solved the problem of settlements with debtors and creditors. The practice of financial result calculation in Datini's company is dated 1366. The calculation algorithm assumed, on the one hand, inventory and registration of stocks of goods in the house and shop, cash, implements, drinks and food stocks, accounts receivable. On the other hand, the accounts payable were inventoried and registered, invested capital (in subsequent periods reinvested capital as well) was registered.

Since 1398, Datini's company in Avignon applied combined accounting, when information on the movement of accounts receivable and payable was formed in the double-entry system in the accounts of the "Venice" form.

Thus, we have obtained confirmation of B. Yamey's statements, who celebrates his centenary these days, that during the period of capitalism development, accounting processes were maintained by both a dynamic and a static model. At the same time, in the given paper we did not set the objective to introduce the obvious advantages of the model based on double-entry bookkeeping, on which W. Sombart insisted in his works.

It is worth noting that the study would not have been possible without the use of modern digital technologies. The authors, during several years of studying the history of accounting, collected their own digitized archive of medieval accounting books, including more than sixty thousand copies. Digitized archival documents allow conducting research remotely, without a permanent presence in the Archives. The method of logical-analytical reconstruction used by the authors allows modeling medieval accounting procedures using digitized materials.

References

Primary Sources

Archivio di Stato di Prato
Prato, A.S., Fondaco di Avignon, D.: Company Francesco di Marco Datini, Tieri di Benci and Tommaso di Ser Giovanni, Quaderni di ragionamento, 1408–1409, No. 178/17
Prato, A.S., Fondaco di Avignon, D.: Company Francesco di Marco Datini, Tieri di Benci and Tommaso di Ser Giovanni, Entrata e Uscita, 1409–1412, No. 130

Secondary Sources

Basu, S., Waymire, G.: The economic value of DEBITS = CREDITS./electronic copy (2017). https://ssrn.com/abstract=3093303
Bensa, E.: Francesco di Marco Datini: discorso detto nell'Aula maggiore del Comune di Prato il dì 21 agosto 1910. Genova, ARTI GRAFICHE CAIMO & C (1923)
Bensa, E.: Le forme primitive della polizza di carico: ricerche storiche con documenti inediti/Enrico Bensa. Genova, Caimo (1925)
Bensa, E.: Francesco di Marco da Prato. Notizie, e documenti sulla mercatura italiana del secolo XIV, Treves, Milano (1928)
Besta, F.: La Ragioneria, 2nd edn., vol. 3. Rirea, Rome (1909). Facsimile Reprint, 2007
Carradori, A.: Francesco di Marco Datini, mercante pratese del sec. XIV, Prato (1896)

Carruthers, B.G., Espeland, W.N.: Accounting for rationality: double-entry bookkeeping and the rhetoric of economic rationality. Am. J. Sociol. **97**(1), 31–69 (1991)

Ceccherelli, A.: I libri di mercatura della Banca Medici e l'applicazione della partita doppia a Firenze nel secolo decimoquarto. Bemporad, Firenze (1913)

Ceccherelli, A.: Le funzioni contabili e giuridiche del bilancio delle società medievali. Rivista Italiana di Ragioneria **14**(8), 371–378 (1914a)

Ceccherelli, A.: Le funzioni contabili e giuridiche del bilancio delle società medievali (Continuazione e fine). Rivista Italiana di Ragioneria **14**(10), 436–444 (1914b)

Ceccherelli, A.: l linguaggio dei bilanci. Formazione e interpretazione dei bilanci commerciali. Le Monnier, Firenze (1939)

Corsani, G.: I fondaci e i banchi di un mercante pratese del Trecento. Contributo alla storia della ragioneria e del commercio. Da lettere e documenti inediti, La Tipografica, Prato (1922)

De Roover, R.: The development of accounting prior to Luca Pacioli according to the account-books of Medieval merchants. In: Littleton, A.C., Yamey, B.S. (eds.) Studies in the History of Accounting, pp. 114–174. London (1956)

Guasti, G.: Ser Lapo Mezzei, lettere di un notario a un mercanto del secolo XIV, Florence (1880)

Kuter, M., Gurskaya, M., Andreenkova, A., Bagdasaryan, R.: The early practices of financial statements formation in medieval Italy. Account. Historians J. **44**(2), 17–25 (2017). https://doi.org/10.2308/aahj-10543

Kuter, M., Gurskaya, M.: The unity of the early practices of financial result formation: the cases of Alberti's, Datini's and Savary's works (1675). In: Théories comptables et sciences économiques du XVe au XXIe siècle: Mélanges en l'honneur du professeur Jean-Guy Degos - Mélanges en l'honneur du Professeur Jean-Guy Degos. Editions L'Harmattan (2018)

Kuter, M., Gurskaya, M., Andreenkova, A., Bagdasaryan, R.: Asset impairment and depreciation before the 15th century. Account. Historians J. **45**(1), 29–44 (2018). https://doi.org/10.2308/aahj-10575

Martinelli, A.: The Origination and evolution of double entry bookkeeping to 1440. ProQuest Dissertations & Theses Global, p. n/a (1974)

Melis, F.: Storia della Ragioneria. Cesare Zuffi, Bologna (1950)

Melis, F.: L'archivio di un mercante e banchiere trecentesco: Francesco di Marco Datini da Prato. Moneta e Credito **7**, 60–69 (1954)

Melis, F.: Aspetti della Vita Economica Medievale. Studi nell'Archivio Datini di Prato. Florence (1962)

Mises, L.: Human Action: A Treatise on Economics. Yale University Press, New Haven (1949). Reprinted 1998, Ludwig von Mises Institute, Auburn, AL

Nicastro, S.: L'Archivio di Francesco di Francesco di Marco Datini in Prato. Rocca S. Casciano, L. Cappelli (1914)

Nussbaum, F.: A History of the Economic Institutions of Modern Europe: An Introduction to Der Moderne Capitalism of Werner Sombart. F. X. Crofts, New York (1933)

Sapori, A.: Una Compagnia di Calimala ai Primi del Trecento. Florence (1932)

Sapori, A.: Libri di Commercio dei Peruzzi. Milan (1934)

Schumpeter, J.: Capitalism, Socialism, and Democracy, 3rd edn. Harper & Row, New York (1950)

Sombart, W.: Der moderne Kapitalismus. Duncke & Humblot, Leipzig (1902)

Yamey, B.S.: Scientific bookkeeping and the rise of capitalism. Econ. Hist. Rev. **1**(2/3), 99–113 (1949)

The Structure of the Trial Balance

Marina Gurskaya(⊠) ⓘ, Mikhail Kuter ⓘ,
and Ripsime Bagdasaryan ⓘ

Kuban State University, Stavropolskaya st., 149, Krasnodar 350040, Russia
marinagurskaya@mail.ru

Abstract. The given paper considers the characteristic features of the early Trial Balances formation in Francesco di Marco Datini's companies in Pisa. Special attention is paid to the activities of the First (1392–1394) and Second (1394–1395) companies. The procedure for the formation of the Trial Balance in these companies and its structure are described in detail. It has been noted that, primarily, indicators from Entrata e Uscita and Memoriale were transferred to "Debtors" side of the Trial Balance. Then, balances of personal accounts of debtors were transferred from the Ledger. And the balances of Merchandise accounts were transferred last of all. Consequently, the indicators on the Debit side of the Trial Balance were placed in order of liquidity decreasing—money, accounts receivable, goods for sale. On the Credit side of the Trial balance, the first indicator was presented in the reserve being created out of pre-distribution profit. Then, the balances of personal accounts of creditors were transferred. The last entry (either in the debit or in the credit) presented the difference between "Debit" and "Credit". The research method is archival, using logical-analytical modelling of sets of accounting books, related to the research periods. This method is used in the use of modern digital technologies, which are the most important tool in research in the field of accounting history.

Keywords: Medieval accounting · Archival materials · Trial balance · Companies of Datini

1 Introduction

The description of the history of accounting practice allows to provide insight into the development and formation of modern accounting methods and tools. The given prospect of research has long been highlighted as a separate area of science – Accounting History. Representatives of this scientific field, representing different regions of the planet, conduct research aimed at finding, recovering and storing information that allows to form an idea of the accounting genesis.

One of the most interesting areas of Accounting History is medieval accounting on the territory of modern Italy, since it is here during that period appear such accounting techniques as depreciation and amortization, reservation, correction, the modern procedure for financial result formation is in the making, and first account balances and balance reporting appear as well.

© Springer Nature Switzerland AG 2020
T. Antipova (Ed.): ICIS 2019, LNNS 78, pp. 103–116, 2020.
https://doi.org/10.1007/978-3-030-22493-6_11

Most of the studies of medieval accounting have been carried out in the 20th century. The foundation of the research in this field was laid by the Italian scientists Alfieri (1891, 1911, 1922), Besta (1922), Ceccherelli (1913, 1914a, b, 1939) and others.

In the 19th century in the Italian city of Prato, there were found documents relating to the business activities of merchant Francesco di Marco Datini (1335–1410). They amount to about 150,000 personal and business letters, bills of exchange, checks, insurance papers, guarantees, etc. (Melis 1962) But, most importantly, most of the books of Datini's companies located in different European cities, which give an idea of the accounting practice of that time, have been preserved, and their study has allowed to reveal many medieval accounting tools (Castellani 1952; Melis 1950; de Roover 1956; Zerbi 1952 and etc.), the appearance of which has long been considered to relate to later periods. Thus, it is in Francesco Datini's companies that you can find documented double-entry bookkeeping as well as basic accounting methods (Kuter et al. 2017, 2018).

The main aim of the study is to provide insight into the accounting procedure during the period of the end of the 14th century and into the formation order and purpose of the Trial Balance. In addition, one of the aims of the study is to promote interest in further research into the history of accounting and archival research.

2   Prior Literature

Luca Pacioli (1494) in chapter 34 of the Treatise wrote: "And having done this with due diligence you must close your ledger account by account in this way: you shall begin with the cash, debtors, merchandise, and customers, and (the balances on) these (accounts) you will transfer to book A, that is in the new ledger. It is not necessary to enter the transfer of the balances in the journal.

You will add all the entries in debit and credit, always augmenting the smaller side, as I mentioned to you above for carry forwards; because the transfers from one ledger to another are similar. The only difference between them is that in the one the balance is carried forward in the same ledger, whereas in this case the balance is transferred from one ledger to another. And whereas in the first case you make reference to the pages in that ledger, in this case you will refer to the pages of the new ledger in such a way that in the transfer from one book to the other you will make only one entry in each ledger, and this is the peculiarity of the last entries in a ledger in which no further entries can be made" (Pacioli 2009, c. 99; Yamey 1994, p. 84).

"...And so the first ledger with its journal and memorial is closed. To make sure that the closing of the ledger is in order, make this further check. That is, on a sheet of paper write all the debits of the ledger marked with a cross on the left, and all the credits on the right; and then add these and make one sum of all the debits, and this is called summa summarum; and in the same manner add the credits which is also called summa summarum; but the first is the summa summarum of the debits and the second the summa summarum of the credits" (Pacioli 2009, c. 103; Yamey, 1994, p. 86).

"...If these two summa summarum are equal, that is each one is the same amount as the other, namely the debits and the credits, you can infer that your ledger has been properly managed, kept and closed, for the reason given in the preceding Chap. 14; but if one of the summa summarum is greater than the other, it denotes an error in the ledger, which you must find with due diligence with the help of the intellect which God has given you and the art of reckoning which you will have learned" (Pacioli 2009, c. 103; Yamey 1994, p. 86).

As you can see, in this case "father of accounting" recommended a direct transfer balances of the accounts from the closing Ledger to the new one. The Trial Balance was formed on separate sheets of paper based on the balances of closed accounts in the old Ledger. Obviously, it is more logical to form the Trial Balance using the balances transferred to the new Ledger.

In chapter 36 (summarizing) Pacioli considers a different method of forming the Trial balance: "When the ledger is full or old and you wish to transfer it to a new ledger, do as follows. First check if the old ledger is marked on the cover. Let us say it is with A. It is necessary that the new ledger to which you wish to transfer it is marked B on the cover, because merchants' books must be in sequence, one after the other following the letters A, B, C, etc. Then you must compile the balance of the old ledger to see that it is equal as it should be, and from that balance copy ill the creditors and debtors into the new ledger, all in the order in which they are in the balance, and make separate accounts for all the debtors and all the creditors, and leave as much space for each as you udge to be needed" (Pacioli 2009, c. 111–113; Yamey 1994, pp. 90–91).

"...Now, in order to cancel the old ledger you must close each (open) account to the above mentioned balance, that is, if an account in the old ledger is a creditor, as the balance will show, debit the account and say: for so much remainder in credit on this account, transferred to the credit in the new ledger marked B on page... And so you have closed the old ledger completely, and opened the new ledger. And as I have shown you for a creditor, you must do the same for a debtor, except that you make a credit entry and transfer it as a debit entry to the (new) ledger. And it is done" (Pacioli 2009, p. 113; Yamey, 1994, p. 91).

If the first variant of the Trial Balance was formed after opening of the transfer of the closed accounts balances of the old book to the new one, then Pacioli's second variant supposes opening of the accounts of the new book directly from the Trial Balance, which is also formed on separate sheets of paper and acts as an intermediate link between the accounts of the old and new books.

Let us single out one detail in Pacioli's text. He indicates the sequence in which the account balances are transferred from the balance to the new book ("you must compile the balance of the old ledger, ... and from that balance copy ill the creditors and debtors into the new ledger, all in the order in which they are in the balance" (Pacioli 2009, p. 103; Yamey 1994, p. 91)) and leaves without recommendations the order of transferring of the account balances to the balance from the old book. We will try to investigate this issue using F. Datini's books of his Second Company in Pisa (1394–1395).

3   Research Method

The principal research method adopted in this study is archival. It uses material found in the State Archive of Prato. Generally, this research team has been working with the material in this archive for the past decade and many of the records have been recorded and linked together using logical-analytical reconstruction. This is an approach that we developed for the purpose of enabling entries in the account books to be traced visually between accounts and books and from page to page.

Unfortunately, because the material studied related to an early stage in the development of accounting and was not organised systematically, it was impossible to do this. Instead, the authors had to work from the Ricordanze, entry by entry, searching for each one in the Ledger until they were all traced.

4   Statement of Basic Materials

We should make it immediately clear that with the accumulation of knowledge and practical experience the accounting registers developed as well as their unification, which took place according to their purpose, that, respectively, led to a reduction in their number. The transfer from Mercanzie to the Ledger of the multi-page the merchandise account (the prototype of Profits and Losses account) led to the clear functioning of each register and the establishment of a specific structure. The Ledger now included debtors and creditors accounts (pages from 2v to 279r were allocated for them), and from page 280r (and up to the last completed account on folio c. 365v–366r) there were accounts that were part of the system of financial result formation. These accounts should include the accounts of the purchase and sales of goods, the dynamic lines of accounts for operational results accumulation, and Profits and Losses account itself, which was placed on Prato, AS, D. No. 362, c. 364v–365r (1).

Taking into account the fact that, unlike individual enterprises, where Profits and Losses account was formed after the sale of all the consignments of goods and the closing of all the accounts of purchase and sales of goods, companies were closed on the earlier agreed date. In this regard, the duration of the Second Company in Pisa slightly exceeded one year – from July 14, 1394 to August 1, 1395. This circumstance could not but affect the size of the account for financial results. If, for example, in Datini's Second Individual Enterprise (1386–1393) the account occupied four folios, then in the Second Company it occupied one folio and on the largest (credit) side the account included 14 positions – nine varieties of profit, the total of profit and four entries of profit distribution, which the accountant mistakenly put on the credit side of the account.

In the First Company, for the first time, the Trial Balance was formed inside the Ledger, on the pages of which for each position of the debit and credit side it was indicated from which page in the closing book the balance of account was transferred to the Trial Balance, as well as on which page in the opening book this balance of account was placed.

Thus, the Trial Balance simultaneously performed two functions: it was, on the one hand, the Trial Balance, and on the other, the balance of closing of the old Ledger and opening of the new one.

And, which is very important, the presence of the internal balance of such a form would embody Schar's idea (Schar 1914) that all the bookkeeping starts and ends with a balance. In this case, the unbroken chain begins in the Trial Balance of the First Company and ends in the Trial Balance of the Second Company. And further, there is a similar connection between the Trial Balance of the Second company and the Trial Balance of the Third Company (2.08.1395–16.11.1996).

In the First Company, as well as in the Second one, the Trial Balance was placed after all the debtors and creditors accounts. It occupied four folios, starting with Prato, AS, D. No. 362, c. 153v–154r. The Ledger's pages from the Trial Balance to the accounts for financial result were left blank, since the Ledger was in advance divided into two sections. For clarity, Figs. 1 and 2 show the photocopies of the first and the last folios of the Trial Balance.

Fig. 1. Photocopy of the first folio of the Trial Balance of F. Datini's Second Company in Pisa (Prato, AS, D. No. 362, c. 153v–154r)

Fig. 2. Photocopy of the fourth folio of the Trial Balance of F. Datini's Second Company in Pisa (Prato, AS, D. No. 362, c. 156v–157r)

The authors have carried out a significant amount of work on translation of the medieval accounting books. A total of 311 key positions and 3 positions of the transfer of total indicators from one account to another were translated on just four folios.

What is the purpose of such significant work? Firstly, it will allow to inscribe all the accounts in the logic-analytical reconstruction of the medieval accounting complex, which will to a great extent increase the visual aspects of the materials under consideration. Secondly, it is aimed at studying the structure of the Ledger in general and the Trial Balance in particular. And, thirdly, it is quite important for establishment of the sequence, according to which a medieval accountant entered the closed balances into the Trial Balance. A quantitative assessment of the structure of the Trial Balance is introduced in Table 1.

Table 1. Quantitative characteristics of the Trial Balance of the Second Company in Pisa (1394–1395)

Folio	Number of positions		Credit
	Debit		
	Total	Including debtors	
Prato, AS, D. №362, c. 153v–154r	52	49	58
Prato, AS, D. №362, c. 154v–155r	46	46	27
Prato, AS, D. №362, c. 155v–156r	56	56	1
Prato, AS, D. №362, c. 156v–157r	60	17	12
In total	214	167	98

As it follows from the table, the Trial Balance occupies four folios, which are successively located in the Ledger. It is noteworthy that on the first three pages the total number of positions and the number of positions belonging to debtors coincide. The difference in three positions on the first folio will receive a special explanation. As for the fourth folio, where debtors represented by material values are collected, this is one of the main issues of the given study.

The selective translation of "Debtors" side of the first folio (Table 2) allowed us to establish the sequence that the accountant followed when forming this section.

Let us emphasize the fact that the first two entries in the Trial Balance do not refer to the Ledger. The first position was transferred from the book of cash accounting (Entrata e Uscita), and the second one from Memorial "B". At first glance, this seemed logical, since it was convenient for the accountant to enter the Ledger, in the first place the entries from other registers, and then to make a selection from the Ledger, with which he used. However, a detailed study of the account of Giovanni di Lorenzo and Cellino d'Orlando significantly changed our point of view.

The first entry indicates that for August 1, 1395, on page 117r of Entrata e Uscita (Prato, AS, D. No. 407), the balance of cash f. 114 s. 14 d. 9 was introduced. The study of Entrata e Uscita fully confirmed the specified sum. In addition, this sum will be indicated in Entrata e Uscita (Prato, AS, D. No. 407) as the opening balance of account in the Third Company's accounting system.

As for the second entry in "Debtors" side, it turned out that Giovanni di Lorenzo and Cellino d'Orlando were bankers, which was not mentioned in the Trial Balance entry, but it did take place in some entries in the accounts. It turned out that F. Datini conducted non-cash transactions with many suppliers and buyers through bankers as far back as the First Company (1392–1394). In the Trial Balance and post balance reporting there was a debit balance (without specifying the type of activity of correspondents). In the Memorial of the First Company, 33 accounts were opened for transactions with bankers, each of which occupied a whole folio, and in the Second

Table 2. Selective translation of positions in "Debtors" side on the first folio of the Trial Balance of the Second Company in Pisa (Prato, AS, D. 362, c. 153v)

1395		
DU1-1	Francesco di Marco da Prato and Manno d'Albizo and Co. **must give** on 1st of August f. 114 s. 14 d. *a oro* which we found in many currencies put in the book Uscita B at c. 117, we put them in the Entrata C in a new account at c. 2	f. 114 s. 14 d. 9
DU1-2	And on this day we gave them as debtor Giovanni di Lorenzo and Cellino d'Orlando as in the Memorial B at c. 155 in f. 797 s. 3 *a oro* and they **must give** in the Memorial C at c. 2	f. 793 s. 3
911.17.9		
DU1-3	And on this day for *cavas* and carpite and ropes we found that we assigned to them in cash in this book at the gain on merchandise's account at c. 365 and that we put in the new book at the expenses on merchandise's account at c. 209 they **must give**	f. 25
DU1-4	And on this day we assigned to them as debtor Piero di Strenna as in this book at c. 5 and we put in the black book C where he **must give** at c. 2	f. I3 s. 3 d. 9
DU1-5	And on this day we assigned to the as debtor Rosso di Francesco as in this he **must have** at c. 7, in the black book C he **must give** at c. 2	f. 14 s. 5 d. 4
DU1-6	And on this day for Antonio di Domenico Genni at c. 5 and put in the black book C he **must give** at c. 2	f. 31 s. 10
	....	
DU1-49	And on this day for Giovanni Bndini da Vila at c. 99 and put in the black book C he **must give** at c. 17	f. 11
DU1-50	And on this day for Lionardo di Sansone at c. 99 and put in the black book C, he **must give** at c. 18	f. 67 s. 8 d. 6
DU1-51	And on this day for Lazzaro Guinigi himself at c. 99 and put in the black book C he **must give** at c. 18	f. 66 s. 5 d. 4
DU1-52	And on this day, we put in this book they **must have** for balance of this account of this set of accounts f. 9456 s. 1 d. 7 *a oro* as at c. 155	f. 9456 s. 1 d. 7 *a oro*
The sum is f. 12355 s. 19 *a oro*		

Company there were opened 29 accounts. The closing balance in sum f. 797 s. 3 in the accounting system of the Third Company was also placed in the Memorial at Prato, AS, D. No. 373, c. 2r.

Of course, a special study will be devoted to the given finding, about which no one has ever written, including F. Melis, who was principal researcher of F. Datini's archives in Prato.

Thus, we can make another assumption: the accountant of Datini's second company in Pisa in "Debtors" side of the Trial Balance put information about money and transactions with bankers in the first place.

The third entry refers to the adjustment entries (Kuter et al. 2019a).

All other entries in "Debtors" side on the first folio (from DU1-4 to DU1-51) are of the same type. Each sum in the entry indicates the credit of the closing account (*avere* – must have) in the Ledger Bianco (Prato, AS, D. No. 362) and the debit of the opening account (*dare* – must give) in the Ledger Rosso (Prato, AS, D. №363). We should note that the rules for closing and opening accounts used by accountants in Francesco Datini's companies at the end of the 14th century are identical to those described at the end of the 15th century (1494) by Luca Pacioli in Chapter 36 of the Treatise.

Let us temporarily postpone the study of DU1-52 entry – the last position in "Debtors" side and move on to "Creditors" side on the first folio of the Trial Balance, a selective translation is shown in Table 3.

Table 3. Selective translation of the positions on the "Creditors" side on the first folio of the Trial Balance of the Second Company in Pisa (Prato, AS, D. 362, c. 154r)

1395		
CU1-1	Francesco di Marco and Manno d'Albizo and Co. *must have* on the 1st of August 200 florins *a oro* which are for a reserve set aside from the gain on the merchandise account as in this ledger at c. 364 and that reserve was made for wages of the workers and for expenses for Livorno middlemen and other expenses, and for mistakes we found. And this reserve *must have* in the Black Ledger at c. 63	f. 200
CU1-2	And they *must have* on this day we assign to them as creditor Giovanni Schiana di Francesco himself as they must give in this took at c. 3 and in the Black book C we put he *must have* at c. 3	f. 17 s. 13 d.
CU1-3	And on this day we gave to the mas creditor Lodovico Tolosini as they *must give* in this book at c. 4, and in the Black book C we put he *must have* at c. 4	f. 4 s. 14 d. 3
CU1-4	And on this day for Deo Ambruogi and Giovanni Francesco di Montpellier in Montpellier as they *must give* at c. 7 and put in the Black book C he *must have* at c. 4	f. 4 s. 8 d. 3
....	....	....
CU1-56	And on this day for Noferi di Bonaccorso in Pisa as in this at c. 134 and put in the Black book C he *must have* at c. 38	f. 200 s. 18 d. 2
CU1-57	And on this day for Bonaguida di Bagio as in this at c. 136 and put in the Black book C he *must have* at c. 41	f. 9 s. 10 d. 4
CU1-58	And on this day for Nicolò dell'Ammannati and Co as in this at c. 138 and put in the black book C he *must have* at c. 42	f. 80 s. 5 d. 6
Sum is f. 12355 s. 19		

The first entry in the account (CU1-1) indicates the reserve of pre-distribution profits of the finished reporting period. As is known (Kuter et al. 2019b), for the first time a reserve of pre-distribution profits was created in the First Company in 1394. Then, the accountant entered it into the credit in the Merchandise account after calculating the profit, as the difference between the right and left sides of the account. In the Second Company in 1395, the reserve was entered correctly into the debit of the merchandise account (f. 200) and by means of cross-entry was reflected by the first position in the credit of the Trial Balance.

The remaining entries in "Creditors" side of the first folio of the Trial Balance are of the same character: the number of the page on which the closing account is to be entered with the indication of the debit (*must give*) of this account. The entry in the closing account is in the debit of the account, since the account refers to the payable and is balanced by the entry in the debit. The second entry in the Trial Balance indicates the credit (must have) of the account in the new book.

After the last entry in the credit of the account on the first folio (CU1-58), the accountant calculated the Total on the credit of the Trial Balance – f. 12355 s. 19, which, as it turned out, exceeded the Total on the debit of the account by f. 9456 s. 1 d. 7.

Here it should be given explanations. In the early accounting, the Trial Balance, serve concurrently as of the account for accounts closing in the old book and accounts opening in the new book as well as acting as a Trial Balance, was perceived as a regular account, which was balanced according to the general rules for accounts in Venetian form, with parallel debit and credit. When the indicators of "Debtors" and "Creditors" were placed on the same folio, as for example, in F. Datini's First Company in Barcelona (1393–1395), there were no problems with the formation of the Trial Balance (Kuter et al. 2019a, b) – the sum of balances of the debit equaled the sum of balances of the credit. Otherwise, as Pacioli taught: "An error crept into your deeds".

In the cases like ours, it was necessary to balance the totals of the sides of each folio. Generally, in most of the Trial Balances that we studied, placed inside the Ledger, on the side with the smallest Total the sum before balancing of the account was indicated. The careful study of Fig. 1, which shows the first folio of the account, indicates that such a total does not exist. Apparently, such a total remained on the draft copy or abacus, with the help of which the accountant kept records. In the same place, in all likelihood, he identified a balancing indicator (addition to the Totals equality). Accordingly, on the "weak" side of the debit in the position of DU1-52, he wrote: "And on this day we put in this book they *must have* for balance of this account of this set of accounts f. 9456 s. 1 d. 7 *a oro* as at c. 155". Thus, the pair of Totals of "Debtors" and "Creditors" on the second folio of the Trial Balance will start with the incoming balance f. 9456 s. 1 d. 7 on the credit side.

A careful study of all the entries in "Debtors" side of the second folio of the Trial Balance shows their identity. As on the first folio, all the entries have the same structure: "as in this book at c. xxx and put in the Black book C he *must give* at c. xxx". Although, in order to simplify the entries, the accountant omitted the keyword *"must have"* in some positions, implying that it is definitely present. As, for example, position DU2-44 says: "And on this day for Lorenzo d'Ango del Doso as at c. 125 and put in the Black book C he *must give* at c. 5, f. 55".

The situation is similar with "Creditors" side of the second folio. At the beginning, the full format is maintained in all the entries, for example, CU2-3: "And on this day for Francesco and Andrea of Genoa for the company of Pisa as they *must give* in this at c. 139 and put in the Black book C they *must have* at c. 43, f. 786 s. 7 d. 3". In the second part of the folio the format is somewhat simplified. It does not have the key word *"must give"*. For example, the last entry CU2-27 is formed in the following way: "And on this day for Lorenzo di Chresci as at c. 152 and put in the black book C he *must have* at c. 63, f. 276 s. 12 d. 2".

As on the first folio, the credit total exceeds the debit total by f. 12916 s. 8 d. 10. Accordingly, the accountant transferred this sum from the debit side of the Trial Balance to the credit side as the opening balance of new account. The transfer entry (DU2-47) says: "And on this day we put in this book as they *must have* for balance of this account f. 12916 s. 8 d. 10 *a oro,* as in this book at c. 156".

The corresponding cross-entry can be found on the third folio in "Creditors" side (Prato, AS, D. No. 362, c. 155r): "Francesco di Marco and Manno d'Albizo and Co. *must have* on the 1st of August f. 12916 s. 8 d. 10 *a oro* for balance of an account of them and they *must give* at c. 154". As we can see in Table 1, the given entry in the section is the only one. It can be easily explained, remembering that the number of debit entries equals 214, and the number of credit entries of the Trial Balance is only 98.

As for "Debtors" side on the third folio, all the first 56 entries are carried out by the accountant meticulously: each entry contains both key words – *"must have"* and *"must give"*. And, what is worth special attention is that all the entries on the third folio, like the entries on the two previous folios (except the two entries on the first folio), refer to the balances of debtors closed accounts that are persons or companies.

The final entry in "Debtors" side of the third folio on AS, D. No. 362, c. 156v says: "And on this day we put in this account and the balance is f. 5210 s. 3 d. 7 *a oro* as at c. 157". This confirms the priority on the sum of credit entries over the debit ones described on the first three folios. Thus, the entries on the fourth folio of the Trial Balance begin with the opening balance of account in "Creditors" section.

For the given research, of particular interest is the sequence of entries in "Debtors" side of the fourth folio (Fig. 2 and Table 4).

The entry DU4-17 (And for Andrea di Bartolomeo dell'Onno as he *must have* at c. 153 and as in the Black book C he *must give* at c. 61) is the last one in the Trial Balance, which refers to the closed accounts of debtors.

Starting with DU4-18 entry (And for *risaalgallo* as it *must have* at c. 282 and as in the Black book C it *must give* at c. 200, f. 2 s. 10 d. 3), follow the entries of the Trial Balance, referring mainly to the account balances of closed accounts of purchase and sales of goods. A special study will be devoted to drawing up the account balances of these accounts, since securing the requirement of a reliable calculation of the cost of unsold goods stock through its inventory and valuation at the cost of acquisition, preparatory treatment and storage is the most important stage in the history of the development of double-entry bookkeeping, which created the necessary conditions for the formation of the merchandise account (at any time, and not after the sale of all consignments of goods), formation of the Trial Balance and balance financial statements.

The Total of the debit balances complete "Debtors" side of the last folio, not on all the Trial Balance entries, but only on the last page (the sum f. 6117 d. 9 *a oro*).

The translation of "Creditors" side of the fourth folio of the Trial Balance in Table 5 includes only 12 entries, and therefore it is shown in complete form.

Table 4. Selective translation of "Debtors" side on the fourth folio of the Trial Balance (Prato, AS, D. 362, c. 156v) of the Second Company in Pisa

1395		
DU4-1	Francesco di Marco and Manno d'Albizo and Co. *must give* on the 1st of August f. 2102 s. 16 d. 4 *a oro* of which we assigned to the for debtor Leonardo Portinari and Co. for the Pisa firma s they *must have* in this book t c. 149 and as in the Black book C they *must give* at c. 57	f. 2102 s. 16 d. 4
....	....	....
DU4-17	And for Andrea di Bartolomeo dell'Onno as he *must have* at c. 153 and as in the Black book C he *must give* at c. 61	s. 12 d. 4
DU4-18	And for *risaalgallo* as it *must have* at c. 282 and as in the Black book C it *must give* at c. 200	f. 2 s. 10 d. 3
DU4-19	And for 1 barrel of incense as it *must have* at c. 282 and as in the Black book C it *must give* at c. 200	f. 53 s. 15
DU4-20	And for 11 barrels of .... it *must have* at c. 283 and as in the black book C it *must give* at c. 200	f. 156 s. 4 d. 8
DU4-21	And for 1 .... of *oricello* as it *must have* at c. 284 and as in the Black book C it *must give* at c. 200	f. 8 s. 18 d. 4
DU4-22	And for many household goods as they *must have* at c. 287 and as in the Black book C they *must give* at c. 201	f. 155 s. 17 d. 9
....	....	....
DU4-53	And for expenses for 105 sacks of leather as it must have at c. 364 and as in the black book C it must give at c. 208	f. 36 s. 15 d. 8
DU4-54	And for many furs as they *must have* at c. 366 and as in the black book C they *must give* at c. 208	f. 122 s. 3 d. 10
DU4-55	And for 1 sack of wool of San Matteo as it *must have* at c. 366 and as in the black book C it *must give* at c. 208	f. –
DU4-56	And for ½ sack of *cassia* and 1 of *sena* as they *must have* at c. 361 and as in the Black book C they *must give* at c. 207	f. 40 s. 6 d. 6
The sum is f. 6117 d. 9 *a oro*		

The first entry was earlier described as the result of balancing of the previous pair of accounts (on the third folio). Ten subsequent entries were directly referred to the closing of accounts payable for delivered goods or services provided to Datini's company without indication of specific persons who delivered the goods or provided those services and to whom this debt was due to be paid.

And, of course, our contemporaries will be quite interested in the last (CU4-12) entry of the section. Typically, a similar entry occurred when the Total of the account balances on the debit side in all the Trial Balance folios did not match the Total of the account balances on the credit side. In the given case, the sum of cash assets and accounts receivable exceeded the sum of claims by f. 508 s. 11 d. 10. The given sum was added to the reserve from profit and was included in the reserves until the mistakes that led to the formation of these amounts were identified. The reverse situation was extremely undesirable, when claims on property by creditors exceeded cash assets and accounts receivable. In this case, the lack of property is perceived as a monetary debt of the accountant to the company and is listed in "Debtors" side until mistakes are detected.

Table 5. Translation of "creditors" section on the fourth folio of the Trial Balance of the Second Company in Pisa (Prato, AS, D. 362, c. 157r)

1395		
CU4-1	Francesco di Marco and Manno d'Albizo and Co. **must have** on the 1st of August f. 5210 s. 3 d. 7 *a oro* as balance of an account of them, we put in this book they **must give** at c. 156	f. 5210 s. 3 d. 7
CU4-2	And they **must have** on this day we assigned to them as a creditor lb. 400 of Burgunin yarn as in this book it **must give** at c. 331 and it **must have** in the Black Book C at c. 204	f. 19 s. 19 d. 8
CU4-3	And on the same day we assigned to the mas a creditor 11 sacks of Provençal wool as in this book it **must give** at c. 332 and it **must have** in the Black Book C at c. 204	f. 53 s. 9 d. 6
CU4-4	And for 20 *fardelli* of ..... as they **must give** at c. 336 and they **must have** in the Black Book C at c. 204	f. 54 d. 2
CU4-5	And for lb. 25 of *cinnamom* as it **must give** at c. 335 and it **must have** in the Black Book C at c. 205	f. 6 s. 10
CU4-6	And for 1 *pondo* of rice as it must give at c. 347 and it must have in the Black Book C at c. 206	f. 4 s. 3 d. 2
CU4-7	And for paint in grain as it **must give** at c. 346 and it **must have** in the Black Book C at c. 206	f. 33 s. 15 d. 7
CU4-8	And for 24 *veals* (leather) of Barcelona as they **must give** at c. 349 and they **must have** in the Black Book C at c. 207	f. 15 s. 1 d. 8
CU4-9	And for 328 *veals* (leather) and leather as they **must give** at c. 351 and they **must have** in the Black Book C at c. 207	f. 149 s. 14 d. 8
CU4-10	And for 5 .... of *veals* and other leathers as they **must give** at c. 360 and they **must have** in the Black Book C at c. 211	f. 56 s. 10 d. 11
CU4-11	And for the poor of Christ as in this book they **must give** at c. 152 and they **must have** in the Black Book C at c. 62	f. 10
CU4-12	And on this day f. 508 s. 11 d. 10 *a oro* which we put in the Black Book C as a reserve made from this account and we found more debtors than creditors. We think they were put in error and the book should be checked another time to find the error in the aforementioned Black Book at c. 63	f. 508 s. 11 d. 10
The sum is f. 6117 d. 9 *a oro*		

5   Conclusion

As a result of the study, it became possible to reproduce the sequence with which accountants of the end of the 14th century transferred the balances of closed accounts to the Trial Balance.

The balances of money accounts from other accounting books, such as Entrata e Uscita (cash balance) and the Memorial, were initially transferred to "Debtors" side. The debit balance of account of the account of settlements with bankers was transferred from the Memorial.

Further, the balances of personal accounts of real (physical or legal) debtors, mainly buyers of goods, were selected from the Ledger. And, at the end of "Debtors" side from the Ledger the balances of the accounts of purchase and sales of merchandise inventories according to not fully sold consignments were transferred. Consequently, active positions were placed in order of lowering liquidity – cash, receivables, goods for sale.

In "Creditors" side, the reserves created from profit were indicated first. Then, the account balances of identified suppliers of goods, persons providing specific services and other creditors by name were transferred. The list was continued by the balances, in which not specific creditors, but the goods supplied and services provided, which led to the emergence of payables, were indicated.

All the entries transferred from the Ledger were arranged in the order in which they took place in the Ledger, and in the same sequence, the accounts were opened in the new Ledger of the next period, more precisely, of the new company, as a rule, with the same owners.

The rules for keeping accounts of owners did not differ from the rules assigned to other personal accounts. The only difference was that the results of profit distribution were transferred to them.

The sum of discrepancy between the totals of balances on debit and credit sides completed the entries (either to debit or credit).

References

Primary

Archivio di Stato di Prato, Prato, AS, D., Fondaco di Pisa, Company Francesco di Marco Datini e Manno d'Albizo Degli Agli, Libri Grandi, 1392–1394, №361

Archivio di Stato di Prato, Prato, AS, D., Fondaco di Pisa, Company Francesco di Marco Datini e Manno d'Albizo Degli Agli, Libri Grandi, 1394–1395, №362

Archivio di Stato di Prato, Prato, AS, D., Fondaco di Pisa, Company Francesco di Marco Datini e Manno d'Albizo Degli Agli, Libri Grandi, 1395–1396, №363

Archivio di Stato di Prato, Prato, AS, D., Fondaco di Pisa, Company Francesco di Marco Datini e Manno d'Albizo Degli Agli, Memoriale, 1394–1395, №373

Archivio di Stato di Prato, Prato, AS, D., Fondaco di Pisa, Company Francesco di Marco Datini e Manno d'Albizo Degli Agli, Libri minori e speciali/ Saldi di ragione - Quaderni di ragionamento, 1395, №1164

Secondary

Alfieri, V.: La partita doppia applicata alle scritture delle antiche aziende mercantili veneziane. Paravia e Comp, Torino (1891)

Alfieri, V.: La partita doppia applicata alle scritture delle antiche aziende mercantili veneziane. V. Alfieri, Roma (1911)

Alfieri, V.: Commemorazione di Fabio Besta. Tipografia dell'Unione Arti Grafiche, Città di Castello (1922)

Besta, F.: La ragioneria, seconda edizione riveduta ed ampliata col concorso dei professori Vittorio Alfieri, Carlo Ghidiglia, Pietro Rigobon. Parte prima, vol. III. Vallardi, Milano (1922)

Castellani, A.: Nuovi testi fiorentini del Dugento e dei primi del Trecento. T. 2. The ledger of Giovanni Farolfi & Company, 1299–1300, is transcribed, almost entire. Firenze (1952)

Ceccherelli, A.: Le funzioni contabili e giuridiche del bilancio delle società medievali. Rivista Italiana di Ragioneria 14(8), 371–378 (1914a)

Ceccherelli, A.: Le funzioni contabili e giuridiche del bilancio delle società medievali (Continuazione e fine). Rivista Italiana di Ragioneria 14(10), 436–444 (1914b)

Ceccherelli, A.: I libri di mercatura della Banca Medici e l'applicazione della partita doppia a Firenze nel secolo decimoquarto. Bemporad, Firenze (1913)

Ceccherelli, A.: l linguaggio dei bilanci. Formazione e interpretazione dei bilanci commerciali. Le Monnier, Firenze (1939)

De Roover, R.: The development of accounting prior to Luca Pacioli according to the account-books of Medieval merchants. In: Littleton, A.C., Yamey, B.S. (eds.) Studies in the History of Accounting, pp. 114–174. London (1956)

Kuter, M., Gurskaya, M., Andreenkova, A., Bagdasaryan, R.: The early practices of financial statements formation in Medieval Italy. Account. Historians J. 44(2), 17–25 (2017). https://doi.org/10.2308/aahj-10543

Kuter, M., Gurskaya, M., Andreenkova, A., Bagdasaryan, R.: Asset impairment and depreciation before the 15th century. Account. Historians J. 45(1), 29–44 (2018)

Kuter, M., Gurskaya, M., Bagdasaryan, R.: The correction of double entry bookkeeping errors in the late 14th century. Working Paper (2019a)

Kuter, M., Sangster, A., Gurskaya, M.: The formation and use of a profit reserve at the end of the 14th century. Working Paper (2019b)

Melis, F.: Aspetti della vita economica medievale (studi nell'archivio Datini di Prato). Monte dei Paschi di Siena, Siena (1962)

Melis, F.: Storia della Ragioneria, 872 p. Cesare Zuffi, Bologna (1950)

Pacioli, L.: Traktat o schetakh i zapisiakh [Treatise on Accounts and Records]. In: Kuter, M.I. (ed.) Finansy i statistika and Prosveshchenie-Yug, Moscow and Krasnodar (2009)

Schar, J.F.: Buchhaltung und Bilanz (Bookkeeping and Financial Statements), 2nd edn. Berlin. (many editions, later by J.F. Schar and W. Prion, Berlin: Julius Springer, 1932) (1914)

Yamey, B.S.: Luca Pacioli. Exposition of Double Entry Bookkeeping, Venice 1494. In: Yamey, B.S., Gebsattel, A. (eds.) Albrizzi Editore, Venice (1994)

Zerbi, T.: Le Origini della partita dopia: Gestioni aziendali e situazioni di mercato nei secoli XIV e XV. Marzorati, Milan (1952)

Francesco Datini's «Entrata e Uscita» of the Second Company in Pisa

Mikhail Kuter[(⊠)] [iD], Marina Gurskaya[iD],
and Ripsime Bagdasaryan[iD]

Kuban State University, Stavropolskaya st., 149, Krasnodar 350040, Russia
prof.kuter@mail.ru

Abstract. The paper allows to consider the characteristic features of the recording of income and outgoing of monetary resources in the accounting books of Francesco di Marco Datini's company in Pisa in 1394–1395. This allows, along with the study of the features of accounting techniques and other accounting objects, to form a full picture of the organization of accounting procedures in companies and individual enterprises of medieval Europe. This issue was not sufficiently presented in the accounting literature earlier, but the presence of cash registers maintenance in medieval companies was repeatedly discussed as one of the prerequisites for the emergence and development of double-entry bookkeeping.

Keywords: Entrata e Uscita · Income and outgoing · Medieval accounting · Francesco di Marco Datini

1 Introduction

In the context of digital science and information technology, broad prospects for studying unsolved issues in the field of the history of economics in general and accounting in particular are being opened. The greatest success in the study of medieval accounting books is achieved by means of computer-based reconstruction of accounting complexes based on the construction of logical-analytical block-diagram models according to digitized photocopies of bookkeeping accounts. Studies, aimed at description of characteristic features of the accounting of individual segments of the medieval business, are difficult to be called popular at present. Often this is due to the insufficiency of interest among modern colleagues in the field of accounting history, the reason for which is the opinion that the results of previous studies are indisputable and that the need for new research does not arise. The peak of studies of medieval accounting practice fell on the beginning and middle of the 20th century (Castellani 1952; Littleton and Yamey 1956; Littleton 1966; Melis 1950, 1962, 1972; de Roover 1956; Zerbi 1952 and etc.). Particular attention is paid to the study of the Venice line in accounting (1891, 1911, 1922; Besta 1909, 1922; Lane 1944; Sieveking 1906), as well as the Tuscany line (de Roover 1956; Melis 1950; Martinelli 1974, 1977). By 1994, the 500th anniversary of the Luca Pacioli's Treatise, the interest in medieval accounting resumed again (Antinori 1994; Yamey 1994; Hernandez Esteve 1994).

© Springer Nature Switzerland AG 2020
T. Antipova (Ed.): ICIS 2019, LNNS 78, pp. 117–130, 2020.
https://doi.org/10.1007/978-3-030-22493-6_12

It is impossible not to note the importance of the safekeeping of sources that allow to carry out such studies. As for the Venetian accounting books, the earliest survived ones that can give an idea of the early methods of accounting are dated the middle of the 15th century (Besta 1922, 302–303; Melis 1950, 532–533). The Tuscany line in accounting was more fortunate. The earliest source confirming the use of the most progressive form of accounting, the double-entry bookkeeping, was discovered and stored in the archives of Florence and dated 1211 (Lee 1973). That makes it possible to agree on the assumption of the primacy of using double-entry bookkeeping in Tuscany, as many scholars suggest (Ceccherelli 1910, 29–34; 1939, 145; de Roover 1937, 274–275, 1956, 141; Melis 1950).

The archives of the merchant from the city of Prato (20 km from Florence) Francesco di Marco Datini (1335–1410) give the most full picture of accounting practices maintenance in the late 14th and early 15th centuries. His companies and enterprises were spread throughout Europe and were engaged in various types of activities, from trading everyday goods in shops to full-scale wool production (Nigro 2010). Currently, the survived documents include more than 500 accounting books, more than 100,000 documents that give insight into the conduct of business in companies (Nigro 2010), which allow to study in detail all the transactions, operations and their accounting support, carried out more than 600 years ago.

One of the most interesting and most difficult objects of accounting supervision in medieval accounting practice is cash accounting. It is not easy to find literary sources, describing the cash movement accounting, among accounting historians. This issue is more interesting to historians of economics (Cipolla 1967, Goldthwaite 2009; de Roover 1999; Mueller 1997), who pay great attention to the history of medieval money and mention the peculiarities of keeping accounting books with the help of special money systems of "money of account" or "ghost money" (Cipolla 1967).

The authors decided to devote the given study to the peculiarities of keeping special accounting books intended to reflect the income and outgoing of monetary resources in one of Francesco di Marco Datini's most progressive companies – the company in Pisa 1394–1395.

2 Review of Prior Literature

Perhaps there will arise a question: why the study of the accounting system of the second company begins with cash? There is an explanation for this. One of the distinguished researchers of the history of double-entry bookkeeping, Professor Tommaso Zerbi, stated that "none of Datini's books contains a clear example of double entry" (1952, p. 136). As R. de Roover stated: "Professor Zerbi, who carefully studied Datini's books, explains this by the lack of a cash account, although there is a cash book that could be used as a substitute" (1956, p. 141). De Roover does not share this opinion, considering that "Combining the cash book with the General Ledger would allow an accountant to establish a balance, provided, of course, that he observes the rules in other respects. Therefore, this argument is not a governing one" (ibid.). We will try to obtain confirmation of what was said in the following text.

Sangster (2016), considering the causes of double-entry bookkeeping emergence, described the importance of "Quaderni di Cassa" both in the banking sector and its use by merchants. Unfortunately, he did not consider a more important register for money movement accounting – "Entrata e Uscita".

The publication (Aleynikov 2017) examined in detail "Entrata e Uscita" in Francesco Datini's early companies in Avignon, where single-entry bookkeeping was used. As a rule, cash accounting books consisted of two separate sections "Entrata" (income) and "Uscita" (outgoing). These two sections could fit in a book under the same cover (Entrata e Uscita), or each section could have a separate cover. There have been cases when a single book, in addition to the specified sections, contained "Memoriale" section.

At the stated periods of time for each section (Income and Outgoing) the totals were calculated. The total of "Uscita" was transferred to "Entrata" and was recorded below the total of "Entrata". The identified cash balance was entered as the first indicator of "Entrata" in the new period (during the course of the company's duration), the entries in "Uscita" began with a zero indicator.

The authors noticed that the entries in the cash outgoing section (Uscita) were transferred from the Memorial for each transaction, and the section of income (Entrata) reflected the total amount accumulated over a day or several days.

The distinctive feature of Entrata e Uscita bookkeeping in F. Datini's individual enterprises and First Company in Pisa is that both income and outgoing of cash entries successively reflect each entry registered in the Memorial or that had correspondences with other accounting registers.

3   Research Method

The principal research method adopted in this study is archival. It uses material found in the State Archive of Prato. This research team has been working with the material in this archive for the past decade and many of the records have been recorded and linked together using logical-analytical reconstruction. This is an approach that we developed for the purpose of enabling entries in the account books to be traced visually between accounts and books and from page to page. By adopting this approach, we are able to see the entire accounting system electronically, making entries and their sources clear in a way that is not possible if all that you have is the original set of account books. This enables us to consider each transaction in detail, trace its classification, and so explain the bookkeeping and accounting methods adopted without possibility of mis-interpretation. This approach represents a new paradigm in how to analyses and interpret accounting practice for periods when there was no concept of either a standard method or a unified approach to either financial recording or financial reporting.

4   Statement of Basic Materials

Let us try to consider the organization of cash accounting in Datini's Second Company on the basis of examination of the entries in the income and outgoing sections based on the study of Entrata e Uscita book with the archive identifier Prato, AS, D. No. 407.

Figure 1 shows the first page (Prato, AS, D. No. 407, c. 2r) of "Entrata", the translation of the text is given in Table 1, and Fig. 2 shows the diagram of interrelation between the indicators on the page.

Fig. 1. The first page (Prato, AS, D. No. 407, c. 2r) on the income side in "Entrata e Uscita". Datini's second company in Pisa (1394–1395)

As you can see, at the end of page Prato, AS, D. No. 407, c. 2r the total f. 126 s. 3 is calculated, that was transferred to page Prato, AS, D. No. 407, c. 105r, where the intermediate totals of the Uscita as at December 30, 1394 are displayed.

The first entry in account Prato, AS, D. No. 407, c. 2r is the transfer of balance of account from the Trial balance placed inside the Ledger (Prato, AS, D. No. 361, c. 269v) at the time of the closing of the First Company – the sum of f. 44 s. 10 d. 3. The text in the Trial Balance says: "And on this day in this book on c. 268 cash *must have*, and in the Red Book of B on c. 285 it *must give*, f. 44 s. 10 d. 3".

We should note that the accountant transferred to page Prato, AS, D. No. 407, c. 105r the totals from all pages 2r–4v. Moreover, at the bottom of page 4v he calculated the sum total of pages 2r–4v (the sum – f. 916 s. 17 d. 8.), as evidenced by the first ellipse in Fig. 3.

Figure 4 shows the procedure of introducing the cash balance for December 30, 1394 (the intermediate balance – the balance – within the period of the company's operations).

Table 1. Translation of the text of page Prato, AS, D. No. 407, c. 2r Entrata e Uscita of F. Datini's Second Company in Pisa

July 14th 1394	
From cash on this day f. 44 s. 10 d. 3 a oro we found in cash when balancing the account in the form of different coins, as in the White Book B at c. 269	f. 44 s. 10 d. 3
From Bartolomeo di Franc....linen cloths on July 16th s. 17 d. 2 a oro we had in cash, Nanni brought the mas in the ~~Red~~ Book B at c. 5	f. – s. 17 d. 2
From Giovanni di Lorenzo and Cielino on July 18th f. 30 d'oro for us to messer Banducio Bonconti and Co. for grossi we got from them in two time sas in the Memorial A at c. 4	f. 30
From Ciuccio swordmaker on this day f. 2 d'oro we had in cash brought by Nanni, as in the ~~Red~~ Book B at c. 29	f. 5 s. 7 d. 2
From Luca Dartio on this day f. 5 s. 7 d. 2 a oro we got on his behalf for him from Ghirigoro d'°Antonio as in the Memorial B at c. 5	f. 5 s. 7 d. 2
From Cino di Bandino linen cloth on 21th July s. 2 d. 6 a oro we had in cash brought by Nani as in the Red Book B at c. 5	s. 2 d. 6
From ~~Giovanni di Cino~~ Michele del Campane from Prato on July 26 f. 6 s. 17 d. 2 d'oro we got in cash from Manno as in the Red Book B at c. 19, and in f. they lost f. 4 s. 1 d. 5 a oro, what remains is	f. 6 s. 16 d. 9
From Berlinghieri Feriola catalan on July 28 f. 12 s. 5 d. 5 a oro for him from Buondi di Giusepo, Jew, at the Memorial B at c. 8	f. 12 s. 5 d. 5
From Expenses for merchandise on July 31th f. 2 we got from Antonio di Bartolomeo Sostegni for insurance and sindaco of C farde of wool we found in the ol account one year ago through the ship of Stefano Broglo from Lorini di Bardini di Cherichino, and we had to give on his behalf to Francesco di Bonaccorso and for our stabling estimated by Antonio (.........) as in the Memorial A at c. 213 and in the Red Book B at c. 285 at the Expenses account	f. 2
From Borsotto di Bochino shipper on August 13th f. 7 we got on his behalf from Francesco da Riglone as in the Red Book B at c. 29	f. 7
From one deposit from San Marco Door f. 1 on this day, until August 1st as in the Red Book D at c. 5	f. 1
From Giovanni di Lorenzo and Cielino d'Orlando on August 4th f. 7 s. 8 d. 7 a oro, from us to Domenico hempmaker and they are for Lorenzo Ciampeli as in the Cash Notebook B at c. 2 must give in the Memorial B at c. 4	f. 7 s. 8 d. 7
From Bonaiuto del Impeto apothecary on this day f. 7 s. 1 d. 2 a oro we got in cash as in the Red Book B at c. 24	f. 7 s. 1 d. 2
126.8.0	
Put in this book in the bottom of c. 105	f. 126 s. 8

To page Prato, AS, D. No. 407, c. 105r six page-by-page totals of the cash withdrawal from the cash register (Uscita) are transferred: 102r – f. 61 s. 18 d. 6; 102v – f. 74 s. 11 d. 5; 103r – f. 43 s. 18 d. 6; 103v – f. 156 s. 2 d. 7; 104r – f. 28 d. 4; 104v – f. 568 s. 18 d. 7. The sum total of cash given was f. 923 s. 9 d. 11.

Below are the totals of the first six pages of the cash income from Entrata: 2r – f. 126 s. 18 d. 6; 2v – f. 126 s. 18 d. 10; 3r – f. 164 s. 13 d. 11; 3v – f. 117 s. 9 d. 7; 4r – f. 217 s. 1 d. 7; 4v – f. 164 s. 5 d.10. The sum total of cash income is f. 916 s. 17 d. 8.

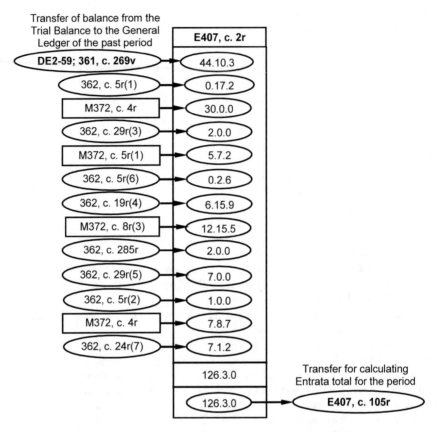

Fig. 2. Linking diagram on the first page (Prato, AS, D. No. 407, c. 2r) of the cash income section in "Entrata e Uscita"

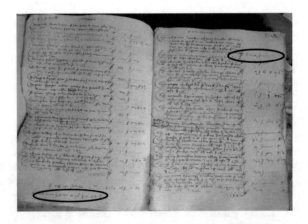

Fig. 3. The page of introducing cash income total (Prato, AS, D. No. 407, c. 4v–5r) on the side of cash income in "Entrata e Uscita"

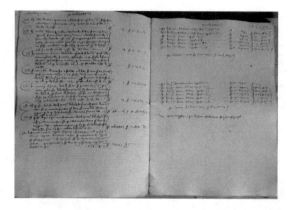

Fig. 4. The introducing of cash outgoing in "Entrata e Uscita" on Prato, AS, D. No. 407, c. 104v–105r.

The last total coincides with the total on page Prato, AS, D. No. 407, c 4v.

Thus, it can be stated that quite an interesting and rare situation arose when the cash register gave f. 6 s. 12 d. 3 too much than it had been entered, as evidenced by the last entry on page: "The Outgoing (Uscita) is higher than the Income (Entrata), f. 6 s. 12 d. 3".

As a rule, the introduced final balance of account (cash balance) is the first indicator in Entrata of the next period. In the given example, such an indicator should have been the first indicator on Prato, AS, D. page. No. 407, c. 5r (Fig. 3, the second ellipse). However, due to the current situation, this did not happen. In addition, the first indicator is the sum of f. 81, the first cash entry in the second intermediate period, and not the sum of the balance of cash introduced.

Let us study the procedure of introducing of the intermediate cash balance in the following period of time – March 23, 1394. Here, an uninformed reader may be surprised: the previous period ended December 30, 1394 and the analyzed period – March 23, 1394. The fact is that in medieval commune-cities, which in future will make up the modern territory of Italy, the year changed in March, respectively, in April the data of 1395 will be studied.

It should be noted that the accountant-manager, who did the bookkeeping and compiled post reporting it the First Company in Pisa (Simone Belandi), received an independent activity sphere and since 1395 headed F. Datini's new company in Barcelona. The new accountant did not possess sufficient knowledge and experience and was often not consistent.

The entries of the new (second) intermediate period were recorded by the accountant in Entrata on pages 5r – 8v, and in Uscita on pages 105v – 109v. If during the previous period the accountant introduced the totals of the income and outgoing of cash in Uscita, then this time he calculated the totals of each section separately.

For that purpose, he summarized the totals for each page: 5r – f. 166 d. 10; 5v – f. 104 s. 10 d. 7; 6r – f. 194 s. 9 d. 8; 6v – f. 157 s. 15 d. 1; 7r – f. 221 s. 13 d. 8; 7v – f. 198 s. 6 d.10; 8r – f. 579 s. 1; 8v – f. 371 s. 8 d. 10. Page Prato, AS, D. No. 407, c. 8v (Fig. 5) displays the total of pages 5r–8v: f. 1993 s. 6 d. 6 (marked with an ellipse).

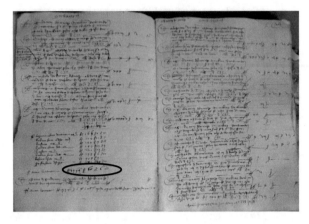

Fig. 5. The page of drawing up the total of cash income (Prato, AS, D. No. 407, c. 8v) on the cash income side in "Entrata e Uscita".

The translation of the last entries on the page recorded below the total:

"From Duccio di Giovanni apothecary on 2th February f. 1 d'oro in cash at the Red Book B at c. 86"

The total sum of all the Entrata (Income) book f. 1994 s. 6 d. 6 *a oro* put at the bottom of the Uscita (Outgoing) in this book at c. 110".

The total is transferred to page Prato, AS, D. page. No. 407, c. 110r.

Further, the accountant entered on page Prato, AS, D. No. 407, c. 110r (the photocopy is in Fig. 6) page-by-page totals of Uscita: 105v – f. 18 s. 8 d. 6; 106r – f. 43 s. 1 d. 2; 106v – f. 75 s. 11 d. 1; 107r – f. 50 s. 17 d. 7; 107v – f. 53 s. 11 d. 9; 108r – f. 80; 108v – f. 237 s. 3 d. 9; 109r – f. 118 s. 1 d. 9; 109v – f. 1271 s. 7 d. 8. The total of cash withdrawal is f. 1998 s. 3 d. 3 (marked with the upper ellipse). The confirmative translation is given in Table 2.

The total of cash income was entered below this total – f. 1993 s. 5 d. 6 (calculated in Entrata on Prato, AS, D. No. 407, p. 8v, lower ellipse).

As in the first case, the sum of cash given exceeds by f. 3 s. 16 d. 9 cash income.

In the second company in Pisa, the accountant drew up the balance in cash two more times: July 4, 1395 and at the closing of the company (August 1, 1395). Figure 7 shows the final pages of Entrata (Prato, AS, D. No. 407, c. 13r) and Uscita (Prato, AS, D. No. 407, c. 114v).

In this case, on the page of Entrata as well as on the page of Uscita page-by-page totals were not listed, but the total of the pages of the section was entered: on c. 13r – f. 2647 s. 15 d. 5; on s. 114v – f. 2648 s. 5 d. 9.

Table 3 shows the translation of the entries that were of certain interest for the research on Prato, AS, D. No. 407, c. 13r, Table 4 shows a similar translation on page Prato, AS, D. No. 407, c. 114v.

Further, the total from the Uscita is transferred and entered on Prato, AS, D. No. 407, c. 13r below the total of Entrata. As in the first two cases, the sum of cash given exceeded the sum of cash received by s. 10 d. 4.

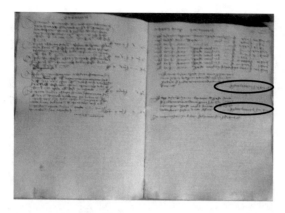

Fig. 6. Drawing up balance in cash in "Entrata e Uscita" on Prato, AS, D. No. 407, c. 109–110r

Table 2. Translation of the text on page Prato, AS, D. No. 407, c. 110r Entrata e Uscita

March 23ʳᵈ, 1394	
Total sum of the first page in this at c. 105	f. 18 s. 8 d. 6 *a oro*
Sum of the second page at c. 106	f. 43 s. 1 d. 2
Sum of the 3rd face at c. 106	f. 75 s. 11 d. 1
Sum of the 4th face at c. 107	f. 50 s. 17 d. 7
Sum of the 5th face at c. 107	f. 53 s. 11 d. 9
Sum of the 6th face at c. 108	f. 80
Sum of the 7th face at c. 108	f. 237 s. 3 d. 9
Sum of the 8th face at c. 109	f. 118 s. 1 d. 9
Sum of the 9th face at c. 109 on the other side	f. 1271 s. 7 d. 8
Sum of all the Outgoing (Uscita) of this account kept by Piero d'Antonio Zampini starting from December 1st 1394 until today, f. 1998 s. 3 d. 8 *a oro*	f. 1998 s. 3 d, 8 *a oro*
Sum of all the income (Entrata) of this account f. 1994 s. 6 d. 6 *a oro* as it can be seen in this book at c. 8, the total sum is at the bottom of the page	f. 1994 s. 6 d. 6 *a oro*
The Outgoing is higher than the Income	f. 3 d. 16 d. 9 *a oro*

August 1, 1395 the second company in Pisa was closed. When closing the company, as in the previous variant of drawing up the intermediate balance of cash, the accountant did not rewrite the page-by-page totals, but only calculated the sum total for Entrata (on c. 14r – the sum of f. 377 s. 3 d. 7) and the sum total for Uscita (on c. 117r – the sum of f. 379 s. 15 d. 6). The translation of the final entries of page Prato, AS, D. No. 407, c. 14v says: "July 31th 1395. It was a sum of all the income from the 13th to the 13th to the 14th century. 13 on the second page until c. 14, they are three pages and f. 377 s. 3 d. 7 *a oro*. This sum is put in this book at c. 117 under the sum of the income of this account".

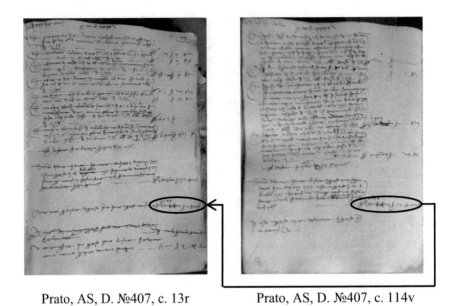

Prato, AS, D. №407, c. 13r   Prato, AS, D. №407, c. 114v

Fig. 7. Drawing up balance in cash in Entrata e Uscita on Prato, AS, D. No. 407, c. 13r and 114v

Table 3. Translation of the text on page Prato, AS, D. No.407, c. 13r Entrata e Uscita

June 30th, 1395	
....	
Total sum of the income (Entrata) from March 22th 1394 to today July 4th 1395, from c. 9 until this c. 13: they are 9 pages and f. 2647 s. 15 d. 5 *a oro*	f. 2647 s. 15 d. 5 *a oro*
They had for the outgoing (Uscita) of this account as in this book at c. 214	f. 2648 s. 5 d. 8
~~This account is unbalanced, the Outgoing is less than the Income~~	~~f. VIIII s. VIIII d. VIII a oro~~
This account is unbalanced, the Outgoing is higher than the Income, the account is kept by Piero d'Antonio Zampini	f. – s. 10 d. 4 *a oro*

Table 4. Translation of page Prato, AS, D. No. 407, c. 114v Entrata e Uscita

July 4th, 1395	
....	
Total sum of all the Outgoing (Uscita) of this account starting from March (NOT READABLE) 1394 until this day July 4th 1395 from c. 210 until c. 214, they are 9 pages f. 2648 s. 5 d. 9 *a oro*	f. 2648 s. 5 d. 9 *a oro*
In this book at the bottom of the Income as in the Memorial at c. 13	

The entries on Prato, AS, D. No. 407, c. 117r are shown in Table 5.

Table 5. Translation of page Prato, AS, D. No. 407, c. 117r Entrata e Uscita

August 1st 1395	
Cash account started on August 2nd of the new *ragione* by Francesco and Manno and Co, f. 114 s. 14 d. 9 *a oro* we found in cash in many coins on this day while balancing the account of the company, as in the Red Book B at c. 153 and in the Income (Entrata) of the new account C at c. 2	f. 114 s. 14 d. 9 *a oro*
The sum is f. 114 s. 14 d. 9 *a oro*	
Total sum of all the Outgoing of this account kept by Piero d'Antonio Zampini starting from July 5 1395 until August 1st 1395 as it can be seen here from c. 115 until c. 117, f. 379 s. 10 d. 6 *a oro*, they are 5 pages	f. 379 s. 10 d. 6 *a oro*
They had for the income (Entrata) of this account as it can be seen here at c. 14 for the amount of	f. 372 s. 3 d. 7 *a oro*
The Outgoing is higher than the Income of f. 2 s. 11 d. 11 *a oro*	
Then, we found the balance of this account – kept by the aforementioned Piero, who got it by Marchacio Rosso transporter – f. 1 s. 18 d. 10 *a oro*, put in the Black Book C at c. 7	f. 1 s. 18 d. 10
In the end, the Outgoing is higher than the Income for f. – s. 13 d. 1 *a oro*	

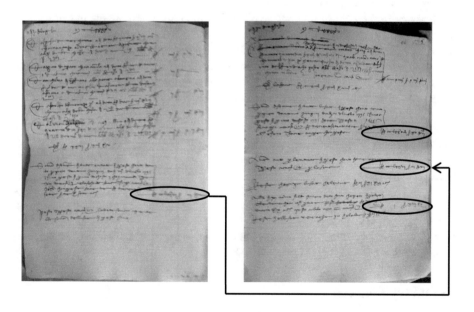

Prato, AS, D. №407, c. 14v   Prato, AS, D. №407, c. 117r

Fig. 8. Drawing up balance of cash in "Entrata e Uscita" on Prato, AS, D. No. 407, c. 117r at the closing of Datini's Second Company in Pisa

As shown in Fig. 8, the sum total of Entrata is transferred to page Prato, AS, D. No. 407, c. 117r and recorded below the sum total of Uscita. The difference is f. 2 s. 11 d. 11, and again it was against cash income.

5 Conclusion

Thus, it can be stated that in Francesco Datini's Second Company in Pisa (1394–1395) there was a logically organized cash accounting system. Entrata e Uscita book consisted of two separate sections: for cash Income (Entrata) and cash Outgoing (Uscita).

Between them there was a reserve of blank pages, which was formed due to the allocation of more pages for Entrata than was actually filled.

It is very important to note that the accountant did not register the accumulated totals registered in the Memoriale, as was the case with single-entry bookkeeping in Avignon. As Fig. 3 confirmed, individual indicators of cash income and outgoing both from the Memoriale and General Ledger were entered in Entrata e Uscita.

Periodically the totals for each section were calculated. The accountant of the Second Company was not consistent. During some periods, he calculated the balance of cash in Entrata, at other times in Uscita.

As the analysis of the reflected business transactions has shown, in each intermediate accounting period the sum of cash given exceeded the sum of cash received. In this respect, the accountant did not give any explanation for this.

Research in the field of accounting history is impossible without the use of modern digital technologies. The possibilities of information technology have greatly facilitated the work of scholars and have significantly increased the level of such research. It should be noted is the possibility of digitized medieval archival materials, which not only contributes to their preservation for future generations but also allows for research using Internet technology not be going to the archives.

References

Primary Sources

Prato, A.S., Fondaco di Pisa, D.: Company Francesco di Marco Datini e Manno d'Albizo Degli Agli, Libri Grandi, 1394–1395, No. 362

Prato, A.S., Fondaco di Pisa, D.: Company Francesco di Marco Datini e Manno d'Albizo Degli Agli, Entrata e Uscita, 1394–1395, No. 407

Secondary Sources

Alfieri, V.: La partita doppia applicata alle scritture delle antiche aziende mercantili veneziane. Paravia e Comp, Torino (1891)

Alfieri, V.: La partita doppia applicata alle scritture delle antiche aziende mercantili veneziane. V. Alfieri, Roma (1911)

Alfieri, V.: Commemorazione di Fabio Besta. Tipografia dell'Unione Arti Grafiche, Città di Castello (1922)

Aleinikov, D., Kuter, M., Musaelyan, A.: The early cash account books. In: Antipova, T., Rocha, Á. (eds.) Information Technology Science. MOSITS 2017. Advances in Intelligent Systems and Computing, vol. 724, pp. 195–207. Springer, Cham. https://doi.org/10.1007/978-3-319-74980-8_18. Accessed 20 Feb 2019

Antinori, C.: Luca Pacioli. E la Summa de Arithmetica. Dopo 500 anni dalla stampa della 1a edizione (1494—1994). La vita, le opere, il Trattato XI de computis et scriptures. Istituto poligrafico e zecca dello stato, Roma (1994)

Besta, F.: La Ragioneria, 2nd edn., vol. 3. Rirea, Rome (1909). Facsimile Reprint, 2007

Besta, F.: La ragioneria, seconda edizione riveduta ed ampliata col concorso dei professori Vittorio Alfieri, Carlo Ghidiglia, Pietro Rigobon. Parte prima, vol. III. Vallardi, Milano (1922)

Castellani, A.: Nuovi testi fiorentini del Dugento e dei primi del Trecento. T. 2. The ledger of Giovanni Farolfi & Company, 1299–1300, is transcribed, almost entire. Firenze (1952)

Ceccherelli, A.: Le scritture commerciali nelle antiche aziende fiorentine. Lastrucci, Firenze (1910)

Ceccherelli, A.: I libri di mercatura della Banca Medici e l'applicazione della partita doppia a Firenze nel secolo decimoquarto. Bemporad, Firenze (1913)

Ceccherelli, A.: Le funzioni contabili e giuridiche del bilancio delle società medievali. Rivista Italiana di Ragioneria 14(8), 371–378 (1914a)

Ceccherelli, A.: Le funzioni contabili e giuridiche del bilancio delle società medievali (Continuazione e fine). Rivista Italiana di Ragioneria 14(10), 436–444 (1914b)

Ceccherelli, A.: l linguaggio dei bilanci. Formazione e interpretazione dei bilanci commerciali. Le Monnier, Firenze (1939)

Cipolla, C.: Money, Prices and Civilization in the Mediterranean World: Fifth to Seventeenth Century. Gordian Press, New York (1967)

de Roover, R.: Aux origines d'une technique intellectuelle. La Formation et l'expansion de la comptabilité à partie double, Annales d'histoire économique et sociale IX, 171–193, 270–298 (1937)

de Roover, R.: The Medici Bank organization and management. J. Econ. Hist. 6(1), 24–52 (1946)

de Roover, R.: New perspectives on the history of accounting. Account. Rev. 30(3), 405–420 (1955)

de Roover, R.: The development of accounting prior to Luca Pacioli according to the account-books of Medieval merchants. In: Littleton, A.C., Yamey, B.S (eds.) Studies in the History of Accounting, pp. 114–174. London (1956)

de Roover, R.: The rise and decline of the Medici Bank: 1397–1494. In: Harvard Studies in Business History. Beard Books, New York (1999)

de Roover, R. The story of the Alberti company of florence, 1302–1348, as revealed in Its account books. Bus. Hist. Rev. 32(1) (Spring), 14–59 (1958)

de Roover, R.: The organization of trade. In: Postan, M.M., Rich, E.E., Miller, E. (eds.) Economic Organization and Policies in the Middle Ages. Cambridge University Press, Cambridge (1971)

Hernandez Esteve, E.: Luca Pacioli: De las cuentas y las escrituras. Titulo Noveno, Tratado XI de su Summa de Arithmetica, Geometria, Proportioni et Proportionalita. Venecia, 1494. Estudio introductorio, trduccion y notas por Esteban Hernandez Esteve. Con una reproduccion fotografica del original. – Editado por Asociacion Espanola de Contabilidad Y Administracion De Empresas (AECA), Madrid (1994)

Goldthwaite, R.A.: The Economy of Renaissance Florence. The Johns Hopkins University Press, Baltimore (2009)

Kuter, M., Gurskaya, M., Andreenkova, A., Bagdasaryan, R.: The early practices of financial statements formation in Medieval Italy. Account. Historians J. **44**(2), 17–25 (2017). https://doi.org/10.2308/aahj-10543

Lane, F.C.: Andrea Barbarigo, Merchant of Venice, 1418–1449. Johns Hopkins Press, Baltimore (1944)

Lane, F.C.: Doubles entry bookkeeping and resident merchants. J. Eur. Econ. Hist. **6**, 177–191 (1977)

Lee, G.A.: The Development of Italian Bookkeeping 1211–1300. Abacus **9**(2) (1973)

Littleton, A.C.: Accounting Evolution to 1900. Russell & Russell, New York (1966)

Littleton, A.C., Yamey, B.S.: Studies in the History of Accounting. Richard D. Irwin, Inc., Homewood (1956)

Martinelli, A.: The Origination and evolution of double entry bookkeeping to 1440. Ph.D. Dissertation, North Texas State University (1974)

Martinelli, A.: Notes on the origin of double entry bookkeeping. Abacus **13**(1), 3–27 (1977)

Melis, F.: Storia della Ragioneria, p. 872. Cesare Zuffi, Bologna (1950)

Melis, F.: Aspetti della vita economica medievale (studi nell'archivio Datini di Prato). Monte dei Paschi di Siena, Siena (1962)

Melis, F.: Documenti per la storia economica dei secoli XIII–XVI, p. 752. Leo S. Olschki, Firenze (1972)

Mueller, R.: The Venetian Money Market: Banks, Panics, and the Public Debt, 1200–1500. Johns Hopkins University Press, Baltimore (1997)

Nigro, G.: Datini Francesco di Marco: the man the merchant. Firenze University Press, Firenze (2010)

Origo, I.: The Merchant of Prato, Francesco di Marco Datini. J. Cape, London (1957)

Sangster, A.: The genesis of double entry bookkeeping. Account. Rev. **91**(1), 299–315 (2016)

Sieveking, H.: Studio sulle Finanze Genovesi nel Medioevo e in Particolare sulla Casa di San Giorgio. Atti della Societa Ligure di Storia Patria. XXXV (1906)

Yamey, B.S.: Luca Pacioli. Exposition of double entry bookkeeping (Venice). Albrizzi editore, Venice (1994)

Zerbi, T.: Le Origini della partita dopia: Gestioni aziendali e situazioni di mercato nei secoli XIV e XV. Marzorati, Milan (1952)

Formation of the Information Space for Audit and Taxation as a Factor for the Improvement of Investment Attractiveness of the Ukrainian Economy

Irina Chumakova[1,2(✉)] [ID], Lyudmila Oleinikova[1] [ID],
and Svitlana Tsevukh[2] [ID]

[1] State Educational and Scientific Institution «Academy of Financial Management», Druzhby Narodiv Boulevard 28, 01014 Kyiv, Ukraine
chumakovaafu@gmail.com
[2] Department for International Economic Relations, University of Economics and Law «KROK», Tabirna Street, 30-32, 03113 Kyiv, Ukraine

Abstract. This article is a research study of supranational and national peculiarities of the formation of the information space for audit and taxation purposes in the context of the digitalization of the world economy, which in turn should contribute to improving the investment attractiveness of the Ukrainian economy within long-term foreign capital inflow. Considerable attention is paid to the supranational framework documents and standards of the financial reporting in the framework of the formation of the information and tax environment in Ukraine and transformation of tax reporting and ways of its transmission through Internet resources, which directly affects the expansion of the information space and facilitates reasonable managerial and investment decision making. The necessity to disclose information for tax purposes and for ensuring the transparency of business processes in the global space strengthens the role of international financial reporting standards, the exchange of tax and financial information in the framework of the development of digital economy. Particular attention is paid to the significance of eXtensible Business Reporting Language as a technology that stimulates and allows each stakeholder to obtain considerable benefits and to get added value from the use of IFRS financial statements throughout the business information chain.

Keywords: Investment attractiveness ·
Information environment for tax purposes ·
Audit of XBRL-based structured digital financial reports

1 Introduction

The process of Ukraine's integration into the European Community presupposes the formation of an attractive investment environment, the determinants of which are the integrity and transparency of business information and its closeness to all regulators and participants of financial markets, which take managerial and investment decisions.

© Springer Nature Switzerland AG 2020
T. Antipova (Ed.): ICIS 2019, LNNS 78, pp. 131–147, 2020.
https://doi.org/10.1007/978-3-030-22493-6_13

The conclusion of the Association Agreement between the European Union and Ukraine and the implementation of reforms in various spheres of socio-economic life have given new impetus to modernization of taxation, accounting and financial reporting and audit related to the necessity of synchronizing the national legislation of Ukraine with the requirements of European directives and principles of international taxation. Declared in the Strategy for the Reforming of the Public Finance Management System for 2017–2020, in the Memorandum of Economic and Financial Policies between Ukraine and the IMF and other legislative acts and normative documents of Ukraine, reforms are aimed at improving tax control, preventing the practice of the tax base blurring and tax evasion as well as improving the quality, completeness and reliability of the data as the basis for managerial decision making in various sectors of economy. In the context of the rapid progress in the field of information technologies (IT), the deepening of the processes of globalization and the digitalization of the world's capital markets and international economic relations, the urgency of solving these issues for Ukraine has increased tremendously and has gained not only internal economic but also transboundary importance for the development of all branches of economic activity.

Theoretical and applied aspects of these issues were considered by a number of Ukrainian economists and practitioners: Iefymenko and Lovinska [1], Kamenska [2], Ozeran [3], Petryk [4], et al. The analysis of the research works prepared by the indicated authors have shown that they have paid considerable attention to the analysis of changes in the concepts and mechanisms of accounting and financial reporting in the field of state and corporate governance, as well as to the scientific substantiation of the application of new paradigms in the development of accounting and auditing systems in Ukraine under conditions of asymmetry of the information space.

Among the studies of foreign scholars on these issues are the research works of Russian scientists Kuter and Gurskaya [5, 6], which are dedicated to the scientific investigations of the evolution of financial accountability for the different groups of the information users in the financial market, in particular those that do not have financial interest.

The great significance for the practical orientation and for value of the results of the submitted scientific article represents the work of Western scholars and practitioners, whose representation in science-computer databases is much larger than that of Ukrainian scholars and practitioners. From the standpoint of the problems within the research study are the following works that are of particular interest: Buckley and Casson [7], Dunning [8], Levitt [9], Tanzi [10], that have been devoted to the disclosure of the impact of globalization on the processes of transformation of economic and public life and the development of cross-border commerce, as well as the benefits and challenges within the globalization processes; Chen, Shu-fan, Thai [11], which have considered the conditions for the transparency of financial markets and the respective challenges; Schwab [12], Manyika [13] and Knickrehm [14], where the authors have analyzed the problem of the expansion of the digital economy, namely its impact on GDP and socio-economic processes at the national and global levels; Baldwin and Trinkle [15], which have concentrated the study on the impact of XBRL [eXtensible Business Reporting Language] on financial reporting and usage of the Delphi technique.

However, the research studies of the Ukrainian and foreign researchers have not focused on analyzing the peculiarities of the global digital economy's influence on the national information space for fiscal and control purposes in order to ensure the transparency of economic processes, inter alia within the investment attractiveness of the country's economy. In this regard, the purpose of this research study is to substantiate the approaches towards the challenges at the national level related to the transparency of economic processes, the prevention of tax evasion and the formation of an attractive investment environment. These challenges are shaped by the globalization processes and enhanced by the development of digital economy; their current consequences are realized through the existing restrictions on the exchange of tax information between countries at the supranational level and information asymmetry on the basis of the incompatibility of reporting formats for financial and tax purposes at the national level. Therefore, the study of the impact of these challenges exerted on the formation of a national information space is predetermined by its actuality as it aims to expand knowledge in the field of existing IT and their application for publishing of the financial reports on the Internet and automation of the control procedures for disclosure of reporting information, and thus creating conditions for generating new ideas for methodological, organizational and technological issues in the field of audit and taxation.

This paper is organized in such a way as to present analysis of supranational regulation of the transparency conditions within tax processes and prevention of tax evasion, as one of the key factors affecting the investment attractiveness of the national economy. The Ukrainian legislation and its scope and directions of changes are examined taking into account its ability to display new technological opportunities for the implementation of supranational norms in order to counter base erosion and profit shifting as well as the issues on exchanging and controlling financial and tax information, ensuring tax fairness and effectively institutionalizing of the process of continuous monitoring and continuity of reporting within enterprises.

2 Materials and Methods

The global economy nowadays is facing drastic changes predetermined by the usage of digital technologies by individuals, companies and governments. These digital transformations influence the every aspects of human economic activity and are based on the digitalization of the economic activity that can be broadly defined as the incorporation of data and the Internet into production processes and products, new forms of household and government consumption, fixed-capital formation, cross-border flows, and finance [16]. So such total digitalization can be regarded as global digital revolution that is characterized by the accelerated pace of digital transformations and technologies' diffusion across countries of the world. These digital transformations result from general-purpose technologies that are characterized by the possession of the power to continually transform themselves, progressively branching out and affecting productivity across all sectors and industries. This leads to the development of the digital economy on the basic of growing interconnectedness of people, organizations and machines that transforms the respective business models and business interaction providing new business opportunities and facilities and carrying additional business risks. Thus according to the IMF definition "digital economy" is sometimes defined narrowly as online platforms, and activities that owe their existence to such platforms and in a broad sense, all activities that use digitized data are part of the digital economy: in modern economies, the entire economy. If defined by use of digitized data, the digital economy could encompass an enormous, diffuse part of most economies, ranging from agriculture to R&D [16].

Ukraine is currently at the transformational stage of the economy digitalization, which constitutes 40–50%, accompanied by qualitative changes in ICT and gives high chances for the accelerated development of the digital society. The development of the digital economy is currently considered by the Government of Ukraine as a factor for increasing production efficiency and competitiveness of the national economy and, accordingly, for economic growth and for foreign direct investments attraction, new jobs creation and for the improvement of the living standards in general. The Concept for the Development of the Digital Economy and Society of Ukraine for 2018–2020, approved at the beginning of 2018, aims to support the transformation of Ukrainian society by stimulating the development of the economy and attracting of investments, overcoming the digital inequalities, deepening the cooperation with the EU in the digital sphere and developing the country's innovation infrastructure and digital transformation. The implementation of the Concept's measures should ensure: stimulation of the economy development and investments attraction; the basis for the transformation of domestic industries into competitive and efficient ones due to their "digitalization"; solving the problem of "digital gap", bringing digital technologies closer to citizens, including by providing citizens with the broadband Internet access, especially in villages and small cities; creation of new opportunities for realization of human capital, development of the innovative, creative and "digital" industries and businesses; export promotion of "digital" products and services (IT outsourcing) [17].

Thus, the digitization of economic activity should become a prerequisite for increasing the competitiveness of national industries and improving the competitiveness of the national economy as a whole in the global environment, as well as for the attracting of foreign direct investments and improving the investment attractiveness of Ukraine. According to the statistic data provided by the State Statistics Committee of Ukraine, as of January 1, 2019, the volumes of the foreign direct investments stock in the Ukrainian economy have amounted to 32.3 billion US dollars within the general trend of the FDI decreasing from 53.7 billion US dollars in 2013 (see Fig. 1).

The inflow of foreign direct investments in Ukraine in 2018 was only 0.686 billion US dollars (in 2017, the inflow of foreign direct investment in Ukraine amounted to 0.376 billion US dollars compared to the overall world inflow of foreign direct investments at 1 430 billion US dollars according to UNCTAD data [18]). At the same time among the major investing countries the largest share is made up of foreign direct investment from the offshore zone of Cyprus (according to the results of 2018, its share amounted to 25.5% of the total volume of foreign direct investments in the economy of Ukraine).

Ukraine has been currently declared unattractive for the long-term capital investments. So, the country now has only 131th place within the International Business Compass (IBC) rating for 2018, according to the results of the seventh IBC study

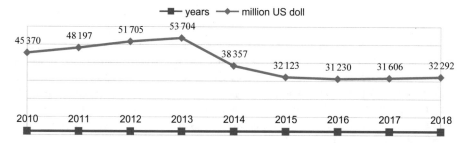

Fig. 1. Dynamics of the foreign direct investments stock volumes growth in the Ukrainian economy from 2010 to 2018.

conducted by the world's auditing and consulting company BDO and the Hamburg Institute of World Economists HWWI, which assess the investment attractiveness of 174 countries the world from all continents with the exception of countries with a population of less than 150,000, as well as excepting Cuba, the West Bank, Jordan, Somalia, Western Sahara, Luxembourg and Syria. According to the results of 2018, Singapore, Hong Kong, Switzerland, the Netherlands and Ireland are among the top five most attractive economies for the FDI in the world. The analysis of the investment attractiveness of the countries of the world within the IBC rating is based on subjective factors, namely economic, political, legal and socio-cultural conditions, the attractiveness of the market and the attractiveness of the allocation of production capacities, etc., which, in contrast to the objective factors, namely peculiarities of the geographical location, the natural climatic conditions etc., are variable in nature and depend on the effectiveness of the relevant national authorities activity within the provision of the competitive investment climate in t. h. the relevant governmental initiatives and decisions.

Thus, among the driving forces of the investment attractiveness of the country are the following factors: macroeconomic and political stability; stability and transparency of the tax system functioning; protection of investors' rights; quality of corporate governance; quality of labor resources, including education, medicine, etc.; quality of infrastructure; level of innovation and technological development of the national economy, etc. Therefore, investment attractiveness is regarded as a model of quantitative and qualitative indicators in the framework of the analysis of the current national conditions for the effective attraction of long-term capital investments. In this respect, the effectiveness of state institutions that ensure stability, transparency of the respective state policies and equality of competitive business conditions plays an important role in the formation of these indicators.

The formation of an institutional environment that ensures the effectiveness and transparency of economic processes and responds to global challenges such as the development of e-commerce, IT technologies requires a methodological and legal regulation at the national level. In this context, the functioning of a stable and transparent tax system is one of the qualitative parameters of the corresponding national investment attractiveness. Thus, the formation of the tax system should be based on clearly formulated set of principles, rules and procedures for their establishment, implementation, control over the timeliness and completeness of tax payment and liability for unpaid taxes.

New challenges that accompany the development of e-commerce and the globalization of economic ties necessitate the change of tax control instruments based on fundamentally new approaches to the formation, processing and application of information for tax purposes. That is why the formation of an effective information space for tax purposes, which involves the international exchange of tax information in order to ensure fiscal transparency and control improving, is an extremely important both at the national and supranational levels.

The main efforts of the countries of the world, including Ukraine, as an integral part of the international information environment for tax purposes, are within the transition of the information exchange process to the "automatic mode" – that is, regular, without separate requests, covering a large number of countries around the world.

The analysis of existing documents of the Organization for Economic Cooperation and Development (OECD) suggests that the exchange of tax information for taxation purposes can be considered as: an administrative method for combating tax evasion; counteracting harmful tax practices and tax base blurring; avoidance of double taxation within the framework of the corresponding investment activity abroad; an instrument for ensuring the transparency of business processes in general, and taxation, in particular [19].

The exchange of information between the tax authorities of different countries plays an important role within the international instruments to combat cross-border tax evasion and distortion of the competitive conditions, while the exchange of information between different authorities within the country is regarded as an effective mechanism for fighting domestic tax crimes.

International cooperation on tax issues is carried out using various legal mechanisms, namely:

- internal: at the national level each country develops its own tax policy, using tax measures as for its residents, that transfer funds to other countries, and foreign investors who are often trying to mitigate double taxation;
- bilateral: the most common mechanism of cooperation remains bilateral treaties of countries, which in practice are divided into: Double Tax Treaties (DTTs) or Double Tax Conventions (DTCs); Tax Information Exchange Agreements (TIEA); Mutual Legal Assistance Treaties (MLAT). As a rule, such agreements regulate the issue of eliminating double taxation of passive investment income and the interaction of the relevant tax authorities;
- multilateral, the most comprehensive of which is the Convention on Mutual Administrative Assistance in Tax Matters. In addition, within the main instruments of the international tax information exchange are, in particular, the International Convention on Mutual Administrative Assistance in Tax Matters, which is open for ratification and signature by all countries, the Convention on the Avoidance of Double Taxation, bilateral treaties on the exchange of information.

The exchange of tax information and the conditions for avoidance of double taxation within the framework of the above mentioned agreements serve as the basis for decision-making in the tax area with regard to resident taxpayers with incomes outside the country. Within the forms of exchange of tax information are the following, namely: the exchange of information on request - the most common method of cooperation; automatic exchange - a systematic transmission of taxpayer information relating to various categories of income and other types of information of the partner countries under the agreement: without prior request and at regular intervals; spontaneous (proactive) exchange is an exchange without prior request, where one country provides to the other relevant information on its own initiative.

The starting point of the global movement of transparency and the exchange of information for tax purposes can be considered the proclamation of the G20 leaders in 2009 intention to limit banking secrecy for tax purposes and the recognition of the need to introduce a single standard for the exchange of information. Over the following years, the international tax environment has changed significantly for increasing transparency and information sharing.

Thus, the FATCA (Foreign Account Tax Compliance Act) adopted in 2010, extended national law to the international level, requiring regular reports to the American Tax Service (IRS) on all accounts and incomes of US citizens in other countries. The main feature of the FATCA mechanisms was the regular automatic sending of reports on all accounts of Americans - that is, without preliminary requests. Having realized the potential of FATCA, an information exchange mechanism was developed within the OECD in an automatic mode - but on a global scale and on the basis of reciprocity. Another important milestone on the road to tax transparency was the adoption of the OECD standard for automatic exchange of financial account information for tax purposes in 2014, which defines the conditions for the bilateral exchange of information on financial accounts in accordance with the requirements of a multilateral agreement between authorities within the framework of the common reporting standards. The standard obliges governments to provide the opportunity to collect detailed account information from their financial institutions and share this information with other jurisdictions automatically on an annual basis.

Information disclosure and access to it, in particular, the exchange of information between the tax authorities of various states, as well as between various authorities within a country, requires a certain standardization of data, unification of presentation forms and digital storage and transmission protection formats. As of July 2016, 54 jurisdictions pledged to make the first exchanges before 2017, and an additional 47 jurisdictions pledged to make the first exchanges before 2018.

From a legal point of view, the CRS standard is not a rule of law and is not legally binding in itself; it is a "model" document. To make binding, the state legally express its consent to comply with the norms of the standard. The document defines the transfer in the framework of international exchange; financial organizations authorized to collect and provide tax-relevant information of their authority for further international exchange; types of accounts and tax residents for which information should be collected and provided; as well as unified methods for checking the integrity of clients when opening accounts (Due Diligence). Thus, according to CRS, all financial institutions that operate in the member countries, such as banks, brokers, investment structures, some types of insurance companies, collect and transfer to their authorized bodies information on accounts for further international exchange. Information should be collected on individual accounts of individuals and corporate accounts of companies, trusts and private foundations. Within the EU, the implementation of the CRS standard was implemented in the form of amendments to the EU Directive on Administrative Cooperation (EU Directive on Administrative Cooperation - DAC2), acting only in relations between EU member states. The implementation of the CRS standard in other countries may take place at the conclusion of bilateral international tax treaties (DTT contracts or agreements TIEA) when concluding multilateral conventions on the exchange of tax information.

In 2013, within the framework of the OECD, a plan of action was developed to counter the erosion of the base and derive profits from taxation BEPS (Base Erosion and Profit Shifting). The plan provides for 15 directions of action in a particular area of taxation, which are proposed for implementation in the national laws of countries and international treaties.

The economic agents of the shadow sector are not interested in data transparency, including those related to transfer pricing control, as well as in organizing the exchange of financial and banking information with other countries for tax purposes. At the same time, transparent companies implementing investment projects around the world are interested in creating transparent business and taxation conditions, because this ensures equal competitive conditions.

Ukraine has joined the program of extended cooperation within the framework of the OECD and has signed a multilateral Convention on the implementation of activities related to taxation agreements [19], in order to counter base erosion and deduction of profits from taxation (Multilateral Convention to Implement Tax Treaty Related Measures to Prevent Base Erosion and Profit Shifting, better known under the short name Multilateral Convention MLI [20], which is a key element in the implementation of the so-called BEPS Plan. The country is committed to implement the Minimum first BEPS Action Plan standard [21] (four mandatory events for fifteen proposed). It is a harmonization of the rules of tax administration in terms of countering financial destabilization due to the erosion of the base and move profits abroad.

According to the analysis of the OESD document "Global Forum on Taxes and Peer Reviews: Ukraine 2016", we can draw the following main conclusions regarding Ukraine:

- the country must guarantee the possibility of identifying information from the relevant structures and its availability to the competent authorities, regardless of the ownership of such information bases. For Ukraine, the issue of the lack of access of the relevant competent authority (Fiscal Office /Ministry of Finance) of information from disparate departmental databases that do not have a single standard for collecting and aggregating information is relevant;
- conditions must be created for reliable, accurate execution and storage of accounting records in relation to all related structures. Yes, there is a certain discrepancy in terms of information storage. Ukrainian legislation does not guarantee that the storage of individual documents after three years of storage of accounting documents, and also does not specify how long the accounting records should be kept after the liquidation of the institution or the closure of the transaction;
- banking information should be available for all account holders, partly provided in Ukrainian legislation. However, the competent authorities should have the right to

receive and provide information within the framework of an agreement on the exchange of information from any organization within their territorial jurisdiction. Today, in order to obtain banking information, the fiscal authorities must go to court. In Ukraine, it is necessary to clarify the legal norms in order to more clearly ensure access to banking information in accordance with the CRS standard, and the rights and guarantees that apply to persons in the jurisdiction of the request must be compatible with the effective exchange of information;

- the information exchange mechanism should be effective and requires clarification of legislation on the amount of banking information provided in accordance with the CRS standard requirements, as well as improving the information exchange process with all available partners and introducing relevant technical protocols, creating conditions for ensuring the confidentiality of the information received and respecting rights and guarantees of taxpayers and third parties;
- banking information should be available for all account holders, partly provided in Ukrainian legislation. However, the competent authorities should have the right to receive and provide information within the framework of an agreement on the exchange of information from any organization within their territorial jurisdiction. Today, in order to obtain banking information, the fiscal authorities must go to court. In Ukraine, it is necessary to clarify the legal norms in order to more clearly ensure access to banking information in accordance with the CRS standard, and the rights and guarantees that apply to persons in the jurisdiction of the request must be compatible with the effective exchange of information;
- the information exchange mechanism should be effective and requires clarification of legislation on the amount of banking information provided in accordance with the CRS standard requirements, as well as improving the information exchange process with all available partners and introducing relevant technical protocols, creating conditions for ensuring the confidentiality of the information received and respecting rights and guarantees of taxpayers and third parties.

It is obvious that the institutional component should take into account methodological peculiarities as well. It is extremely important to employ the best practices in the regulation of international transactions related to the minimization of barriers towards free movement of capital under the condition of the provision of the high institutional level of reflection in the existing rules of the transparency, accountability, compliance principles, etc., which form the basis for the investment attractiveness of the country's economy.

The process of financial globalization has created conditions for international tax planning, and its instruments are widely used by the businesses. The usage of IT, digital platforms and e-commerce has changed significantly the structure and characteristics of business environment and consequently lead to the expansion of the information space. This, in turn, increases the efficiency of the managerial and investment decisions. To a large extent, this will be facilitated by the harmonization of financial reporting currently being implemented in Ukraine by introducing International Financial Reporting Standards (IFRS) for securities market participants and other public enterprises.

Charles Hoffman, who is an inventor of XBRL, says in his blog that "…The fourth industrial revolution is occurring … Three primary enabling technological innovations are driving this significant change to the current accounting practices, processes, and methods: XBRL-based structured digital financial reports, knowledge-based systems and other application of artificial intelligence, and blockchain-based distributed ledgers" [22].

The use of web technologies for publishing IFRS financial statements, as well as the use of XBRL will help reduce information asymmetry due to the incompatibility of reporting formats provided for various regulatory (primarily fiscal) authorities and user groups. However, the process of practical implementation of IFRS poses new issues related to the creation of unified systems within the access to the financial reports of companies and the harmonization of rules for its compilation on the basis of the taxation system and reducing the significant impact of accounting for tax consequences while displaying reported indicators. The solution of these issues is extremely important for Ukraine. Currently, only the state institutions operate employing the electronic reporting standards in Ukraine. Enterprise entities, investors, auditors and software suppliers only react to the introduction of new reporting forms and reporting formats [3, pp. 360–361]. Accordingly, the main obstacles to the implementation of XBRL will, in essence, be the same as those on the way of implementing IFRS, namely: the "immaturity" of domestic financial markets and, as a result, the lack of demand for financial information; lack of interest of owners and management of enterprises in the disclosure of financial information, in particular its publication on the Internet; business opacity; lack of culture and confidence in electronic documents; the existence of distrust of the Internet as a source of information for analysis, forecasting, etc.; lack of sufficient experience and knowledge of the technological aspects of XBRL among representatives of the accounting and financial professions.

Despite these difficulties, Ukraine is increasingly involved in the globalization process, and an increasing number of international exchanges strongly recommend the companies to submit financial statements in the XBRL format. Complications of the corporate structure, the development of e-commerce, digitalization of international capital markets and economic relations determine the need to move to XBRL. As a consequence of these processes, at the end of 2017, the Law "On accounting and financial reporting in Ukraine" was supplemented by changes, according to which, starting from 01/01/2019, entities of public interest should prepare IFRS financial statements and submit them to the government authorities and other users on the basis of XBRL taxonomy in a single electronic format, defined by the Ministry of Finance of Ukraine.

3 Results

Ukraine currently has bilateral double tax avoidance treaties with 71 countries [23] and has no agreement on the automatic exchange of tax information. Obviously, from the declared intentions to practical implementation, a number of measures should be implemented that will bring internal standards and protocols closer to those provided within the framework of automatic exchange of tax information. In order to prepare the implementation of automatic exchange of tax information, it is necessary to introduce international structuring of such information at the national level. This will allow for the automatic exchange between jurisdictions that have introduced the CRS standard. To prepare for joining the automatic tax information exchange system, it is advisable to take into account the findings of the first stage of the review of Ukrainian legislation on the readiness of the system for such an exchange [24]. Thus, in accordance with CRS, tax information that is loaded into the system must meet three criteria: identification by property; accounting records; banking information; access that is characterized by the ability of the competent authority to receive and provide information; reporting procedures to the competent authority to receive information, rights of the competent authorities to receive information as well as security measures for the transfer and use of information; exchange, that provides information exchange mechanisms with all existing partners; confidentiality; rights and security measures of third-party taxpayers; timeliness of responses to requests for the exchange of information.

The implementation of changes within the information tax environment is primarily aimed at ensuring the possibility of introducing the CRS standard for automatic information exchange, but its implementation will lead to streamlining of information flows for internal taxation purposes as well (Fig. 2).

For this purpose, first of all, it is necessary to unify the collection, storage and processing of information about assets, accounts, business activities and individual transactions of individuals according to a single standard. It is necessary to ensure the regulatory and technological readiness of various authorities and responsible institutions to exchange such information, as well as its proper storage and confidentiality.

The national and supranational level of use of tax information is not isolated from each other. On the one hand, the completeness, relevance and timeliness of providing information from interstate databases (national level) to interstate (supranational level) is ensured, and on the other hand, it provides the possibility to quickly and fully provide tax inspectors with the information from interstate databases (national level).

Moving from the supranational level of information circulation and automatic information exchange for tax purposes, in which various stakeholders are interested, it is necessary to take into account the standards of financial and business reporting that should also be in line with a specific format, which will determine a wide range of its use and universality. In this context, the capabilities of XBRL help to improve financial analysis and risk management. This requires not only methodological readiness, but also the development of appropriate software and the availability of software agents.

Taking into account that XBRL's capabilities allow to improve financial analysis and risk management, software suppliers are required to provide the development of software agents and artificial intelligence software informers that can extract XBRL-encoded data from relevant sources for forecasting, for example, for financial performance of the enterprises, but in the perspective for the exchange of tax information between different jurisdictions. The most important for XBRL application in finance is that it is open and accessible to all users (XBRL is an open source technology). Such flexibility compared to deficient solutions is not available within the framework of the usual accounting, ERP systems, since they are not open (in most cases, such systems were based on a closed source software). The entire technological part of XBRL is built on the principles of XML - markup language for the transfer of information, each

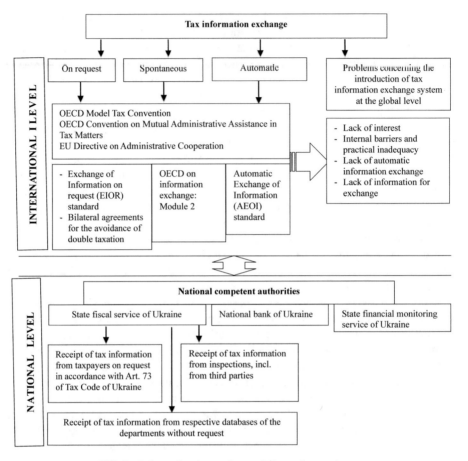

Fig. 2. Information space for tax information exchange

semantic part - based on accounting standards and principles of financial reporting in this or another jurisdiction. This is the uniqueness of XBRL for finance and financial officers and simple clients of financial information. Just as in the United States, Europe and Asia, the fact is that representatives of the accounting and financial profession do not have enough and the main thing - knowledge of the technological aspect of XBRL. This problem can be solved simultaneously by the specialized education and the specialized software products that are taught and already presented within the Ukrainian market.

The codification of the accounting data in XBRL should promote the effective institutionalization of the interrupted reporting process at enterprises. This, in turn, should improve the effectiveness of business's performance and help in the realization of business goals. All business process participants - accountants, regulators, financial analysts, investors, lenders, government, software suppliers - and many others will benefit from the use of XBRL.

In order to achieve the goals of all participants within the economic process, it is vital to address the issue of a single format for the compilation and submission of financial electronic reporting. Such a format should allow reporting to become more comparable and suitable for analysis by software, taking into account the various tasks that regulators, fiscal authorities, investors, and others are assigned.

Due to the fact that the accounting software packages continue to include the XBRL tags, it will be easier to employ it. More and more regulatory authorities currently have the mandate to use XBRL in their surveys and that is the reason for its widespread usage. The main contribution of this study is to provide evidence of the prospects of XBRL usage in accounting, corporate reporting and regulation. This study provides support for some general concepts of XBRL exposure, but distorts expectations of other impacts. Within this study, we develop a framework (see Table 1), which will include more likely consequences and exclude the least likely effects.

The business information network includes a wide range of stakeholders, including: investors, counterparties, partners, regulatory and fiscal authorities. The above mentioned scheme allows to distinguish between several links of the business information supply, which are united by the same processes of preparation, interpretation, submission of IFRS financial statements and guaranteeing their high quality and participants in these processes. For each link (process) of business information supply, one or another XBRL tool is used. Users who are most interested in the purpose of this research study are divided into following groups, namely: (1) Corporation and Compliance, (2) Financial Reporting, (3) Users of Financial Reports, and (4) Audit.

Thus, within the 1st link concerning business operations XBRL for segment reporting is employed. It is the level of internal corporate governance and control, which must ensure the effectiveness of business decision making and compliance with the respective legislature.

Within the 2nd link IFRS financial statements are prepared with the application of XBRL formats for systematic accounting that facilitate continuous reporting and availability of financial statements, including through approximation to the generally accepted accounting principles, that make it possible to plan audits.

Within the 3rd and 4th links of the business information supply chain the external financial statements are prepared and further published using IT technology on the Internet. The appropriate XBRL formats are equally convenient for generating and aggregating of financial statements for further submission to regulatory authorities (including fiscal authorities) and for publishing on business web sites that allow to increase the volume and usefulness of information that is of interest to investors, financial analysts, lenders and the state as a whole.

Therefore the results of the research study serve as the explicit evidence of the importance of XBRL as it stimulates the technology transfer and allows each interested party to receive significant benefits and to get added value throughout the corporate reporting business chain. Despite the overall positive impact and convenience of XBRL-based structured digital financial reports, such a format is unlikely to eliminate the necessity for convergence of generally accepted accounting principles. The arguments of most experts on these issues are still perceived with some skepticism regarding the remarkable effects of business disclosures that some XBRL advocates and accounting futurists demonstrate in their speeches, writings, and Internet blogs.

Table 1. The business information supply chain and the role of XBRL.

Link	Processes	Participants	Framework of likely XBRL impacts
1	Business operations	Company Partners Contractors Software developers	**Corporation and Compliance** • XBRL increases the efficiency of business decision making • XBRL allows for easier regulatory compliance
2	IFRS financial statements	Company Management Accountant Software developers	**Financial Reporting** • XBRL facilitates continuous reporting • XBRL enhances the availability of financial reports. • XBRL eliminates the need for convergence of generally accepted accounting principles
3	Publishing of IFRS financial statements on the Internet with the use of IT technologies	Company Auditors Software developers	**Users of Financial Reports** • XBRL provides more accessible financial reports to users • XBRL increases the use of un-audited information by investors • XBRL increases the ability of financial analysts to perform cross-sectoral analysis within industries • XBRL allows more efficient investment decision-making by users of financial reports
4	Investment-credit analysis regulation	Aggregation and publication of financial statements Regulators Software developers	**Audit** • XBRL facilitates continuous auditing
5	Macroeconomics economic and monetary policy	Investors Central banks Regulators Software developers	

This table presents the analysis of the role of XBRL and its potential impact on the business information delivery network. Existing XBRL research studies have been synthesized and analyzed in order to determine the conceptual framework for corporate reporting and disclosure of potential XBRL impacts.

The fact that experts do not suggest that the XBRL-based structured digital financial report will contribute to its comprehensibility or reduce the cost of conducting its audit indicates that XBRL is not a technological leap for all interested users of financial statements. In contrast, panelists uniformly agree that XBRL is very unlikely to eliminate the need for convergence of generally accepted accounting principles. Most panelists' rationales were tempered with some skepticism regarding the near miraculous impacts that some XBRL adherents and accounting futurists have proposed. The fact that the panelists do not predict that XBRL will promote understandability or reduce financial statement audit costs suggests that XBRL is not the be-all and end-all technology leap that some would have us believe.

Audit of XBRL-based structured financial reports refers to "Audit 4.0", which «strongly relies on a mirror world representation of processes and a strong analytical interlinking of not only financial but especially nonfinancial to financial linkages…. will be applicable to many types of assurances (external, internal, specialized), and will be mainly automated» [25, p. 9].

The active inclusion of Ukraine in the process of the global financial integration, changes in the reporting system, its accessibility via the Internet make it necessary to further reform the accounting and financial reporting system and its audit. At the end of 2017, the Verkhovna Rada of Ukraine adopted a new Law on Audit of Financial Reporting, which brought the norms of national legislation in line with EU Directives 2006/43/EC and EU Regulation 537/2014. At the same time, the introduction from 01.01.2019 of the XBRL-based structured digital financial reports as a mandatory format for entities of public interest and its further audit requires the adaptation of knowledge, skills and talents of a professional auditor in the direction of assimilating and mastering technological innovations, that have caused the significant changes in current accounting and reporting practices.

4   Conclusion

In this study, the authors analyzed the supranational regulation of the transparency conditions in the framework of tax processes and the prevention of tax evasion, as one of the key factors affecting the investment attractiveness of the national economy. The legislation of Ukraine, its scope and directions of changes were considered taking into account its ability to display new technological capabilities for implementing supranational standards in order to counter base erosion and redistribution of profits, as well as exchanging and controlling financial and tax information to ensure tax equity and effectively institutionalize the ongoing process of monitoring and continuity of reporting in enterprises. Special attention is paid to the availability of information to the regulatory authorities, such as the Fiscal Service, the National Bank of Ukraine, the State Financial Monitoring Service.

The possibility of introducing XBRL for the formation and presentation of financial statements is being considered, which allows reducing the information asymmetry caused by the incompatibility of financial and tax reporting formats. At the same time, the transparency of the financial reporting within the business environment in the country strengthens the investment attractiveness and country's reliability in a global environment.

In order to form relevant institutions and to solve a number of problems in obtaining and processing tax information at the international level, it is necessary to simultaneously solve the relevant issues on information standardization, reliability of storage and confidentiality at the national level.

Prospective research studies should focus on the methodological readiness of financial and non-financial reporting standards implementation as well as usage of the latest approaches to the processing, analysis and consumption of information. The specified methodological readiness should form the conditions for the application of CRS. In addition, special attention in the framework of digitalization of the economy and the acceleration of information flows should be paid to extending the possibilities of applying XBRL to various sectors of the economy. At the same time, the following important questions remain open: How does the implementation of XBRL affect the comparability of analytical data from different sectors of the economy? How does the integration of XBRL into accounting software affect the quality and efficiency of companies' financial information? How does this business format help ensure transparency at a supranational level for tax purposes?

References

1. Iefymenko, T.I., Zhuk, V.M., Lovinska, L.H.: The information in crisis management: the global dimension of standardization of accounting and financial reporting. SESI "Academy of Financial Management", Kyiv (2015). (in Ukrainian)
2. Kamenska, T.O.: Internal audit. Modern look. DP "Inform-analyst. agency", Kyiv (2010). (in Ukrainian)
3. Ozeran, A.V.: Theory and Methodology of Formation of Financial Reports of Enterprises. KNEU, Kyiv (2015). (in Ukrainian)
4. Petryk, E.: Ukrainian audit in an era of change: the paradigms of the past and the new time. In: Paradigm of Accounting and Auditing: National Realities, Regional and International Trends (2016). http://www.ase.md/files/catedre/cae/conf/conf_aprilie_2016.pdf. Accessed 28 Mar 2019. (in Russian)
5. Kuter, M.I., Gurskaya, M., Andreenkova, A., et al.: The early practices of financial statements formation in medieval Italy. Acc. Historians J. 44(2), 17–25 (2017). https://doi.org/10.2308/aahj-10543. Accessed 28 Mar 2019
6. Kuter, M.I., VY, Sokolov: A Global History of Accounting, Financial Reporting and Public Policy: Eurasia, the Middle EST and Africa, pp. 75–106. Emerald Group Publishing Limited, Sydney (2012)
7. Buckley, P.J., Casson, M.: The internationalization theory of the multinational enterprise: past, present and future. Br. J. Manage. (2019). https://doi.org/10.1111/1467-8551.12344. Accessed 28 Mar 2019
8. Dunning, J.H., Lundan, S.M.: Multinational Enterprises and the Global Economy, 2nd edn. Edward Elgar Publishing, Cheltenham (2008)
9. Levitt, T.: The globalization of markets. Harvard Bus. Rev. 61(3), 92–102 (1983)
10. Tanzi, V.: Social protection in a globalizing world. Invited Policy Paper, pp. 25–45 (2005). https://core.ac.uk/download/pdf/6500111.pdf. Accessed 28 Mar 2019

11. Chen, H.K., Shu-fan, H., Tai, M.: Who wins and who loses in transparent markets? Daily and intraday analysis of Taiwan stock market. Taiwan Econ. Forecast Policy **41**(2), 127 (2011). http://www.econ.sinica.edu.tw/UpFiles/2013092817175327692/Periodicals_Pdf2013093010104847832/EC412-4.pdf. Accessed 28 Mar 2019

12. Schwab, K.: The fourth industrial revolution. Foreign Affairs (2015). https://www.foreignaffairs.com/articles/2015-12-12/fourth-industrial-revolution. Accessed 28 Mar 2019

13. Manyika, J., Chui, M., Bisson, B., et al.: Big data: the next frontier for innovation, competition, and productivity. Report. McKinsey Global Institute (2011). https://www.mckinsey.com/business-functions/digital-mckinsey/our-insights/big-data-the-next-frontier-for-innovation. Accessed 28 Mar 2019

14. Knickrehm, M., Berthon, B., Daugherty, P.: Digital disruption: the growth multiplier. Optimizing digital investments to realize higher productivity and growth. Accenture (2016). https://www.accenture.com/t00010101T000000__w__/br-pt/_acnmedia/PDF-14/Accenture-Strategy-Digital-Disruption-Growth-Multiplier-Brazil.pdf. Accessed 28 Mar 2019

15. Baldwin, A.A., Trinkle, B.S.: The impact of XBRL: a Delphi investigation. Int. J. Digit. Acc. Res. **11**, 1–24 (2011)

16. International Monetary Fund.: "Measuring the digital economy" IMF report, International Monetary Fund, Washington (2018). https://www.imf.org/en/Publications/Policy-Papers/Issues/2018/04/03/022818-measuring-the-digital-economy. Accessed 28 Mar 2019

17. Cabinet of Ministers of Ukraine.: On approval of the concept for the development of the digital economy and society of Ukraine for 2018–2020 and approval of the plan of measures for its implementation (Decree No. 67-p, January 17) (2018). https://zakon.rada.gov.ua/laws/show/67-2018-%D1%80. Accessed 28 Mar 2019. (in Ukrainian)

18. UNCTAD. World Investment Report 2018. https://unctad.org/en/pages/PublicationWebflyer.aspx?publicationid=2130. Accessed 28 Mar 2019

19. Ukraine signs landmark agreement to strengthen its tax treaties, OECD (2018). http://www.oecd.org/tax/ukraine-signs-landmark-agreement-to-strengthen-its-tax-treaties.htm. Accessed 28 Mar 2019

20. Multilateral convention to implement tax treaty related measures to prevent base erosion and profit shifting, OECD (2016). http://www.oecd.org/tax/treaties/multilateral-convention-to-implement-tax-treaty-related-measures-to-prevent-BEPS.pdf. Accessed 28 Mar 2019

21. Ministry of Finance of Ukraine.: Recommendations for Implementing the BEPS Action Plan (minimum standards) (2017). https://www.minfin.gov.ua/uploads/redactor/files/2017_Roadmap_BEPS_UKRAINE_ua.pdf. Accessed 28 Mar 2019. (in Ukrainian)

22. Hoffman, Ch.: Accounting and Auditing in the Digital Age. http://xbrl.squarespace.com/journal/2017/6/28/accounting-and-auditing-in-the-digital-age.html. Accessed 28 Mar 2019

23. The Exchange of Tax Information Portal. http://eoi-tax.org/jurisdictions/UA#agreements. Accessed 28 Mar 2019

24. Peer Review Report, Phase 1. Legal and Regulatory Framework. Ukraine. http://www.oecd-ilibrary.org/taxation/global-forum-on-transparency-and-exchange-of-information-for-tax-purposes-peer-reviews-ukraine-2016_9789264258716-en. Accessed 28 Mar 2019

25. Improved Business Process Through XBRL: Use Case for Business Reporting. https://xbrl.us/wp-content/uploads/2007/12/20060202FFIECWhitePaper.pdf. Accessed 28 Mar 2019

Organizational Aspects of Digital Economics Management

Margarita Melnik[1] and Tatiana Antipova[2(✉)]

[1] Financial University, Moscow, Russian Federation
[2] Institute of Certified Specialists, Perm 614015, Russia
tatiana462@yahoo.co.nz

Abstracts. The digital economy opens up new horizons for improving of Digital Economics management at all levels of state. The improvement of this management at the meso-level, where effective methods of cooperation and coordination of the activities of legally independent economic entities, united in a single process of producing end-use products, are of particular importance. Considering the state's attention to accelerating and improving the research and development, special attention should be paid to the digital economy to forms of coordinating research & development work (R & D) and preparatory production processes and accelerating the innovative development of all fields of activity based on them. Bearing in mind this requirement, authors have elaborated Conceptual Model of Digital Economics Management. Under these conditions, new communication methods created in the digital economy provide solutions to the tasks of improving strategic planning and solving global problems of modernizing the country's economy. It is important taking into account the new opportunities opened up by the digital economy and turn production organization into the most important driver of its successful development.

Keywords: Digitization · Organization of production · Production process ·
Organizational structure · Centralization · Standardization ·
Holding formations · Effectiveness · Improving organization ·
Meso and macro level

1 Introduction

In modern conditions, much attention is paid to the impact of the digital economy development on the organization of industrial production management, the formation of new factors of its development. However, the majority of specialists, determining the importance of the digital economy, pay great attention to the processes of distribution, exchange, forms of interaction between economic actors in different spheres of activity. The problems of direct changes in the organization of production processes, which should be included as traditional processes, i.e. processes directly related to the transformation of raw materials into finished products (the so-called transformation processes) and on innovative processes. Most importantly, it is in the digital economy that most traditional business processes acquire several other forms, taking into account

© Springer Nature Switzerland AG 2020
T. Antipova (Ed.): ICIS 2019, LNNS 78, pp. 148–162, 2020.
https://doi.org/10.1007/978-3-030-22493-6_14

the high level of their automation, and in some cases even the transition to implementation using modern "smart" technology, so-called Industry 4.0.

Started in Germany, "the Fourth Industrial Revolution", Industry 4.0 pays much attention in modern literature [1]. The basic concept was first presented at the Hannover fair in the year 2011 [2]. Industry 4.0 is defined as "the integration of complex physical machinery and devices with networked sensors and software, used to predict, control and plan for better business and societal outcomes" [3] or "a new level of value chain organization and management across the lifecycle of products" [4] or "a collective term for technologies and concepts of value chain organization" [5].

Industry 4.0 mainly contains from following elements: digitization; information and communication technology (ICT); cyber-physical systems (CPS); network communications; big data and cloud computing, modelling, virtualization and simulation; improved tools for human-computer interaction and cooperation. CPS is a system of collaborating computational entities which are in intensive connection with the surrounding physical world and its on-going processes, providing and using, at the same time, data-accessing and data-processing services available on the Internet. CPS improve the capability of controlling and monitoring physical processes, with the help of sensors, intelligent robots, drones, 3D printing devices [6, 8–10]. Industry 4.0 is the transformation of industrial manufacturing through digitization and exploitation of new technologies. Industry 4.0 is a promising approach to business and manufacturing integration. Application of the generic concepts of CPS and industrial Internet of Things (IoT) to the industrial production systems are one of the technical aspects of these requirements. That is, the Industry 4.0 is based on the connections of CPS blocks [6, 7].

Availability and use of the internet and Internet of Thing (IoT), integration of technical processes and business processes in the companies, digital mapping and virtualization of the real world, 'Smart' factory including 'smart' means of industrial production and 'smart' products are all concepts that are used in the Industry 4.0 [6, 7] as a new part of Digital Economics.

Also, Industry 4.0 characterizes the transition to implementation using modern "smart" technology, i.e. robotics. Anyway Industry 4.0 developing is possible in the digital economy conditions. It is important that there is a significant change in the structure of the business process itself in any field of activity, because the preparatory processes associated with a very high level of knowledge-intensiveness, the formation of new types of organization of the production process, most focused on the use of modern technology and information technologies for performing creative processes that significantly accelerate research and development, and especially design development. As a result, the change of the market (from standardized to diversified) with the production of customized products. Machines and robots are able to communicate each other, to take decisions and to self-update. The production lines are automated: control and maintenance tasks can be performed remotely [8–10].

This, on the one hand, contributes to the acceleration of the renewal of production, and on the other hand, implies a fairly frequent fundamental changes in the organization of production processes. It is clear that it is from this work that the main effect of the digital economy must be obtained, since it must be associated precisely with changes in production processes, first of all, with their acceleration and with a corresponding reduction in production cycles. This will contribute to a significant reduction in fixed costs, which can currently occupy a significant share of total production costs.

To bringing in new technologies and digitization, it finds out the profit-generating and cost-cutting opportunities. According to some sources, Industry 4.0 factory could result in decrease of production costs by 10–30%, logistic costs by 10–30% and quality management costs by 10–20% [7, 9, 10].

Considering that the digital economy is developing in all countries of the world, including at a fairly high rate and in previously backward, and now rapidly developing countries, the use of its tool will help expand economic ties between countries, which will most fully meet the requirements of sustainable development estimated not only by economic results, but also ensuring the economic and social sustainability of the world economy as a whole. This is especially important in the implementation of global programs: the environment program, the rational use of natural resources, the provision of employment as the most important factor of social sustainability.

2   Research Methodology of Digital Economics Management in Digital Age

Most experts as an essential characteristic of the digital economy emphasizes an active role in the virtual environment, which currently complements economic reality and changes the essence of economic relations through the use of the Internet, cellular communications, etc. However, if the basis of the selection of positions and features of the digital economy is the system of relations, then it should be more clearly defined that it is not the essence that changes, but the form of this relationship. In this case, first of all, the approaches to spatial conditions change; the approach to the location of selected partners is changing depending on the expansion of contacts and the specifics of the implementation of work cycles in the context of new forms of communication between economic actors. But at the same time, it is very important to note that in principle, the digital economy helps to accelerate the exchange of experience in the provision of services (primarily consulting), which make it possible to identify the most effective and efficient ways of production's development. However, this is primarily due to the acceleration of information exchange, which expands the possibilities of contacts, but the requirements for rationalizing material flows do not change. It is very important to remember that when choosing the forms of interaction between economic actors, it is important to take into account the economic essence of the production management relations that determine their content.

Evolution on production paradigm. A wide discussion about the nature and significance of the digital economy's improvement allows us to conclude that the majority of specialists pay special attention to the forms of interaction between economic entities and to a lesser extent consider the significance of the influence of the digital economy on the real change in production, primarily production processes. This is a very serious risk that arises when evaluating and anticipating the effect of the introduction, development and widespread use of the techniques and methods of the digital economy. In this regard, when it comes to improving the organizational and production structures, the creation of fundamentally new production positions on the formation of organizational and legal forms, the regulation of their interrelations, the creation of standards defining the restrictions that are set for the parent company and the obligations which the management organization must perform. It is in the combination of interests, while maintaining a certain legal freedom, that such types of organizations are created as elements of Digital Economics Management.

So, it is very important that the unity of change in management processes is based on the real improvement of the main business processes that implement the production of finished products. At the same time, of course, the problem in the ratio of different stages of the business process significantly changes, including much more laboriousness arises in the process of preparing for the creation, developing the basic elements of production, including the creation of more advanced fixed assets, a slightly different requirement for the organization of material support processes, and above all, a fundamental change in the role of a person in a business process, when all intellectual capital, intellectual groups of workers are and will focus on the preparation, i.e. conducting research, design, regulatory and economic stages of preparation of production.

This process should cover all levels and functions of management - from the macro level, including meso-levels, to individual business processes within economic entities. However, it is necessary to take into account that institutional reform, the emergence of a large number of fundamentally new organizational and legal forms, within which approaches to assessing the effectiveness of the activities of individual economic entities within their composition, as well as the peculiarities of value added formation in a number of industries require a specific approach to the further development of the organization of production and management. New approaches to the distribution of the productive forces were actually already defined when a market economy was established, when large economic structures were created, both focused on capital consolidation and associated with the establishment of stable economic relations, which formed stable groups of independent legal entities, but organically linked in a single reproduction process.

At present, attention to the regulation and deployment of forces according to the sectoral principle is changing significantly again, because the meso-level that has been formed in our time, and having a huge amount of funds, has independent capital aimed at further improving industrial capital ensuring real production efficiency. It is the economic entities themselves, each independently or in groups for coordinating interests, allocate funds for the further development of production, including the formation of research departments that choose the most effective methods for developing and improving production and management.

At the heart of the formation of centers for the development of certain industries lies the process of producing products of increased consumption, and all other production processes are formed around the parent enterprise. The choice of the latter is directly related to the features of the production processes. Within the framework of analytical processes, this is usually an organization that provides the entire chain of interrelated enterprises with an initial raw material resource. In synthetic processes, this enterprise forms the final product, connecting and assembling the main components made at other enterprises, and providing the composition in the product that meets the needs of specific consumers. In a straight-line process, a group of enterprises is formed. Its formation occurs, as a rule, less clearly. This approach to the formation of centers within specific territories formed the basis for the formation of clusters, in which the sectoral and territorial approach to the distribution of productive forces was effectively combined. Clusters are based on a group of enterprises implementing the main business process, but at the same time around them are formed production structures that accompany the production processes: auxiliary, service, processing waste of the main industries.

Logistic improvement. In this regard, we must recognize that the digital economy, the expansion of the use of basic information technologies in all areas of activity and the emergence of new types of communication links significantly change the possibilities of organizing business processes and relies on further improving organizational structures, the formation of new growth factors and increasing production efficiency. That is, again, it should be noted that for the wide use of modern organizational and legal forms, the emergence of new communication links between independent economic entities and entities belonging to the group. Great attention should be paid to the regulations of their interaction; it is at the level of regulatory documents that the tasks and development directions that are assigned to each legal entity that has specific tasks within the group of united enterprises must be approved.

KPI system optimization. The specificity of digital economics business processes implemented in individual organizations implies a significant differentiation of key performance indicators, some of which may be related to the organization of production (for example, the task of accelerating the turnover of resources, reducing production cycle activities, working in sync with accessory services, lean organization). Depending on the specifics of the basic business process, interrelated groups of industries can be formed in which the principles of sectoral and territorial management of productive forces are actively and effectively combined. This provides a synergistic effect from the full use of the sectoral approach and the specialization of individual economic actors, while using local resources. This approach concerns, first of all, labor resources by increasing employment, attracting production capacities of already operating organizations and reducing costs based on infrastructure facilities or newly created enterprises tied to a specific territory.

Control increasing. The development of a network of subsidiaries and the formation of consolidated units around the latter next level allows us to create a multi-tier structure of economic entities that include legally independent enterprises whose financial and economic activities are under the control of the head units. This causes new phenomena in the management system, first of all, an understanding of the real assessment of the effectiveness of accounting and control over the activities of the system of the organization as a whole and of its individual production groups. As a result of new types of organizational and legal forms of production, the socio-economic relations between their participants are changing to a certain extent. Hence, the role of audit and control over the activities of each participant in the system increases significantly, which should ensure the unity of approaches to the management of the system of distribution of resources, rights and responsibilities of each link in the system in order to achieve harmony of their interests.

In the accounting policy of the organization, internal audit/control, which determines the basic methods of interaction between members of a society or a network of interconnected external partners, forms requirements for the main management regulations that are fixed in internal regulatory documents. It is this that significantly speeds up the maximum standardization of all the main stages of production management in the first-place accounting and control and analytical types of work. At the same time, there are not only internal production standards, but also those documents that regulate the development of the organization and the relationship between group members, including relations between parent and subsidiary enterprises, the emergence of so-called management companies, the formation of special management units responsible for the activities of the system of enterprises in general, and documents regulating the activities of participants in such groups.

Control over the use of such regulations is, of course, a function of internal control, and internal audit controls and to a certain extent coordinates the methods used by internal control, i.e. provides real coordination of work of all interconnected divisions.

Consequently, the overall control methodology is determined at the level of the parent organization, agreed and approved by the Board of Directors, which, as a rule, includes the heads of subsidiaries, and the internal control system is focused on ensuring that tasks are accomplished and supporting methods of rational use of resources, improving product quality and production efficiency.

Currently, it is very important to define the role of the research and development unit in the organizational and production structure. It is the interest of the research and development unit in the final results of manufacturing enterprises that can significantly accelerate the final implementation processes of scientific and technical developments and their direct access to the system of new types of production and, accordingly, the creation of the finished product, ensuring the most rational use of all production resources and complete customer satisfaction of these products.

Big Data & analytics. Analyzing and evaluating management regulations of large corporate structures, it is advisable to pay attention to a number of management aspects. First, it is necessary to clearly calculate the raw materials indicators, reflecting the tasks and taking into account the specifics of the business processes of each enterprise included in the corporate structure. In this regard, many enterprises have raw materials indicators that are not directly, but indirectly related to the characteristics of the general financial results of the corporation (production volume, profit, profitability, etc.), but are factors in achieving them, shaping the cost of production, ensuring the acceleration of the turnover of resources and their economic development, saving of some particularly scarce resources, etc.

Subdivisions that are aimed at mastering new types of products, introducing modern technologies, etc., have substantial specificity of tasks. They are of particular importance with the increase of innovative production, when in a number of divisions in the period of development of innovations and after effective ways of their use, additional costs occurred. They can be attributed to the costs associated with the formation of intangible assets that will be in demand for further use in many departments.

Secondly, when creating corporate structures, you can clearly define the rights and responsibilities of separate organizations, the ability to independently solve specific functions, such as purchasing material, selling additional products, attracting additional labor based on a contract, etc.

One of the most pressing issues is the distribution of profits and participation in the use of funds allocated for the development of production. Objectivity of priorities should be confirmed by a strategic audit of the projects proposed by individual participants in the corporate structure. The procedure for the distribution of profits should be transparent and clear to every participant in the company. These issues should provide an appropriate level of control within each organization and internal audit within the group of companies.

Thirdly, the organization of the internal audit and control system becomes an important condition for successful cooperation in the system of uniting legally independent enterprises. This is especially important in cases when a number of participants are not connected with the parent enterprise by capitalization of capital, but is under its control in terms of the choice of the product range (provision of services, performance of work), establishing communication with partners that are not part of the corporate structure (joint stock company, network) and development strategy development. In this case, it is the well-founded, taking into account the specific role of each member of the society or network, the accounting policy to a large extent becomes the key to preventing possible conflicts and clearly defining the role of each link in the overall reproduction process, which ensures the sustainable development of the corporate structure as a whole.

Thus, in digital economics the organization of production at the meso-level is implemented to a large extent, the organizational aspect of management is becoming increasingly important.

In the digital economy, the organization of corporate structures is significantly less affected by the spatial distribution of their members, since fundamentally new communication links, remote control capabilities and monitoring of production progress are used. At the same time, in these conditions, the importance of clarity of regulation of management significantly increases based on the standardization and diversification of both production and management processes.

The organizational aspect of improving digital economics management becomes relevant at the micro level, i.e. when organizing production and management processes directly in the enterprise. This is due to a number of reasons—changes in the structure of operations in the production process system, including an increase in their automation and mechanization, the growth of intellectualized labor associated with managing the system of mechanisms, the need to quickly switch production from one type of process to others, and the corresponding construction of a flexible infrastructure (serving and auxiliary operations), individualization of a number of tasks at specific workplaces due to an increase in the share of individual unit production.

Under these conditions, it is necessary to clearly plan the organization of production processes, taking into account their proportionality, time synchronization (logistic scheme), consistency of quality, etc. It should be noted that in recent years, attention to the organization of production has decreased significantly due to the lack of regulatory internal planning and production control.

Staff training. In the conditions of digitization of the economy in terms of the use of machines and mechanisms that implement most of the production process, the role of workers is largely associated with the organization of rational use of production resources, i.e. with the implementation of preparatory processes, technological and regulatory and economic training, which determines the uniqueness of the decision on loading equipment, providing the process with the necessary materials and ensuring timely measures to maintain equipment operability, etc. The formation of the digital economics improvement, in which highly qualified people are very often engaged in performing quite complex labor processes, ensure the importance of highly skilled groups of the population. In addition, they provide for the creation of related industries in which women and younger people who do not have high qualifications are employed, which prevents the occurrence of unemployment or dissatisfaction with the employment of second and third family members, and ensures social balance in the society.

This task is activated with the growth of innovation in production and the rapid turnover of the tangible elements of business processes.

Consequently, in the conditions of digitalization of the economy, the role of man in production is increasingly connected not with his direct participation in the execution of operations, i.e. in a transformational economy, and with the preparation of conditions for their successful implementation of machines and mechanisms. Thus, in the conditions of constant change in all spheres of activity, the organization of management becomes more and more important and becomes one of the main factors for increasing the efficiency of economic development. Since the Internet of Thing (IoT) is based on Cyber-physical Systems (CPS), the introduction/implementation of these systems is impossible without ensuring a high level of Cybersecurity.

Considering above mentions, the approach to developing Digital Economics is presented in brief on Fig. 1.

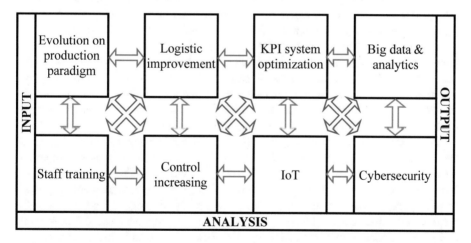

Fig. 1. The conceptual model of Digital Economics Management. *Source: own elaboration*

The model that shown on Fig. 1 consists of following elements: Evolution on production paradigm; Logistic improvement; KPI system optimization; Big data & analytics; Staff training; Control increasing; IoT; Cybersecurity.

3 Discussion and Some Russian Lessons Learned

As the most important tendency to improve the organization of production and management, its shift towards the macroeconomic level should be noted, i.e. The main issues related to the development of the organization of production processes and organizational structures of production and management are now being solved practically at the meso-level, and it is by large economic systems that include enterprises located in different territories of Russia, and sometimes even abroad, working in a different mode of standard time. But at present, with the wide development of remote control, the intensification of the development of new methods of communication, which practically provide instant transfer of the necessary information, including visual information, problems and their solution are associated primarily with the presence and attraction of a large number of capital investments for which advanced economic structures spend fairly large amounts of money, forming the appropriate investment resources and financing the implementation of individual projects through them or system activities for modernization, change of production, etc.

In this case, it is necessary to pay tribute to large economic systems, primarily working within the framework of the synthetic production process, in which special divisions are created to form new types of products, i.e. those innovations are formed that are the base of product information, and as a result, projects are being developed to improve the technology and attract the technology that makes it possible to manufacture relevant types of products that are in demand and meet the requirements to increase their competitiveness not only in Russia but also in the global market.

However, this process is still not properly regulated in our country, and most importantly, we have lost the core of interaction between the state and these large economic structures, because when it comes to the formation of funds to support business development. To the greatest extent, regulatory documents are oriented towards small and medium businesses, which are helped by mastering new production methods, new types of production, new types of products, etc.

At the same time, we are well aware that serious fundamental technical changes are taking place precisely within the framework of large economic structures. Problems of interaction between government and business should be addressed more actively. Moreover, this cannot be limited to the creation of public private partnerships, which, as a rule, help to implement projects that have already been tested, and on innovative areas, coordinated actions of those research and design institutes that are funded by the budget are needed with commercial organizations they can and must realize the projects they are developing, and it is natural to bear with it certain risks, assuming them and often carrying certain losses in connection and with the development of new types of products, the need to rework production, etc. In this case, the state can also take part using various economic instruments, concluding various royalty agreements with the subsequent deduction of a share of profits in the case of state funding, etc.

These questions are seemingly outwardly purely economic. In fact, they are deeply productive, as they ensure the pace, real possibilities and systematic implementation of projects, supporting the rational interaction of the main elements of production, increasing the profitability of business activities and promoting research and development work directly into production. This guarantees the growth of the Russian economy, increasing its competitiveness and creating the possibility of using the developments made by Russian scientists in other countries.

Traditionally in Russia there was a strong scientific school of production organization. Research focused on the organization of business processes directly within the production facilities. It should be emphasized that the presence of these scientific developments made it possible to promptly raise the question of the need to centralize a number of production processes and create a single service center for a group of interested enterprises. In the 60s–70s of the last century, this was manifested in the form of creating centralized procurement and auxiliary workshops and, above all, workshops for initial production processes, the technologies of which are based on processing raw materials and form the basis of future elements of finished products. This applies to the foundry and forging shops in the framework of mechanical engineering in broad terms. In this case, in a number of regions, the so-called centralists were created, the centrifuges were those large productions that allowed the use of more modern methods for the initial processing of raw materials, which saved material resources and, in some cases, switched to more accurate production [11–16].

This applies to the cut-off processes in relation to a number of high-tech productions - cutting of especially scarce materials in aircraft manufacturing, in the manufacture of spacecraft, and other similar industries. In material-intensive industries, this actively influences the total cost of production due to the most rational use of materials.

It should be emphasized that in this case, when it comes to creating centralized industries for the initial use of raw materials, there is a large indirect environmental effect, because, moreover, it makes possible a more rational use of raw materials, it is possible to more efficiently process waste, their full use. To implement this work, it is necessary to create a special environmental accounting, internal control methods at each enterprise and in groups of enterprises of the same type. When creating powerful production of this type, the relative costs of environmental treatment facilities are relatively reduced, i.e. per unit of output, these costs are reduced.

In this case, the issue of environmental efficiency. Such centralized auxiliary production, as a rule, are created in places remote from settlements, i.e. in the suburbs, in a special industrial zone, a number of social problems are being solved.

In this regard, we still have a lot to do, because it is precisely at present that such processes can be triggered or initiated by the digital economy that led to Industrial 4.0 implementation. There will again be questions about the understanding of the optimal capacity of individual production processes and, as a result, the choice of new methods of communication between separate business processes. At the same time, the Russian scientific school of the organization of production was aimed at improving production within the enterprise and its general production process. Currently, there are many new issues that have to focus the attention of the state and individual corporate structures.

It seems that one of the very important areas of a more complete description of the activities of economic entities in modern conditions is the need to develop indicators of the structure of economic objects. This direction is actively implemented in connection with the development of the stakeholder theory of analysis and management, which allows us to more fully present and evaluate the effectiveness of the economic relations of any legal entity. It should be remembered that for a long time in Russia, during the period of the centralized economy, the course was taken to create complex enterprises that have essentially a closed production cycle, and sometimes reproduction, since those units whose main task was to repair and maintain equipment in working condition, engaged in a significant measure of modernization of fixed assets obtained in a centralized manner, conducted research on the possibility of using additional devices on the observed and regulation of the equipment, increased the possibility of additional support and service automation stages of operations. That is, practically within the framework of a number of economic enterprises, the repair shop performed to a large extent the task of further developing, modernizing equipment and directly adapting some complex equipment systems to the specific conditions of a particular organization.

In a number of industries, even special types of devices and mechanisms were created that helped to mechanize and improve the organization of certain types of business processes.

In the period of the 60–70s, when creating industrial associations, these issues were largely resolved and regional authorities actively intervened in their solution, in particular, in Leningrad and in the Leningrad Region, a great deal of work was carried out to create centralized support services. Productions on the organization of the primary processing of materials. These works to a large extent carried out research institutes of economics and organization of production relevant industries. In this regard, several were distinguished, but the essence was not even in industries classified by main types of products, but by main types of business processes, which often accumulated the same methods of organizing production, albeit for the production of different types of products [11–16].

This process concerns discrete productions, where synthetic processes prevail, productions with a predominance of analytical processes in which fundamentally different types of products and productions were produced from a single source of raw materials, which had fairly simple business processes for which the end-to-end approach to developing technological chains with the same rhythm and with a high degree of synchronization.

In order to implement approach that shown on Fig. 1, great attention must be paid to Evolution on production paradigm; Logistic improvement; KPI system optimization; Big data & analytics; Staff training; Control increasing; IoT; Cybersecurity. In the meantime, we need to revise and improve our views on standardization, modification, typing of production processes, standardization of individual parts of finished products, typing and modification of individual business processes, formation of batches of products simultaneously launched into production and shipped to the consumer, evaluation of other characteristics of production organization.

Currently, this problem arises again as one of the most important, since the right of enterprises to form an appropriate organizational structure in a market economy is expanding significantly. Stakeholder management theory is aimed at finding the leading, and sometimes global, stakeholders in a certain direction. And already within the group of individual legal entities, on the basis of the chosen forms of cooperation, the positions that allow creating an effective holding are determined. It may be fundamentally different - formed in the form of a joint-stock company, which unites enterprises vertically or horizontally, but in principle it provides for interaction between various economic entities, regardless of the variety of forms. Such holdings can combine not only production, but also management business processes. In a number of cases that are defined in the format of an organization's sustainable development, a holding may include subdivisions that solve social and environmental problems common to a group of enterprises that, in the production of business processes, may be tied insignificantly or simply unrelated.

The stakeholder theory makes it possible to form relevant non-financial information characterizing, on the one hand, the methods of interconnection of individual legal entities, and on the other hand, requiring appropriate digitalization and subsequent monitoring of their change and execution of the tasks set and the assessment of efficiency.

At present, the problem of developing Key Performance Indicators (KPI system) based on in-depth marketing analysis is of tremendous importance when consumer requests are considered and for each of them groups of consumers are formed that have different requirements for the finished product. In the first case, as a rule, the integration of enterprises involves the formation of joint capital, in the other case - linking the organization's activities through production processes, since certain classes of parts and assemblies and elements of products used by this legal entity are used.

In this regard, there is a need for new relations when it comes to monitoring the use of commitments for the production of standard parts, the development of projects of certain product groups that are part of the finished product. At the same time, certain representatives of the group of enterprises are required to produce them. In this case, an additional mechanism can be used associated with the allocation of appropriate funds to increase the production of standard parts, which are less marginal for the manufacturer. Specially we have to emphasize this direction, because in a market economy, each company decides on its own which product range is more profitable. If it is necessary to secure him as a mandatory production of certain types of products with a lower marginal income, it implies the adoption of appropriate measures.

The next moment, which is also of great importance and concerns to a certain extent the defense complex of enterprises, is the maximum use of the developed modern technologies, their transfer and use in other industries that do not produce defense products, but are oriented towards the use of high technologies. This is the so-called dual use of technology. In a centralized economy, this was typical mainly for the defense complex, and now it can be extended to almost any economic entity that produces high-tech products and does not have a developed research and development and engineering unit.

It should be borne in mind that users of scientific research may conduct additional research, but they will also need less capital investment and therefore many enterprises are willing to do so. In this regard, it is very important to disseminate the experience of individual enterprises in the field of production organization, improvement of technology and modernization of equipment, but in a market economy we have to take into account the problems of trade secrets, cost competitiveness, product quality and, consequently, they must be based on certain mechanisms adequate to the requirements of a market economy.

At the same time, it is necessary to actively consider how to evaluate such technologies, how to track their transfer to interested parties. This work should be deployed in the light of remuneration to the enterprise developer or the pioneer enterprise in the use of new technologies in the event that they are transferred to other objects. In this case, the franchising system may work, when instead of creating an additional division in innovation factories, innovation pioneers can assume the consulting and control function of using appropriate technologies at specialized enterprises of a commercial type. Innovation developers will also be interested in this, because this will increase their income and, in addition, allow them to learn from the experience of other enterprises, which may be an accelerator of further changes that they are conducting as part of research and design work. Control over these processes should be carried out on the basis of dividing the resulting effect between the enterprise-developer and the first-time user of technological processes and the enterprise-consumer of a specific technological process.

This position certainly deserves attention, and it can also be associated with the need to develop a system of indicators, quantify them and substantiate methods for monitoring the effectiveness of the use of innovations.

This question in Russia was developed quite well and fully. In a centralized economy, based on regulatory methods for accounting and monitoring the implementation of established standards, indicators have been well developed that characterize the organization of a business process in an enterprise and methods for monitoring their strict observance [11–16]. This refers to the formation of the main activities aimed at the economical construction of the production cycle. The production cycle of each product is considered from the moment the material is put into production until the finished product is obtained. The regulation of financial relations between enterprises in the production cycle is associated with the purchase of raw materials and materials, the necessary equipment, and the organization of the sale of end-use products is often not included in this process. Probably today, when it comes to a market economy, when each particular enterprise has a certain influence on the form of acquiring material resources, an extended interpretation of the production cycle, starting from the moment of using the funds to purchase materials, and ending with the receipt of funds on the company's settlement account for the realized products would have to be included in the general concept of the production cycle.

With this formulation, issues related to the need to improve not only accounting, but also management are raised. It is possible to dwell only on one side related to the sale of products, when the enterprise itself, by entering into appropriate agreements, selects accounting methods associated with large risks, which are taken into account in treasury control, which is becoming increasingly important not only in state but also in commercial organizations i.e. to a certain extent, the study of accounting for revenue when funds are received - revenue based on the accrual and shipment of products.

If you consider the problem not as a purely accounting, but as a control and management, it is important to find those control points and issues that need to be taken into account when evaluating the overall characteristics of the business process, and determine the influence of the main decisions on the overall activity of the production cycle, the reduction of which provides an acceleration of turnover assets. Consequently, in this case, firstly, it is possible to get along with smaller volumes of working capital per one ruble of the products sold. Secondly, there is a certain reduction in domestic stocks between individual business processes and individual operations.

It should be noted that from the analytical side of the question one has to constantly refer to the characteristic of the activity of the use of resources. However, when it comes to financing any production, to proving the effectiveness of an enterprise, the problem of organizing production is taken into account significantly less. When technological processes are developed, the problems of organizing business processes are left behind. They are practically charged with the obligations of the user of the new technology, which is hardly legitimate if it comes to the use of budgetary funds. Apparently, in this regard, when transferring relevant technologies in financing production, it is very important to have those indicators that characterize the duration of the production cycle, i.e. directly related to the organization of production.

4 Conclusion

At present, as the study shows, the issue of improving production organization at the meso- and macro-level, which should be developed taking into account new realities of Economics and the digitization of production, is becoming increasingly important. Improving the organization of production processes and organizational and management structures at all levels in a digital economy can be a real driver for its successful development. To do this we have developed Conceptual Model of Digital Economics Management.

The importance of this Conceptual Model in general has always been a definite locomotive that united and led Digital Economy along the path of efficient technical and technological renewal of industry, i.e. the organization of production has always been the most important prerequisite for further technical development. At present, the organization of production is divided and included in different areas of knowledge. Some of the questions need to be restored using the development and experience of Russian enterprises and merged in order to obtain a truly synergistic effect from further improving the organization of production. Other areas require additional basic research. It seems that at present there are all the prerequisites for talking about the digitalization of precisely this aspect of the activity of economic entities.

First, it is very important to trace the genesis of the development of the organization of production and its influence on the final result of the activity of the economic subject.

Secondly, it is necessary to trace the evolution of the development of accounting and control methods for changing the forms of an organization, including the development of a system of indicators, the structure of specific measures to change the organization of production, labor and the effectiveness of research and development and design work in an enterprise.

Thirdly, the new opportunities that provide information technology, including new methods of monitoring and controlling the activities of individual production processes in order to carry out a real digitalization of relevant groups of indicators – KPI system.

References

1. Maresova, P., Soukal, I., Svobodova, L., Hedvicakova, M., Javanmardi, E., Selamat, A., Krejcar, O.: Consequences of Industry 4.0 in business and economics. Economies **6**(3), 46 (2018). http://dx.doi.org.aucklandlibraries.idm.oclc.org/10.3390/economies6030046
2. Kinzel, H.: Industry 4.0–where does this leave the Human Factor? J. Urban Cult. Res. **15**, 70–83 (2017)
3. Industrial Internet Consortium: A Global Industry First: Industrial Internet Consortium and Plattform Industrie 4.0 to Host Joint IIoT Security Demonstration at Hannover Messe 2017. https://www.iiconsortium.org/press-room/04-20-17.htm. Accessed 28 Apr 2019
4. Kagermann, H.: Chancen von Industrie 4.0 nutzen. In: Industrie 4.0 in Produktion, Automatisierung und Logistik, pp. 603–614. Springer, Wiesbaden (2014). ISBN 978-3-658-04681-1
5. Hermann, M., Pentek, T., Otto, B.: Design principles for industrie 4.0 scenarios. Paper presented at 2016 49th Hawaii International Conference on System Sciences (HICSS), pp. 3928–37, Koloa, HI, USA, 5–8 January 2016
6. Senvar, O., Akkartal, E.: An overview to industry 4.0. Int. J. Inf. Bus. Manage. **10**(4), 50–57 (2018)
7. Nienke, S., Frölian, H., Zeller, V., Schuh, G.: Energy-management 4.0: roadmap towards the self-optimising production of the future. In: Proceedings of the 6th International Conference on Informatics, Environment, Energy and Applications, pp. 6–10. ACM, March 2017
8. Baldassarre, F., Ricciardi, F., Campo, R.: The Advent of Industry 4.0 in Manufacturing Industry: Literature Review and Growth Opportunities. University of Dubrovnik, Dubrovnik (2017). http://0-search.proquest.com.www.elgar.govt.nz/docview/2068860672?accountid= 40858. Accessed 29 Mar 2019
9. Ivančić, L., Vukšić, V.B., Spremić, M.: Mastering the digital transformation process: business practices and lessons learned. Technol. Innov. Manage. Rev. **9**(2), 36–50 (2019). http://0-search.proquest.com.www.elgar.govt.nz/docview/2197277088?accountid=40858. Accessed 28 Apr 2019
10. Gavriluță, A., Nițu, E.L., Gavriluță, A., Anghel, D.C., Stănescu, N.D., Radu, M.C., CreȚu, G., BiriȘ, C.M., Păunoiu, V.: The development of a laboratory system to experiment methods to improve the production flows. Proc. Manuf. Syst. **13**(3), 127–132 (2018). http://0-search.proquest.com.www.elgar.govt.nz/docview/2137835746?accountid=40858. Accessed 28 Apr 2019
11. Kamenitser, S.E.: Organization and planning of industrial enterprises, 591 p. M.: Politizdat (1967)
12. Razumov, I.M., Glagoleva, L.A., Ipatov, M.I., Ermilov, V.P.: Organization, planning and management of the machine building enterprise, 544 p. M.: Mashinostroenie (1982)
13. Letenko, V.A.: Turovets OG organization of engineering production: theory and practice, 208 p. M.: Mechanical Engineering (1982)
14. Stepnov, I.M.: Theory and methodology of the use of innovative potential in the industry of the region: dissertation for the degree of Doctor of Economic Sciences. St. Petersburg (2001)
15. Kovalchuk, Y.A.: The concept of advanced production organization. Production organizer, no. 1, T.24, from 5–9 (2005)
16. Glukhov, V.V.: Organization of production and marketing, 369 p. SPb.: Polytechnic University (2012)

Digital Accounting and Auditing

Accounting Policies, Accounting Estimates and Its Role in the Preparation of Fair Financial Statements in Digital Economy

Denis Lugovsky[ID] and Mikhail Kuter[(✉)][ID]

Kuban State University, Krasnodar 350040, Russia
prof.kuter@mail.ru

Abstract. The paper considers the main problems and limitations of the reliable presentation of reporting information. Ensuring the reliability of accounting information is objectively hampered by its conventionality associated with the need to give a numerical estimate of accounting objects. Subjective limitations are connected with a variety of information needs of users of reporting data, which determine differences in accounting methodology. The accounting conventionality is most evident when calculating the financial result, therefore one of the priority problems in forming reliable financial statements is the accuracy problem of financial result calculation. Equally important is the problem of recognition of contingent and evaluation assets as well as liabilities. The choice of accounting methodology at the level of accounting standards or at the level of an economic entity has a decisive influence on the reporting data presented to users: the same business transactions can be presented in completely different ways. The choice of the degree of freedom provided to economic entities during the preparation of reporting lies in the area of the distinction between global and local accounting policies. In addition, reliability of financial statements is influenced by many other factors related to the choice of accounting and depreciation policies, changes in accounting estimates, legal justification of accounting entries and many others that are also considered in the paper.

Keywords: Accounting policy · Global accounting policy ·
Local accounting policy · Professional accounting judgement ·
Accounting estimates · Contingent assets · Contingent liabilities ·
Financial statements

1 Introduction

The formation of reliable information in financial statements is recognized as the main purpose of its compilation. The provisions of international and national standards, as well as accounting policy that is being formed by economic entities, are aimed at this. At the same time, the concept of "fairness" is very conditional. Reporting information, which is formed on the basis of the rules established by accounting standards, is considered to be formally reliable. However, the question of choosing an accounting methodology always remains open for the developers of standards, regulations, instructions as well as accountants who are required to ensure compliance with the established requirements, taking into account the specifics of the company. Accounting

T. Antipova (Ed.): ICIS 2019, LNNS 78, pp. 165–176, 2020.
https://doi.org/10.1007/978-3-030-22493-6_15

legislation is not able to foresee all possible economic situations, to give definite answers to all the questions arising in practical activity. The accountant is required to apply constantly professional judgment in choosing accounting policies, determining accounting estimates, identifying, evaluating and classifying accounting objects. The desire to improve the accounting accuracy leads to the expansion of its information space, filling it with an increasing number of virtual categories such as profit, capital, reserves, contingent and estimated assets and liabilities.

On the one hand, this contributes to the increase in completeness and efficiency of data formation, on the other hand, reduces its accuracy and reliability. As part of the most pressing problems of forming reliable financial statements, the accuracy of calculating profit and comprehensive income indicators, the problem of presenting estimated and contingent assets and liabilities in financial statements, the problem of recognizing and disclosing information about estimated and contingent income and expenses, the problem of correlation of global and local accounting policies, as well as the ratio of accounting policy and accounting estimates should be highlighted.

The problems of profit calculation are touched upon in the works of many authors. In determining the financial result, as Needles, Powers and Crosson note, one of the most difficult problems of accounting is considered to be the attribution of expenses and income to one or another accounting period. When calculating profit, an accountant should proceed from the assumptions of going concern and matching rule of income and expenses [1]. Limperg and Zappa can be considered among the most distinguished researchers of the issues of formation of financial result, evaluation and accounting policy. The former in his research progressed from expenses to income, based on the statement: expenses are obvious, income is regulated. The latter, on the contrary, hold the view that income is obvious, and expenses are regulated [2]. The problems of subjective estimate during the formation of professional judgment, the choice of accounting policy, the recognition and evaluation of contingent assets and liabilities were analyzed by many scientists. In particular, Zack draws attention to the fact that the conventionality of the accounting data creates the prerequisites for the manipulation of financial statements indicators [3]. In this aspect, he emphasizes the estimate at fair value, which, in his opinion, is of subjective character and lacks the advantage of the initial (historical) estimate, such as validity. In turn, the validity and confirmation of the estimate of an accounting indicator or a business transaction are aimed at increasing of the reliability of financial statements. Sher also singled out the evidence of books before a court as one of the mandatory requirements for accounting [4]. Paying attention to the correlation between global and local accounting policies Hendriksen and van Breda will highlight the accounting policy formed at the company level, and the accounting policy formed above the company level [5]. The application of fair value as an accounting policy instrument was studied by Karniz, Alexander and other authors [6, 7].

In the first part of the work, the listed problems are analyzed from the standpoint of their influence on the accuracy of digital financial reporting data, the possible areas of their solution based on the application of fundamental accounting principles and finding the optimal balance between the qualitative characteristics of financial reporting are identified. In the second part of the study, the main conclusions and results are presented, and areas for further work in this area are identified.

2 Problems of Fair Financial Statements Preparation

It is known that the main purpose of financial statements is the formation of reliable information that is useful when its users make economic decisions. At the same time, it is believed that information generated on the basis of current accounting and reporting rules established by international or national standards is considered to be reliable.

However, standards establish only general accounting principles, giving the opportunity of choosing the accounting policy and determining the accounting estimates, thus allowing taking into account the peculiarities of the functioning of a particular economic entity and, as a result, improving the quality and usefulness of reporting information. At the same time, the choice of accounting policy is largely subjective and requires the use of accounting professional judgment. The basis of the subjectivity of the choice of accounting policy is the conventionality inherent in most modern objects of accounting observation. To the greatest extent it affects such virtual objects as profit, capital and reserves.

The amount of profit is influenced by many accounting methods and means related to the choice of accounting policy, accounting estimates and the use of professional judgment in recognizing the amount of income and expenses of the reporting period. The determination of the concept and criteria for the recognition of current and deferred income and expenses, reservations, recognition of asset impairment, depreciation policy and a lot more are among them.

Thus, the reliability of financial statements is largely associated with the methods of financial result calculation. It aggregates in itself the whole complex of elements of accounting policy related to the calculation of income and expenses of an economic entity, including the choice of the method of calculating the depreciation of long-term assets, writing off the cost of raw materials, final products, etc. Of course, the amount of income, expenses and the financial result is influenced not only by the methods of accounting and evaluation of accounting objects, enshrined in the accounting policy, but also by the real efficiency of the business, the rationality of the implementation of financial and economic activities. But such concepts as "efficiency" and "result" are also conditional. It is possible only to increase the relative accuracy of their calculation, but it is impossible to achieve the absolute accuracy. In part, this can be achieved if we look at the history of accounting: it already occurred in the early stages of the development of double-entry accounting, when there was a practice of calculating the financial result from the moment of establishment to the moment of liquidation of the enterprise, and there were no such modern objects as intangible assets, reserves, financial instruments and many other things.

Certainly, financial result, calculated for the entire period of the company's operation, turns out to be much more accurate than the periodic one, since it eliminates many conventions, for example, from depreciation as such, from choosing the method of settlement of inventory values, from assessment of work in progress or determining the procedure for writing off overhead expenses. In addition, the exclusion from the list of accounting objects of everything except property and monetary resources will significantly increase and bring the accuracy of determining the financial result closer to the absolute accuracy. However, for some reason this does not happen. The price for accuracy is low efficiency and information capacity of the reported data. The governing body of modern enterprises requires a lot of information for making timely management decisions in the shortest possible time, otherwise in the fierce competitive environment, it may be too late. In such a situation, it is important for the users to be aware of the entire degree of conditionality of the financial statements and use it only in close connection with the accounting policy applied in the organization and due attention to the system of accounting estimates. Otherwise, such information will be of not much benefit.

The reverse is also true: price for the speed of information is the decrease in its accuracy. Today, the period for which the financial statements are prepared and the financial result is determined is a calendar year (a quarter – for interim reporting). In the accounts the financial result is presented even more often – monthly. Technically, it is not difficult to calculate the financial result for a shorter period, but this does not occur, since the error of such a calculation may be unacceptable.

In addition, the fair, at first glance, desire to obtain more and more diverse digital data on the financial position and financial performance of the enterprise is also associated with the decrease in its absolute and relative accuracy. Absolute accuracy is possible only with respect to monetary resources (and even then, with certain reservations). As soon as the object of accounting, proclaiming the cost measurement as one of its fundamental principles, becomes something other than money, an objective basis for the evaluation's conditionality arises. Moreover, it arises even when it comes to material objects – the property of the organization, such as fixed assets, materials or goods, which are widely represented on the market. Because even the market value of identical or interchangeable assets may vary significantly depending on the seller, region, point in time and other circumstances. If the situation is complicated by the absence of analogs and (or) material and physical form, then the spread of accounting estimates and the subjectivity of the estimate increases. In addition, market valuations are far from the only possible one to use in accounting. Different circumstances necessitate the use of the original, replacement, discounted and amortized cost.

The distribution of the initially conditional value of assets and liabilities over time, the formation of income and expenses using such assumptions as depreciation and capitalization, reinforces the subjectivity of their initial valuation.

And finally, the culmination is the calculation of the financial result, which requires the correlation of income with expenses, which is also based on accounting, virtual methods and means. In particular, the principle of conformity (linking) of income and expenses, procedures for deferral and trans-formations, determining the procedure for deferred income and expenses accounting, as well as reservation and the associated with it recognition of estimated liabilities.

Thus, profit, being the main purpose of the activities of a commercial organization and the most important object of modern accounting, has absorbed all the conventions inherent in it. The reason for this is the appearance of more and more new virtual accounting objects, which contribute to the reduction of the relative accuracy of digital data. Therefore, an important direction in the development of modern accounting theory and practice is finding the optimal ratio between the relevance of information, on the one hand, and the reliability of its assessment, on the other hand. A balance is needed between both the real and virtual data in accounting and reporting.

In these circumstances, it is impossible to overestimate the importance of information concerning the accounting policy used, accounting estimates and changes occurring in them for assessing indicators of the financial position and financial performance of an enterprise, without which such indicators have no sense.

Therefore, paradoxically, disclosure of non-financial information promotes reliability of financial statements. According to Efimova and Rozhnova, structured financial and non-financial information helps to improve the accuracy and quality of information disclosure in financial statements [8].

When preparing financial statements, companies are faced with the necessity to choose accounting policy and determine accounting estimates. Enterprises develop accounting policy taking into account the specifics, scope and types of economic activity that they carry out, existing regulatory restrictions and other factors.

The main difficulty in developing accounting policy is to ensure its focus on achieving the main goal of financial reporting – to form a reliable understanding of the financial position and financial performance of a company. Rejection from one method in favor of another is a change in the balance of priorities of various groups of users of financial statements. Increasing accuracy in one position means losing it in another. This is an objective inevitability inherent not only in accounting.

In addition, subjective factors, such as optimization of taxation, unification of accounting and tax accounting, dividend policy and others, can also have a negative impact on the accuracy of accounting data.

Formation of the accounting policy is also associated with technical difficulties caused by the disclosure of what does not relate to it as part of the information on the accounting policy applied. First of all, the imperative instructions of accounting legislation, for instance, the procedure for the initial assessment of fixed assets. In addition, variable regulations of financial or economic legislation, for instance, the rules for distribution of profits among shareholders, are erroneously attributed to the accounting policy. And finally, in the composition of information on accounting policies there may be elements that determine the procedure for evaluating a digital indicator of financial statements in the framework of the adopted method. Such information should be introduced as part of information on changes in accounting estimates.

Close association of accounting policy with an economic entity is not accidental, since in practice it is its officials who choose from among the alternatives provided by the accounting regulators and draw it up in the appropriate order. The imperative requirements of accounting standards have nothing to do with the accounting policy of the company due to the lack of any ability to influence them. However, from the position of the regulator that establishes mandatory accounting rules for all economic entities, such rules are the accounting policy.

As Hendriksen and van Breda note, the form and methods of forming the financial statements of the company constitute its accounting policy. It is recognized as the basis of accounting standards, recommendations, interpretations, rules and regulations used by companies during the preparation of financial statements [5].

According to IAS 8, accounting policies are specific principles, grounds, agreements, rules and practices adopted by an enterprise for the preparation and presentation of financial statements. The Russian Accounting Regulation 1/08 treats accounting policy as a set of accounting methods – primary observation, cost measurement, current grouping, and a final summary of the facts of economic activity.

The main difference between the presented interpretations lies in the fact that the Russian standard defines the accounting policy from the point of view of accounting, and the international one from the point of view of financial statements. Accordingly, in Russia, the accounting policy covers all the elements of the accounting method, and in the international practice it covers only those elements that affect the digital indicators of financial statements. In this case, the definitions are unified in the interpretation of the essence of the accounting policy as the choice of methods of accounting (reporting), as well as the entity carrying out its formation – the enterprise.

In a broad sense, the choice of accounting is carried out not only by economic entities, but also by the draftsmen of accounting standards that establish common principles and rules for accounting and reporting. Such a choice is nothing but a national (international) accounting policy.

National (international) accounting policy is the establishment of uniform general principles (rules) of accounting for all economic entities within a single state (group of states) [9].

In other words, the choice of the authors of accounting standards lies between the use of single or double-entry bookkeeping, the concept of ownership right or economic control, the priority of substance over form or form over substance, static or dynamic accounting, the use of depreciation and reservation mechanisms, etc. In turn, within the framework of the national accounting policy of each particular organization, its internal (local) accounting policy is formed. In this case, the choice is made by the chief accountant between specific permitted methods of assessment, classification, reflection in accounting and reporting of certain objects of accounting supervision, for instance, between the authorized methods of fixed assets depreciation.

In a general sense, accounting policy is determination of the rules for accounting. Such a choice, made at the state (international) level, is implemented in the code of published laws, standards, instructions and other acts of standard that determine general rules of accounting and its principles. At the level of an economic entity (organization), on the basis of general principles, it is possible to choose between permitted accounting options [10].

Thus, today it is appropriate to speak about the presence of at least two levels of accounting policies – global (national, international) and local (accounting policies of the organization).

Moreover, as can be seen from the above, as part of the global accounting policy, it is possible to distinguish two separate levels – international and national. From the point of view of the declared fundamental principles, the differences between them are minimal and continue to reduce as national standards adapt to IFRS, but they remain in the peculiarities of their interpretation and practical application.

For instance, the principle of prudence declared in IAS 1 and Russian Accounting Regulation 1/08 is treated equally, but is implemented differently in the content of the accounting standards system in Russia and abroad. The concept of prudence is very relative. For instance, you can recognize as income only earned and actually received sums. You can reduce prudence and include into the income from the reporting period sums in respect of which the right of claim arose. So, IFRS include in the income of the reporting period not only sums in respect of which a legal right to claim repayment from a debtor arose, but also sums in respect of which such a right is expected, showing minimal prudence.

Bloom notes the declining influence of prudence in private companies. In his opinion, conservatism is a perennial issue in accounting practice and accounting standard setting [11].

The practical application of the principle of priority of substance over the form raises no less questions. And such fundamental principles (methods) as double entry and documentation remained completely outside the framework of regulation by the system of international standards.

Therefore, it is fully justified to single out the national accounting policy as an independent level, separate from the global accounting policy, but very close to it.

Similarly, within the framework of the national accounting policy, it is possible and purposeful to form a corporate accounting policy, that is, the accounting policy of a group of companies. In accordance with the current accounting rules, enterprises in the group are not obliged to do this at all: they can carry out various activities, pursue different strategic and tactical goals, implement one or another financial policy, etc. And even if required, there is not always technically possible to unify accounting policies. The situation is complicated by the changes in the structure of the holding, the acquisition and sale of new companies and shares in them, loss of control and other factors. The need to level the differences in the accounting policies of companies belonging to the group based on the principle of consistency arises only in the preparation of consolidated financial statements. At the same time, reporting, rather than accounting data, is adjusted, with the mandatory disclosure of all related information in the explanatory note.

Nevertheless, corporate accounting policies still have the right to exist. The possibility of its formation is provided for by Russian Accounting Regulation 1/08, according to which, if the main company approves of its accounting standards, which are obligatory to use by its subsidiary company, such a subsidiary company chooses the accounting methods based on these standards.

In this regard, it is important to pay attention to two circumstances – the voluntary formation of the corporate accounting policy and succession (consistency) of accounting policies at a lower level to the accounting policy of a higher level.

Therefore, in the context of stable and long-term economic relations between the companies belonging to the group, integration of the management system, planning and coordination of ongoing activities, the presence of the system of internal management regulations (standards), etc., and the main thing is that if there is a need on the part of the parent company, it can form and approve of corporate accounting policy.

Thus, it is possible to classify the accounting policy in relation to the entities engaged in its formation as global (IFRS level), national (national accounting standards level), corporate (group of companies' internal standards level) and local (enterprise accounting policies level).

The use of such a classification will contribute to the succession and consistency of the formation of accounting policies at all levels, will ensure the rationalization of the content of orders on the accounting policies of economic entities. In addition, a clear distinction between the levels of formation of accounting policies will reduce the probability of the most common mistake when drafting an order on an organization's accounting policy—including in it the elements related to higher levels of accounting regulation, manifested in the duplication of non-competitive regulations of accounting legislation.

The depreciation policy of an economic entity is often referred to as one of the priority areas of the accounting policy.

The presence of a variety of views on the economic nature of depreciation is due to the existence of various goals, tasks and functions attributed to it, which, in turn, are predetermined by the interests of individuals whose concern is accounting [12].

The choice of the concept of depreciation, as well as the choice of accounting policies, is directly related to the economic interests of financial statements users. So, behind the legal concept, aimed at determining the sufficiency of property for repayment of debt obligations and, as a result, treating depreciation as the degree of asset value loss, are the interests of creditors: banks, suppliers, insurance companies, etc. The economic concept is based on the interests of the company's management, for which the treatment of depreciation as an element of prime cost, and, consequently, the management of expenses and financial result, is priority. The financial concept is close to the economic one, seeing in depreciation the transfer of value, but seeing in it not so much the consumption of the asset value but the emergence of a reserve for restoration of depreciable property, which is a priority for the owners, since it is aimed at solving their most important task – preserving and increasing their capital. The tax concept, reflecting the interests of the state, allows to include depreciation in expenses that reduce taxable profit, thereby freeing the cost of renovation of depreciable property from taxation.

Each of the presented concepts is aimed at achieving its inherent priorities, based on the interests of persons capable of influencing the choice of depreciation policy, however, making it difficult to apply alternative concepts.

The economic concept of an enterprise is most closely related to the financial concept, since it is focused on calculating profits as the main goal of its activities. And in the long term, based on the concept of maintaining capital. The funds of the depreciation fund should be sufficient to carry out the modernization, replacement and repair of machinery, equipment and other long-term assets, in such a way as to not only preserve, but also increase the property and production potential of the company.

Modern accounting is being recently replenished with new, more and more virtual categories, in connection with which the question arises concerning their correlation with the objects of real property and property rights.

The inclusion in the financial statements along with the property and unconditional liabilities of leased assets, rights to use software, deferred taxes, estimated liabilities and all sorts of reserves, on the one hand, increases its information value, on the other hand, reduces comparability of indicators and in a way "litters" accounting information system. Thus, the recognition of valuation and contingent assets and liabilities significantly influences the formation of reliable financial statements.

Recall that comparability is recognized as one of the key principles of the formation of reliable financial statements [13].

In the most general and simple sense, conditional assets and liabilities should be understood as any objects other than property (property rights), the subsequent transformation of which into real values (rights) is probabilistic in nature. However, in modern accounting, in order to strengthen the user's perception of the reliability of its data, only unlikely events are called conditional. At the same time, highly probable operations, with the possibility of a reliable assessment of their consequences, are no longer considered conditional, but are called evaluative ones. For instance, decrease in the market value of commodity stocks the selling transaction of which has not yet been conducted, or increase in the amount of debt to workers by the amount of the upcoming vacation pay, the monetary obligation the payment of which has not yet occurred, however, its amount can be determined, the probability of the given events is high. The value of such assets and liabilities is nothing other than accounting estimate. In the first case, the term "estimated reserves" is used to identify them, and in the second, "estimated liabilities". Their formation is aimed at implementing the concept of prudence and serves as a tool for minimizing and insuring risk – the risks of evaluating accounting observation objects.

The presence of risks of non-payment of income implies the need of protection against them, i.e. insurance of the sums of recognized incomes; the main event in this direction may be the establishment of estimated reserves for doubtful debts [2].

Assets and liabilities, information about which is enough to identify them at least somehow, but not enough to recognize them as estimated (for example, guarantees received and issued, participation in court proceedings) are assumed to be called contingent in modern accounting.

All other, conditional, unlikely, in their essence, events are not formally related to conditional and are beyond the scope of modern accounting practice, however, taking into account the emerging tendency to deviate from the principle of prudence, in the future it is possible that the boundaries shift in the identification of objects from estimated to real, from conditional to estimated and from unidentifiable to conditional.

Conditional, that is, probable, in a broad sense, can be not only assets and liabilities, but also income and expenses. Moreover, it is the income and expenses expected to be received in the future, and not the upcoming incomings to pay off already existing receivables and payables, i.e., the objects that still replace the content of these concepts. It is this circumstance that caused the measures aimed at refusing to use them in the Russian financial statements and bringing it in line with the provisions of IFRS.

It is quite reasonable if we proceed from the composition and content of the objects included in the composition of the income and expenses of future periods, in most cases, representing receivables or payables, which for some reason did not have a place in the accounts for settlements, or income and expenses of not the future, but the reporting period. However, if you take a fresh look at the concepts of income and expenses of the future periods as the results of events that are of an estimated, probabilistic nature, then they have every chance of being recognized as items of financial statements.

Obviously, under the income and expenses of future periods, as opposed to income and expenses of the reporting period, in a broad sense, it should be understood the potential increase or decrease in the economic benefits of the organization, planned or expected in the future. At present, such amounts are not recognized in the financial statements, including those prepared according to IFRS. This is a prerogative of economic planning and management accounting.

However, as noted earlier, the emerging tendency to abandon accounting prudence, the emergence of new virtual accounting facilities, the shift of formal boundaries between real and conditional facts of economic life, allows to identify income and expenses of future periods as conditional facts of economic life. And if there are contingent assets and liabilities, then there should be contingent income and expenses.

Based on modern boundaries between real, estimated and conditional facts of economic life, the expected increase or decrease in economic benefits due to received or issued guarantees, legal disputes, etc. can serve as an example of conditional income. In the future, they may also include objects of more virtual character. If, on the contrary, to make a start from the principle of accounting prudence, then the income and expenses from the implementation of long-term construction work should be considered conditional, for example. This is, in essence, income and expenses of future periods, which, however, in accordance with current international practice, are included in the income and expenses of the reporting period. But even in this case, in our opinion, it would be more correct to determine them as estimated income and expenses, which should also include exchange differences, the result of the revaluation of material values and all other incomes and expenses arising from changes in accounting estimates.

Thus, all the above mentioned, as well as many others, questions of fair financial statements presentation in the digital economy that remain outside the scope of this study are predetermined by the very essence of the business, which is always associated with the risk of uncertainty of the financial result. And they, for the most part, do not have a simple and unambiguous solution due to the high degree of conventionality and uncertainty of the digital information generated in modern financial statements. However, in any case, such a conventionality should be minimized by all available means, and all the circumstances related to it should be introduced in the explanatory notes to the financial statements.

3   Results

The choice of accounting policy, changes in accounting estimates and professional accounting judgment have a significant impact on the formation and presentation of information in financial statements, especially in the context of digital economy and its inherent technical means of generating, processing, presenting and analyzing of accounting data.

Modern information technologies allow to raise financial reporting to a new level, to bring into the accounting information system a lot of new virtual and derived indicators. At the same time, the conventionality of the generated digital data significantly increases, the need arises to choose global and local accounting policies, to change accounting estimates, and to apply professional judgment. The given circumstances have a dual impact on the reliability of financial statements, on the one hand, contributing to its completeness, on the other hand, reducing its reliability.

One of the most important problems in the formation of reliable financial statements is the problem of the accuracy of calculating virtual reporting indicators, primarily profit. The solution to this problem lies in the area of finding a compromise between the relevance of information and its reliability.

The development of accounting policy is also associated with finding a balance between the priorities of various groups of users of financial statements. It is recognized as part of the general economic policy of a company and forms a whole along with depreciation, dividend and tax policies. Subjective factors also influence the choice of accounting policy, which adversely affects the reliability of financial statements.

The lack of a clear distinction between the levels of accounting policy and also accounting policy and accounting estimates may result in the omission or misrepresentation of essential information. In relation to the entities engaged in the formation of accounting policy, it can be classified into global (international), national, corporate and local. Such a distinction will reduce the likelihood of errors associated with decision-making and introducing in the accounting policy of the company the field outside its competence, or not at all related to accounting policy.

The presence in the financial statements contingent and estimated objects along with the real ones, increases its information capacity, but reduces the comparability and consistency of information. In addition, not only assets and liabilities can be contingent and estimated, but also income and expenses, that is, income and expenses expectable in the future.

The obtained results indicate an increment of available knowledge. They can be used as a basis for further research into this area, as well as in choosing of accounting policy, determining accounting estimates and justifying professional judgment during the formation and presentation of financial statements in the context of digital economy development.

4 Conclusion

The purpose of the research was to determine the main problems and limitations of the formation of reliable financial statements in the context of digital economy and to develop ways of their solution. The research was based on the analysis of modern accounting methodology, historical prerequisites for its formation, fundamental accounting principles, searching for cause-and-effect relationships, studying the works of scientists and experts in accounting, comparing Russian and international approaches to the formation of accounting policy.

The main results of the work are in the formulation and systematization of the main problems of preparation of fair financial statements under current conditions. Among such problems we should call the following:

- the problem of the accuracy of calculation of profit and other virtual reporting indicators;
- the problem of excess of virtual indicators in financial statements;
- the problem of formation of accounting and depreciation policy and determination of accounting estimates;
- the problem of distinguishing between global and local accounting policies;
- the problem of recognition of estimated and contingent assets and liabilities;
- the problem of recognition of estimated and contingent income and expenses.

The solution of these problems in most cases requires finding a compromise between compliance with accounting principles (quality characteristics of financial statements). For instance, between the relevance of information and its reliability, between its accuracy and timeliness. Another direction should be an integrated approach to the development of accounting, financial, tax, dividend and depreciation policies of an enterprise. It is also necessary to ensure the consistency of the accounting policy of a lower level to the higher one: local accounting policy in relation to corporate one, corporate accounting policy in relation to national one and national accounting policy in relation to international accounting policy.

Of course, the carried out research is not able to cover all the existing problems, however, consistent work on overcoming the determined contradictions and understanding of the existing limitations of the preparation of fair financial statements will allow to advance significantly in this area.

A promising direction for further research is the use of information technologies and artificial intelligence in the formation of accounting policy and the determination of accounting estimates, including fair value.

References

1. Needles, B., Powers, M., Crosson, S.: Financial and Managerial Accounting. Houghton Mifflin Company, Boston (2007)
2. Kuter, M., Lugovsky, D., Ponokova, D.: The reserving as a way of risk hedging. Financ. Credit **34**(238), 40–46 (2006)
3. Zack, G.: Fair Value Accounting Fraud. New Global Risks and Detection Techniques. Wiley, Hoboken (2009)
4. Sher, I.: Accounting and balance. Economic Life. Moscow (1926)
5. Hendriksen, E., van Breda, M.: Accounting theory. Richard D. Irwin. Homewood, IL (1992)
6. Cairns, D., Massoudi, D., Taplin, R., Tarca, A.: IFRS fair value measurement and accounting policy choice in the United Kingdom and Australia. Brit. Acc. Rev. **43**(1), 1–21 (2011)
7. Alexander, D., Bonaci, C., Mustata, R.: Fair value measurement in financial reporting. Procedia Econ. Financ. **3**, 84–89 (2012)
8. Efimova, O., Rozhnova, O.: The corporate reporting development in the digital economy. In: Antipova, T., Rocha, A. (eds.) Digital Science. DSIC18 2018. Advances in Intelligent Systems and Computing, vol. 850, pp. 71–80. Springer, Cham (2019)
9. Kuter, M.: Introduction to accounting. Education-South. Krasnodar (2013)
10. Lugovsky, D.: Accounting policies and estimate standards: essence, content, interconnection. Bull. Adyg. State Univ. Episode 5 Econ. **2**(141), 98–106 (2014)
11. Bloom, R.: Conservatism in accounting: a reassessment. Acc. Hist. J. **45**(2), 1–15 (2018)
12. Kuter, M., Lugovsky, D., Mamedov, R.: Depreciation policy is an accounting policy element to ensure the financial strategy of the owner. Econ. Anal. Theory Pract. **29**(158), 17–23 (2009)
13. Grishkina, S., Sidneva, V., Shcherbinina, Y., Dubinina, G.: Comparability of financial reporting under different tax regimes. In: Antipova, T., Rocha, A. (eds.) Digital Science. DSIC18 2018. Advances in Intelligent Systems and Computing, vol. 850, pp. 88–93. Springer, Cham (2019)

Production Accounting System at Breweries

Mizikovsky Effim Abramovich$^{(\boxtimes)}$ ®,
Zubenko Ekaterina Nikolaevna$^{(\boxtimes)}$ ®,
Sysoeva Yuliya Yuryevna$^{(\boxtimes)}$ ®, Ilyicheva Olga Valerievna$^{(\boxtimes)}$ ®,
and Frolova Olga Alekseevna$^{(\boxtimes)}$ ®

State Budgetary Educational Institution of Higher Education «Nizhny Novgorod
State Engineering and Economic University», Knyaginino, Russia
zubenkoen@yandex.ru

Abstract. Construction of production accounting at industrial enterprises in a market economy pays close attention, since the amount of profit received from final products sale directly depends on proper organization of cost accounting and production costs calculation. At the same time, the significant part of brewing enterprises profits, however, like other enterprises of food industry of the Russian economy, often grows as a result of not always an objective increase in selling prices, even when antitrust sanctions are imposed. The article presents a fundamentally new approach to cost accounting, which fully meets industry specifics and peculiarities of breweries technological process.

In the paper, the authors gave a description of an important area of accounting for brewing organizations, such as the system for recording production costs and calculating costs of production by trade names of products. The article discusses the basics of building cost accounting system in detail, providing a procedure for their management. Emphasis is placed on key points that determine formation of production costs of products (works, services): the manufacturing enterprise specifics, the classification of costs.

It was determined that production enterprise costs should be considered for technological operations, and it was also revealed that management of brewing products cost price directly depends on the proper organization of cost grouping according to calculation items.

Keywords: Costs · Production process · Technological operations · Cost price · Breweries

1 Introduction

Russian brewing (large-scale production of brewing and non-alcoholic beverages, mineral and drinking water) is a predominantly stable, highly profitable industry of production activities, finished products of which are officially included in the government list of food products of population.

Domestic and foreign economic practice shows that successful functioning of economic management system of enterprises that systematically implement cost saving programs (and on this basis, systematic reduction of production and other costs) is largely predetermined by the quality of primary and accounting data cost.

© Springer Nature Switzerland AG 2020
T. Antipova (Ed.): ICIS 2019, LNNS 78, pp. 177–183, 2020.
https://doi.org/10.1007/978-3-030-22493-6_16

Studies conducted at a number of large brewing enterprises for 2010–2017 showed that production accounting is usually conducted according to the methodology established by the USSR Ministry of Finance in 1970 and outdated due to known circumstances. Accounting reform, carried out in Russia at the beginning of the two thousandth, did not practically influence production accounting at brewing enterprises, and its functions in a fundamentally new economic environment have not changed in essence.

Production accounting at breweries currently continues to perform only accounting functions, reduced to an accounting summary ("summary") of the production and sales cost for the whole enterprise, at best distributed over unreasonably enlarged product groups, impersonal with respect to trade names. Ultimately, accounting information presents an impersonal financial result from sale of brewing products, taken into account while completely ignoring one of fundamental domestic and foreign accounting principles - comparability of sales revenue of a unit of products of labor with their total cost. Such information cannot be recognized as valuable for the managing system of economy of any type of production activity of the industry in general, and brewing in particular.

The success of the reform will be real only if traditional impersonal ("boiler") production accounting, which has been practiced at breweries for dozens of years using the same rules, increasingly lagging behind realities of the modern brewing economy, is replaced by standard accounting in its full methodological scope.

It is advisable to start reforming the existing production accounting system by examining composition of costs for each production cycle process of brewing and other products, which will determine sectoral (specialized) nomenclature of calculation items.

2 Main part

Significant changes in the main approaches to the manufacturing enterprise management occurred after Russia's transition from a planned to a market economy. So in a planned economy, enterprises did not have the ability to set, as well as to influence a specific price level, since the priority direction for setting prices for products, taking into account the pricing specifics, belonged to the state.

With the transition to the market economy conditions, managers of industrial enterprises had the opportunity to make decisions independently on such major issues as: planning an output; formation of assortment and pricing policies, etc. [6, pp. 117–118].

Correct, adequate management decisions taken in a competitive environment affect functioning and further development of specific economic relations. Successful economic decision making is influenced by objective factors, which are due to new technologies, government regulation and enterprises growth, the structure of a production enterprise, users of financial information, and various areas of production activity.

In connection with what, under the market economy conditions, as well as with industrial enterprises emergence, primarily those oriented to market demands, a new phenomenon appeared in economics - "management accounting" [4, pp. 141–143].

3 Methodology

The Russian food industry includes groups of enterprises producing the following types of products: confectionery, dairy, meat, wine-making, cereals, and also beverages and tobacco. Enterprises engaged in production of food products, including beverages and tobacco, have a complex structure as the main ones among them: baking and flour milling, dairy, meat, wine-making, cereal production. Brewing and non-alcoholic beverages industry is one of the branches of food industry, which is engaged in beer, soft drinks, and mineral water production.

The brewing and non-alcoholic beverages industry is one of the branches of food industry, which is engaged in a wide range of beer, soft drinks, kvass and soft drinks production. Production at breweries is regulated with the basic National Standards of the Russian Federation: GOST 31711-2012 – "Beer. General technical conditions", GOST R 55292-2012 – "Beer drinks. General technical conditions", GOST 31494-2012 – "Kvasy. General technical conditions" and GOST 28188-2014 – "Non-alcoholic beverages. General technical conditions".

According to the Russian Brewers Union ("Report of Brewing Industry"), there are 157 breweries in the Russian Federation, including independent regional companies and branches of major brewing companies (Table 1).

From Table 1 it follows that in the Russian Federation 87 (55.4% of the total) breweries are small, with a production capacity of up to 1 million daL per year; 29 plants (18.5%) are average, with a production volume from 1 to 6 million daL; 41 (26.1%) are large ones with a production capacity of over 6 million daL a year.

Table 1. The composition of brewing enterprises, depending on production capacity, mln daL.

Production capacity	2006		2010		2014		Absolute deviation
	Total number	%	Total number	%	Total number	%	
Up to 1 million daL	125	59,2	98	58,7	87	55,4	−38
From 1 to 6 million daL	47	22,3	31	18,6	29	18,5	−18
Over 6 million daL	39	18,5	38	22,7	41	26,1	+2
Total	211	100	167	100	157	100	−54

Production process is an integral element of the cycle of organization funds consumption. As a result of production process, an enterprise uses material, labor and financial resources that form cost price of goods produced (work performed, services rendered) [3, pp. 177–179].

Production process of breweries has an impact on the cost accounting system. Brewing production process includes the following process steps, shown in Fig. 1.

As can be seen from Fig. 1, finished half-stuff products of brewing technological processes are served in a regulated working time as ingredients to conveyor of the next technological process, until the final technological process, which results as a final product of brewery's work.

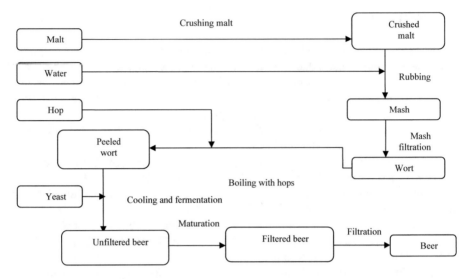

Fig. 1. Production cycle by technological operations at breweries

Beer production technological process begins with production of malt from barley and preparation of wort. In a malting workshop, as a result of crushing barley, crushed malt is obtained (malt grinding), which is then mixed with water and mash is obtained. Mash obtained in a brewhouse is heated to the temperature of 76 °C and then filtered. Mashed malt is in a brewhouse until beer pellet settles in boilers, after which prepared mash is filtered. Wort obtained in the previous step is heated, brought to a boil and hops are added. Next, boiled wort is filtered from remnants of hops and allow it to settle. After wort is completely cooled, yeast is added to vat. After ripening, beer goes through another filtration. At the final stage of beer production, it is poured into a container of various types.

Operational structure of technological processes in production of their own half-stuff products by cost centers in brewing is shown in Fig. 2.

In the beer production costs structure, malt, hops, yeast and water costs prevail, which are the main components of beer formulation. Among the mentioned components, malt in technological process of malting plant accounts for 43 to 49%. Costs of bottling shop in total production costs of brewery occupy from 3 to 7%. Costs of brewing, fermenting and fermentation workshops are formed under the internal factors influence, the external factors influence is insignificant. The emphasis on regulatory accounting of brewing production costs by technological processes necessitates initial registration and subsequent summarization of costs by cost centers. The scientific definition and their economic and reference content is well known and, in our opinion, does not need revision. Along with this, it is obvious that it is necessary to specify the cost centers composition and their production purpose in relation to technical and economic features of an economic entity.

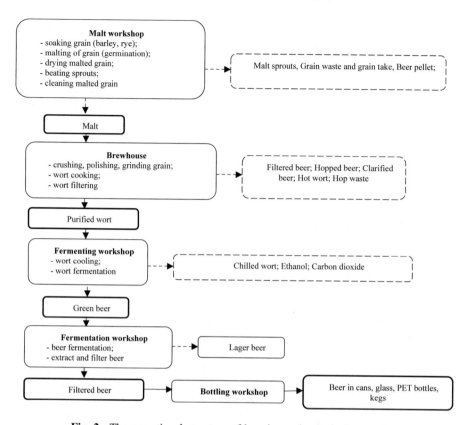

Fig. 2. The operational structure of brewing technological processes

The main conditions for effective management of a manufacturing enterprise are completeness, reliability and efficiency of information about costs that form production cost (works, services), so production cost, their structure and dynamics are always the focus of experienced managers [2, pp. 215–221].

Therefore, the cost accounting system for production of brewing products should be built on an integrated basis, which in turn increases timeliness and analyticity of information focused on making management decisions in the present and future [5, pp. 75–82].

4 Results

An important prerequisite for the proper organization of cost of production accounting and calculating of brewing products is a clear classification of production costs [2, pp. 215–221].

For the purpose of cost control in brewing enterprises accounting, the classification of costs according to costing items is used. The list of calculation articles is established by an enterprise independently in accordance with industry peculiarities [1, pp. 447–450].

Currently, brewing enterprises practice the use of a standard nomenclature of calculation items, which does not fully reflect industry-specific features of production process of these enterprises.

Production costs are a significant source of economic savings due to a more rational organization of brewery production activities. Taking into account the arguments set forth above, we consider the following nomenclature of cost items acceptable for breweries:

- "Raw materials and basic materials";
- "Costs of acquiring material stocks" (Debit of account 10 "Materials", Credit of account 60 "Calculations with suppliers and contractors" in the assessment adopted by an accounting policy);
- "Returnable waste" (deducted);
- "By-product" (deducted);
- "Auxiliary materials for technological purposes";
- "Fuel and energy for technological purposes";
- "Chill for technological purposes" (Credit of account 23 "Auxiliary production" in the assessment adopted by an accounting policy);
- "Basic and additional wages of production workers";
- "Expenses for social needs and social security";
- "Expenses for preparation and mastering of production";
- "Costs for the sewage treatment plants maintenance" (Debit of account 26 "General economic expenses" Credit of accounts 23 "Auxiliary production", 60 "Settlements with suppliers and contractors" in the assessment adopted by an accounting policy);
- "Expenses for labor protection and safety engineering" (Debit of accounts 25 "General production expenses", 26 "General expenses" Credit of accounts 10 "Materials", 23 "Auxiliary production", 60 "Calculations with suppliers and contractors", 70 "Calculations with wage personnel" in the assessment adopted by an accounting policy);
- "Costs of maintaining administrative and managerial personnel" (according to analytical accounting data to account 26 "General business expenses", if such expenses are included in production cost according to the accounting policy of an enterprise);
- "General workshop expenses";
- "Plant costs";
- "Other production costs";

Total: Production cost (Credit turnover on account 20 "Primary production", taken into account during the reporting period, minus cancellations for defect, shortages, etc.).

5 Conclusion

The above list of cost items is flexible for adaptation in relation to the cost accounting construction and their generalization by technological processes and periods of brewing production, hierarchical levels of operational control, analysis and brewing production costs accounting. The recommended nomenclature of articles, in our opinion, makes it possible to form blocks of information in the production accounting system for monitoring and controlling costs.

References

1. Evdokimova, Y.I., Asfandiyarova, R.A.: Production costs accounting at the food industry enterprise. Alley Sci. **3**(8(24)), 447–450 (2018)
2. Kalcheva, T.F.: The main aspects of the cost accounting construction for production. Scientific works of the Southern branch of the National University of Life and Environmental Sciences of Ukraine "Crimean agrotechnological university". Series: Economic Sciences, no. 147, pp. 215–221 (2012)
3. Kolesnikova, E.N.: Modern approaches to cost accounting organization for production. News of the Orenburg State Agrarian University, no. 3(27), pp. 177–179 (2010)
4. Thamokova, S.M.: Management accounting of material and production stocks at food industry enterprises. New Science: Theoretical and Practical View, no. 2–1(63), pp. 141–143 (2016)
5. Uspenskaya, G.A.: Production costs accounting of an industrial enterprise on the example of Ltd. Petroremmash. In the Collection: The Economic System of Modern Society: Economics and Management, a Collection of Materials of the II International Scientific Practical Conference, pp. 75–82 (2017)
6. Khodyrevsky, D.I.: The urgency of the problem of cost accounting management in an enterprise. Current Trends in Economics and Management: A New Look, no. 26, pp. 117–118 (2014)

Digital Economics

Tourism Export Potential: Problems of Competitiveness and Financial Support

Elena Dedkova⬤ and Aleksandr Gudkov^(✉)⬤

Orel State University, Orel 302020, Russia
sashaworld777@gmail.com

Abstract. Tourism is not only big business but also tourism should be seen and protected as perhaps one of the world's premier export products. An export industry is one that sells a significant share of its goods or services outside of the country, thus bringing new money into the local economy. If we, then, assume that exports refer to money going from place X to place Y due to the sale of product Z, then tourism easily meets this standard. In this connection, consideration is given to the possibilities of developing the tourism industry as the most important direction of Russia's non-commodity exports in the medium term in digital age, capable not only of bringing significant revenues to the budgets of various levels, developing domestic tourist infrastructure, but also characterized by inexhaustible resources and forming a favorable image of the country on world markets. The purpose of this article is to consider the problems of competitiveness and financial support of the tourism industry from the standpoint of increasing its export potential in digital age and ensuring a high level of attractiveness of tourist destinations.

Keywords: Tourism · Export · Economy growth

1 Introduction

The development of any economy ultimately rests on the limited domestic consumption market, the impossibility of further development and expansion within one state, the gradual degradation of positioning and management systems, and a decrease in the competitiveness of business entities. International trade is the most important driving force behind the development of the world economy. Organizations in the global economy import and export a large amount of goods, works, services in order to expand production, get more profit, create a favorable world image of the state, strengthen competitive advantages and the economy of their country as a whole. Currently, exports to the state are the most promising and revenue source of the budget [7, 11, 17, 31].

The inbound tourism's characteristic is the creation of an advanced comfortable infrastructure inside and around the traveler attractions, which allows us to talk not only about stimulating the tourism exports development, but also about the formation of favorable living and recreational conditions inside the exporting country. This sets tourism apart from other export destinations, because its development brings a twofold positive effect to the state and its population [7, 8]. In this aspect, the search for

T. Antipova (Ed.): ICIS 2019, LNNS 78, pp. 187–202, 2020.
https://doi.org/10.1007/978-3-030-22493-6_17

effective measures for the development of tourism as a direction of non-commodity exports in Russia becomes relevant, including in terms of GDP growth. There is still no Russian federal program to promote tourism, Russian hospitality and diversity. Russia does not have the kind of advertising campaigns that can be seen in neighboring countries such as Azerbaijan, Macedonia, Ukraine, etc. These campaigns invite tourists to come and experience the culture and related attractions [26]. For example, the share of tourism in the structure of Ukraine's GDP is 7.8% (in Russia – less that twice). Despite the fact that Ukraine, having a considerable tourist potential for the formation of tourist flows, is ranked 85th among 139 countries of the world in terms of tourism attractiveness [18]. In this regard, an objective need arises to consider tourism as the most promising direction for the development of Russian exports, taking into account the possibilities of increasing its competitiveness in world markets and the formation of an adequate system of industry incentives.

2   Analysis of the Recent Research and Publications

The increase in the implementation of internationalization strategies by firms has led to the rapid growth of international trade and the severity of trade deficit pressures in many countries. This has attracted considerable interest from various researchers, managers, and public-policy makers into understanding the determinants of export choice and degree. In the literature, exporting is considered the most frequently used strategy of internationalization due to it being a flexible and cost-effective mode [14]. Krammer advanced a two-stage theoretical model which contends that the export performance of emerging economy firms (EEFs) will depend both upon their firm-specific capabilities and their home institutional environments. Specifically, they argue that EEFs will be more likely to export when facing more uncertainty at home from greater political instability, substantial informal competition, and high corruption [15].

Kaliappan regarded services trade as a new source of income, especially for developing countries. The findings imply that the developing countries should focus on formulating appropriate policy measures to enhance the performance of services sector and service export to stimulate the economic growth [13]. Hjerpe explored outdoor recreation as a sustainable export industry and its regional economic impacts [11]. Wilderness attracts tourists and generates visitor spending in proximate communities as people enjoy wilderness for outdoor recreation. Wilderness also attracts amenity migrants and out-of-region investments into surrounding regional economies. Designated Wilderness areas in the U.S. collectively provide for substantial national economic contributions, estimated to be over $700 million in total output [10]. Nationally, outdoor recreation services have been estimated to be an $887 billion annual industry in the U.S. with increasing trends expected in both participation and total recreation-related expenditures [36]. Mahmoodi examined the causal relationship between foreign direct investment, exports and economic growth in two panels of developing countries (eight European developing countries and eight Asian developing countries). There is evidence of long-run causality from export and FDI to economic growth, and long-run causality from economic growth and export to FDI for both of the aforementioned panels [17].

Paraskevaidis and Andriotis in their study adopted an interdisciplinary approach by combining theoretical and empirical evidence drawn from evolutionary biology, and social sciences to theoretically substantiate altruistic behavior in tourism. To explore altruism in tourism from the perspective of host communities and more specifically from the viewpoint of voluntary tourism associations twenty-one interviews were conducted. Tourism's development potential refers to community members who earn their living through tourism and tourism impacts greatly on the development of their community. With that in mind, members of voluntary tourism associations, need to believe that the tourism sector has the potential to contribute to the development of their community [22].

Tourism research has yet to confirm whether an integrated destination image model is applicable in predicting the overall destination image and behavioral intentions of local residents. Study examines whether the cognitive, affective and overall image – hypothesized to be predictors of behavioral intentions – are applicable to residents and tourists. The findings support the applicability of the model to local residents and also showed that among tourists, the affective component exerted a greater influence than the cognitive on overall destination image and future behavior. As such, the study helps researchers understand how differences in the overall image and future behavior of the two groups (residents and tourists) develop. The model also assists destination practitioners by providing recommendations for the development of different marketing strategies to achieve a suitable positioning for each stakeholder group [29]. There is a plethora of studies segmenting the lucrative tourism market, limited attention has been given to identifying potential segments of local residents based on their image of the place they live in as a tourist destination. Stylidis study aims to address this gap by clustering local residents of a tourist destination based on their images of that place; and identifying whether those image-based resident groups share similar/different levels of place attachment and intentions toward tourism (support for tourism, intention to recommend it to others) [30].

Russia has a significant potential for tourism export development. Russia remains hindered by numerous issues such as destination image, infrastructure development, workforce training and education, quality management, and sustainable management [1]. One of the main issues in Russia's efforts to enhance tourism competitiveness: to educate a qualified workforce at the university level. Better education at universities enhances students' employability at the time that supports tourism firms to perform better. Both together help to boost tourism destination competitiveness and sustainability, favoring progress and socio-economic development [2].

Strategic management has become an important subject in university-level tourism and hospitality education since it aims to prepare and train future managers to develop a holistic management point of view. Okumus critically reviewed and evaluated different methods of teaching the subject. In particular, he provided discussions about the aims of teaching strategic management and referred to the challenges and difficulties of doing so in tourism and hospitality programs. A number of key conclusions and recommendations for practice and future research are also provided. However, due to contextual factors, changing teaching methods or adapting new methods may not be very easy since doing so may have important cultural, structural, and resource implications for tutors, students, and the institution [21].

Tourism cannot be concentrated only in megacities, and therefore the study of the development of small cities, as well as family business of different generations, is actualized, since this type of business has great chances to enter international markets. Berezka analyzed the modern approaches to the development of tourist value proposition and to discussed relevant cases of small cities in Russia. She used a case study approach based on the experience in Vladimir Region with respect to government and its support for tourism development. Several recommendations are offered for developing and improving the tourist value propositions of small cities [5]. Pavlov tried to present some conceptual foundations of the small family business as a strategic entity for the future of the non-capital regions. Sustainable territorial development based on the activity of small and medium-sized businesses is one of the research fields currently being intensively developed in the world economic science [23].

Sheresheva sheds light on the major shift to domestic tourism that is partly due to the turbulent economic environment that is forcing Russia to search for internal sources of development. People began actively visit the Crimea, the Caucasus, the Republic of Tatarstan, the Far East, as well as numerous small Russian cities that collaborate and cluster on the basis of unique cultural heritage and unique identity. At the same time, the decision to rely on diversity of opportunities and sustainable regional tourism make it possible to have a broad variety of tourist value propositions in almost all parts of Russia. This also makes Russia more attractive to foreign tourists. Still, to make Russia a popular destination internationally, improvement in almost all elements of the Russian tourist product is needed. The Russian tourism market suffers from the poor development of the country's tourism infrastructure, including a shortage of accommodation and entertainment resources and the poor state of many local attractions and road networks. The infrastructure is also poorly adapted for people with disabilities, as well as for foreign tourists, because of a lack of supporting information in English. The discrepancy between outbound and inbound expenditure suggests more can be done for travelers visiting the country [26–28].

For the development of inbound tourism, it is first necessary to create a tourist infrastructure, objects of increased attention of tourists - for example, theme parks and recreation areas. Nowadays theme parks generate billions of dollars in revenue, have a substantial effect on local economies and therefore, are considered a significant driver of the hospitality industry [19]. They will continue to have an impact on society and their roles in destination development and sustainability will continue to grow and affect a variety of demographic and socio-economic sections of the communities in which they operate. Developing new theme parks or adding new attractions into the current theme parks require major capital investments. Therefore, their development in Russia at the present stage is impossible without financial incentives.

Tourist evaluation criteria developed by scientists Lopes, Munoz and Alarcon-Urbistondo on the example of Portugal. To measure the basic offer at each destination, they used three sub criteria: the number of businesses at each destination within the hospitality industry (accommodation, catering and similar), and the lodging capacity at each destination, subdivided into hotel establishments and local accommodation (rural tourism and housing tourism). The complementary offer was measured by examining both the natural environment and the cultural offer [16].

Based on the analysis of various studies of the development of tourism exports in the world, it can be concluded that this problem has many subsystems related to potential, competitiveness, infrastructure, safety [4, 6], investment, education and behavioral intentions of residents and tourists. This article attempts to outline the range of problems of improving competitiveness and financially stimulating the export of tourism in Russia in digital age.

3   Goal of the Research

The Russian economy which is highly dependent on oil and gas exports is now faced with a decline in petro-dollars because of falling oil prices. Under new circumstances, Russian leaders have to decide how to reduce the energy resource dependence of the economy and, therefore, pay close attention to tourism and hospitality as potential drivers of the national economy [26]. Given Russia's significant tourism potential in the world market, it is necessary to divert the country's economy from product dependence, identifying the main problems of competitiveness and financial support for the tourism industry in digital age in terms of increasing its export potential and ensuring a high level of attractiveness of tourist destinations in the formation of an effective program for the development of inbound tourism.

4   Methodology

The authors used such methods as synthesis, generalization, theoretical modeling, analysis, graphical construction, formalization, forecasting. The positions of the research are argued also with the help of private scientific methods of research (formal, comparative, functional, concretization, etc.). Official statistics and analytical reports of UNWTO [32–35] and Russian federal authorities [24] served as the information basis of the study.

5   Results

Tourism is already the largest industry in the world, generating 10% of global GDP. In 2016, according to the UNWTO, 30% of international tourists visited the following 5 countries of the world: France, USA, Spain, China, Italy. The number of foreign tourists visiting Russia in 2016 decreased to 24.6 million people, although in 2015 Russia was in 9th place out of 189 countries of the world with a total attractiveness of 33.7 million people per year. In 2030, UNWTO predicts an increase in international arrivals to 1.8 billion people, with about 40% of those arriving in Europe [32–35]. At the same time, the number of domestic trips per year exceeds 5 billion people, which underlines the direct link between domestic and inbound tourism, based on the country's common tourism infrastructure [8].

Russia is an excellent choice for many different kinds of tourism. It offers a blend of Eastern and Western culture with a wide variety of historic places, diversity of climatic zones and beautiful scenery, outstanding cultural heritage and rich natural resources, including spa resources. There are more than 1,000 cities in Russia, of which around 70% are smaller cities, and any of them have their own peculiarities and unique historical value. As to the big Russian cities, all of them can be a destination for business tourism and big events. Moscow is traditionally the top urban tourism destination in Russia, together with St. Petersburg, with its acclaimed masterpiece of world architecture and culture [26–28]. However, over the past five years since 2013, the export of tourism services in Russia has significantly decreased from the level of 20.2 to 12.8 billion US dollars in 2016. At the same time, the United States, Spain and the Great Britain are steadily increasing their income from foreign tourism. Table 1 shows the main factors of tourism growth and the associated risks.

Table 1. Fundamental factors of growth of world tourism and related risks (UNWTO data)

Drivers of growth	Main risks
Favorable economic environment	Economic slowdown
Strong outbound demand from major source markets	Unstable fuel prices
Consolidation of the recovery in key destinations affected by previous crises	Geopolitical and trade tensions
Enhanced connectivity	Brexit uncertainty
Increased visa facilitation	

Exports of tourism in the world in terms of incomes are approaching such categories as "fuel" and "chemicals", already ahead of "food" and "automotive products". In this regard, we can talk about a high degree of attractiveness of the industry for development and investment. The development of inbound tourism is one of the priorities for the development of non-primary exports to the Russia, along with education, the chemical industry, the automotive industry, etc. Recently, the Russian Export Center, a state institute for supporting non-commodity exports, was established and successfully operates in Russia, providing Russian exporters with a wide range of financial and non-financial (institutional, legal) support measures. The main problems in the development of tourism exports are presented in Fig. 1.

Effective development of the export component of tourism is possible only in the context of a comprehensive solution to all the problems identified, which can be combined into two large groups - problems of ensuring the competitiveness of a destination and problems of financing the industry. Using the methodology of the Tourism and Travel Competitiveness Index (TTCI) and experience in assessing the competitiveness of tourism in Portugal [16], one can present the criteria for the competitiveness of Russian tourism in Table 2.

However, the potential of inbound tourism is far from being fully used. The incoming tourism is mostly limited to visits to Moscow and St. Petersburg or quite traditional routes that are the most famous cultural destinations in Russia (i.e. the Golden Ring). Many Western experts and potential foreign guests are convinced that Russia is not a «suitable» place for tourists. The reason is partly due to the negative information flow about Russia in Western media [25]. At the same time, as Andrades

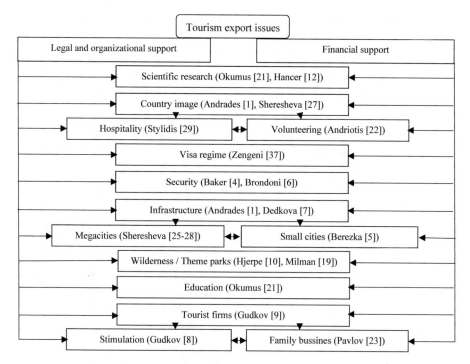

Fig. 1. Actual problems of tourism export development in scientific literature

and Dimanche underlined, tourism development in Russia remains hindered not only by destination image but also by such issues as infrastructure development, workforce training and education, quality management and sustainable management [1]. There are also difficulties of visa processing for foreigners and insufficient development of road networks in many Russian regions, as well as an acute shortage of entertainment and accommodation resources in almost all attractive regional destinations, especially in small towns that are actually «hidden gems» of Russia [27].

Given the multidimensionality and multifunctionality of tourism, as well as the main trends in modern tourism in digital age, it is possible to offer the main directions of its perspective development in Russia:

(1) The organization of flights with stop-overs for tourists traveling from Europe to Asia, from Asia to Europe, from the USA to Asia. In this case, the experience of Istanbul and Lisbon is very indicative, which allow you to get information about the city and country during transfers, ensuring the return of tourists for a longer period.
(2) The creation and promotion of mobile applications for tourist attractions in various cities, which significantly increases the awareness of tourists and improves perception. Im and Hancer in their study enhances the understanding of travel mobile application usage behavior by investigating interrelationship of utilitarian/ hedonic motivation and self-identity on attitude toward using travel mobile application [12].

Table 2. The main criteria for the competitiveness of Russian tourism

Criteria	Sub criteria	Sub index TTCI	Priority	Weight
Tourism	Number of accommodation and catering for tourists	C «Infrastructure»	Max	0.2
	Environmental costs	B «Policy and enabling condition»	Max	0.05
	Number of cultural heritage sites, including museums and art objects	D «Natural and cultural resources»	Max	0.1
	Amount of entertainment	C «Infrastructure»	Max	0.15
Environment	The proximity of the international airport	C «Infrastructure»	Min	0.05
	Extensive road network	C «Infrastructure»	Max	0.1
	The development of ICT	A «Enabling environment»	Max	0.1
	Number of university graduates	A « Enabling environment»	Min	0.05
	Crime rate	A «Enabling environment»	Min	0.1
	Level of medical care	A «Enabling environment»	Max	0.1

(3) Travel cards for transportation and discounts that are widely used in Europe, for example, the Berlin Welcome Card or the Lisboa Card.

(4) Development of projects for the introduction of carpal lanes or the allocation of bus lanes for tourist vehicles in metropolitan areas, especially in Moscow.

(5) Coordinated change in the visa policy of the state. Given the diversity of Russia and the territorial fragmentation of tourist sites, tourists should be given at least annual multivisas with a streamlined filing process, including the possibility of issuing electronic visas, which will significantly increase tourist flow, even if the high level of consular fees remains. The advantages of visa reforms are historically visible in the form of an observed increase in attendance after its implementation (and, conversely, a decrease in attendance while maintaining restrictive measures).

(6) Financing and development of theme parks attracting a large number of international tourists, for example, in France, Spain and the USA. It has been scientifically and practically proven that theme parks and attractions significantly affect the socio-economic stability of a tourist destination and its competitiveness in the world [19].

(7) Modernization of the education system for the training of specialists in the field of tourism and hospitality with the introduction of mandatory annual international internships.

(8) Positioning of cities, tourist attractions of Russia not only within the country, but also on the world stage.

(9) Development of federal and regional programs for direct and indirect financial incentives for tourism firms and enterprises. Including tax incentive tools designed specifically to support the tourism industry [8, 20].

A number of federal and regional programs and strategies for the development of the tourism industry are currently operating in Russia, including the Strategy for the Development of Tourism in the Russian Federation until 2020 and the Federal Target Program "Development of Domestic and Inbound Tourism in the Russian Federation (2019 – 2025)". The program is expected to be allocated from the federal budget in the amount of about 1 billion US dollars. In addition to this, a passport is currently being developed for the national project "International Cooperation and Export", within the framework of which the federal project "Export of services" is provided, including, among other things, services in the field of tourism. From the point of view of financing, much attention is paid to the megacities of Russia, which is due to the possibility of a quick return on investment and the absence of the need for significant advancement on the international market due to good publicity (Table 3).

Table 3. The main criteria for competitiveness of the leading tourist megacities of Russia

Moscow	Saint-Petersburg
Advantageous geographical position	Advantageous geographical position
Transport accessibility (including 4 international airports) and good infrastructure	Transport accessibility and good infrastructure
Financial center	Cultural center
Conduct major world-class sporting events	Conduct major world-class sporting events
The high degree of development of domestic tourism	The high degree of development of domestic tourism
Reputation of the largest scientific and educational center	Reputation of the largest scientific and educational center
Prospects to become one of the leading megacities of the world	
State capital	

The development of a tourism development strategy in the modern metropolis implies an understanding of the specifics and principles of functioning of large urban agglomerations, as well as the main trends in the development of the global tourism market and key factors that contribute or impede the expansion of the contribution of the tourism industry to the economy of the megalopolis. In this regard, it is important to carefully analyze the activities of the leading world cities, which are generally recognized tourist destinations, identify and compare the best practices used by them, and evaluate their applicability in the conditions of the Russian megacities. In terms of the use of advanced foreign experience, it seems appropriate to consider the leading megacities – benchmarks presented in Table 4.

Table 4. The experience of France, Italy and China of tourism stimulating in megacities

Megacities – benchmarks		
Purpose of stimulation	Incentive tools	Incentive results
Paris		
1. Increasing the number of rooms for tourists	Adoption of state programs, creation of conditions for the development of tourism, co-financing of socially significant projects, an increase in tourist facilities in order to increase the number of overnight stays of tourists	The number of rooms – 84,473 thousand; The number of nights spent – 44 million; Occupancy rate of tourist accommodation – 69,4%
2. Increase in spending of foreign tourist	Providing tourists with different categories of accommodation, the formation of an expanded supply of various indicators to best meet the demand	Costs - $12,03 billion
3. Improving of tourist service	Development of hotel and transport tourism infrastructure based on the Tourism Development Plan in Paris	Creating new tourist attractive areas. Ensuring a comfortable environment and safety of tourists. Increased control and attention of the city administration to tourist accommodation facilities
4. The growth of foreign tourist flows	Formation of a favorable tourist image of the city, development of accommodation facilities in the average price range	Foreign tourist flow – 14 million people; Number of rooms in hotels 3–4* – 58001
5. Optimization of the average room rate	Control by the authorities, fiscal incentives, informational support to balance the cost of rooms, the needs of tourists and their capabilities	Average room rate – 230 euros
Rome		
1. Increasing the number of rooms for tourists	Adoption of state programs, creation of conditions for the development of tourism, co-financing of socially significant projects, an increase in tourist facilities in order to increase the number of overnight stays of tourists	The number of nights spent – 26,9 million; Occupancy rate of tourist accommodation – 69,3%
2. Increase in spending of foreign tourist	Providing tourists with different categories of accommodation, the formation of an expanded supply of various indicators to best meet the demand	Costs – $4,5 billion

(*continued*)

Table 4. (*continued*)

Megacities – benchmarks		
Purpose of stimulation	Incentive tools	Incentive results
3. Expansion and diversification of the offer for tourists	Strategy for the development of tourism in Italy. Identify opportunities to use the seaside area to create tourist offers non-beach recreation	Transformation of the southwestern part of the city into a seaside recreational zone
4. The growth of foreign tourist flows	Formation of a favorable tourist image of the city, development of accommodation facilities in the average price range	Foreign tourist flow – 9 million people
5. Optimization of the average room rate	Control by the authorities, fiscal incentives, informational support to balance the cost of rooms, the needs of tourists and their capabilities	Average room rate – 148 euros
Beijing		
1. Increasing the number of rooms for tourists	Adoption of state programs, creation of conditions for the development of tourism, co-financing of socially significant projects, an increase in tourist facilities in order to increase the number of overnight stays of tourists	The number of number of accommodation facilities – 2236; Occupancy rate of tourist accommodation – 63,7%
2. Increase in spending of foreign tourist	Providing tourists with different categories of accommodation, the formation of an expanded supply of various indicators to best meet the demand	Costs – $4,6 billion.
3. Higher degree of satisfaction of the population with the tourist industry	General state strategy reflected in five year plans	The comprehensive contribution of the tourism industry to the national economy should exceed 12%
4. The growth of foreign tourist flows	Formation of a favorable tourist image of the city, development of accommodation facilities in the average price range	Foreign tourist flow – 4 million people
5. Optimization of the average room rate	Control by the authorities, fiscal incentives, informational support to balance the cost of rooms, the needs of tourists and their capabilities	Average room rate – 66 euros

*4 stars hotel

Taking into account international experience, we can see how high the level of development of tourism is in the leading destinations of the world. From the other hand modern-day world faces a hostile climate, depleted resources and the destruction of habitats. The dream that growth will lead to a materialistic utopia is left unfulfilled by a lack of ecological and economic capacity. The only choice is to find alternatives to increased growth, transform the structures and institutions currently shaping the world, change lifestyles and articulate a more credible vision for the future and lasting prosperity. As a reaction to the problems accrued by capitalism, new development approaches such as the concept of degrowth have evolved [3]. However, the use of this concept, in our opinion, in Russia at this moment is impossible.

Russia is still a country with elements of a planned economy, market processes are not sufficiently developed, therefore the importance of the state in the processes of regulating the activities of various industries is quite important. Strict legislative regulation of activities on the one hand pursues the goals of ensuring security, preventing financial fraud and deceiving consumers, but on the other hand in practice leads to the formation of too intricate and very restrictive development of norms and rules, which do not allow to appear new businesses. The same applies to the tourism industry, despite the prospects, at present, since 2014, there has been a significant decrease in the profitability of enterprises in the industry, as evidenced, for example, by travel agency surveys [9]. In this regard, a way out of this situation is possible only through state intervention. As already noted, the most significant are the measures of financial and tax support. Figure 2 presents the proposed hierarchy of financial support for the export of tourism in Russia.

The smallest value of the indicator "image" is explained by the fact that the image of the country as a tourist destination is formed by a combination of subordinate factors that increase its export orientation. It is important to create the Tourism Development Fund. It will invite for applications for grants to support events or projects that will attract additional visitors in Russia and help stimulate the tourism industry even more. In particular, from small businesses and start-ups. Organizations wishing to apply to the Fund should be able to demonstrate that their events or projects will attract more visitors, can attract additional funding from sources in addition to the Fund and create sustainable enterprises in the coming years. Many funds all over the world already have supported numerous successful events and infrastructure development, covering a wide range of tourism initiatives.

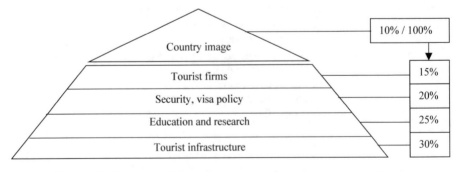

Fig. 2. The hierarchy of financial support for the tourism export in Russia

6 Conclusion

Tourism is important for the development of the country, as it contributes to the preservation and development of historical, cultural, natural, resort and recreational potential of various subjects of Russia, as well as smoothing the disparities in the socio-economic development of regions. In addition, the tourism industry stimulates the growth of about 50 other dependent sectors of the economy. At the same time, additional jobs are being created, investments are being attracted, small and medium-sized businesses are developing, the amount of tax revenues to the budgets of the budget system of Russia is increasing, and international relations are improving. In Russia, there has been a certain growth trend in domestic and inbound tourism. Domestic tourist flow has increased significantly since 2015. However, at the moment, the tourism industry in Russia has almost exhausted the resources of domestic growth and development, which were given during the annexation of the Crimea, the Olympic Games and the World Cup, the closure of popular outbound areas, which actualizes the solution to the problems of inbound tourism. Demand from Europe that declined dramatically in 2015 and was partly replaced by Asian demand, began to recover in 2016–2017. An important factor in this development relates to depreciation of the ruble: spending in Russia, including accommodation costs, is now much cheaper for clients from Europe due to the change in the currency exchange rate. The entry flow from Asian countries is also growing, primarily from China. There are now many foreign guests coming to Moscow and St. Petersburg for holidays, including those with relatively high incomes that stay in luxury hotels [28].

Taking into account the goals of the Federal Target Program "Development of domestic and inbound tourism in the Russian Federation (2019 – 2025)" with a 70% increase in the contribution of tourism to GDP before the end of the program, tourism should show growth twice as much as the growth rate of the entire state economy. This is possible only through the effective distribution of financial support for the industry based on the proposed hierarchy, paying close attention to the main areas of its prospective development in Russia, taking into account the criteria of competitiveness and the experience of megacities - benchmarks of the world's leading tourist destinations. Figure 3 summarizes the current problems and conditions for the development of inbound tourism in Russia.

The Russian government is making a huge effort to improve the country's tourism infrastructure and to help the Russian regions in their movement toward sustainable tourism. It is planned to give priority to the cluster approach in tourism and to boost five priority types of tourism, namely, cultural, health, active, cruise and ecotourism. Other key objectives set in the Concept for the next few years are as follows: ICT infrastructure development, advanced training programs for the tourism industry, the promotion of Russia as a tourist destination and plans for subsidizing entrepreneurial and public initiatives in regional tourism [27].

The main results of the study are to comprehensively address the problems of exporting tourism and the main drivers of its global growth, form a set of criteria for the competitiveness of Russian tourism, suggest directions for its future development based on the world's best tourist practices and determine the hierarchy of financial support for

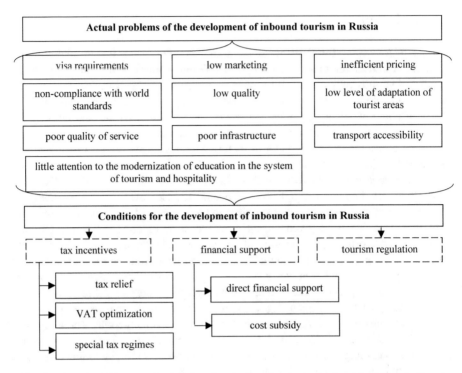

Fig. 3. Actual problems and conditions for the development of inbound tourism in Russia

tourism exports. The practical value of the results obtained is that they can be used at the macro level to form an effective program for the priority development of inbound tourism in accordance with the results provided by the grant of the President of the Russian Federation for the purpose of stimulating and supporting non-commodity exports.

Acknowledgement. The article was prepared within the framework of the President of the Russian Federation grant for the state support of young Russian scientists – Ph.D. MK-404.2018.6 "Tax incentives of non-primary exports in Russia".

References

1. Andrades, L., Dimanche, F.: Destination competitiveness and tourism development in Russia: issues and challenges. Tour. Manag. **62**, 360–376 (2017)
2. Andrades, L., Dimanche, F.: Destination competitiveness in Russia: tourism professionals' skills and competences. Int. J. Contemp. Hosp. Manag. **31**(2), 910–930 (2019)
3. Andriotis, K.: Degrowth in Tourism: Conceptual, Theoretical and Philosophical Issues. CABI, Oxfordshire (2018)
4. Baker, DMcA: The effects of terrorism on the travel and tourism industry. Int. J. Relig. Tour. Pilgr. **2**(1), 58–67 (2014)

5. Berezka, S., Kolkov, M.Y., Pereskokova, E.V.: The development approaches to tourist value propositions of small cities: the case of the Vladimir Region. Worldw. Hosp. Tour. Themes **10**(4), 498–509 (2018)
6. Brondoni, S.M.: Global tourism and terrorism. Safety and security management. Symphonya. Emerg. Issues Manag. **2**, 7–16 (2016)
7. Dedkova, E., Gudkov, A., Dudina, K.: Perspectives for non-primary export development in Russia and measures of its tax incentives. Probl. Perspect. Manag. **16**(2), 78–89 (2018)
8. Gudkov, A., Dedkova, E., Dudina, K.: Tax incentives as a factor of effective development of domestic tourism industry in Russia. Probl. Perspect. Manag. **15**(2), 90–101 (2017)
9. Gudkov, A., Dedkova, E., Dudina, K.: The main trends in the Russian tourism and hospitality market from the point of view of Russian travel agencies. Worldw. Hosp. Tour. Themes **10**(4), 412–420 (2018)
10. Hjerpe, E., Holmes, T., White, E.: National and community market contributions of Wilderness. Soc. Nat. Resour. **30**(3), 265–280 (2017)
11. Hjerpe, E.E.: Outdoor recreation as a sustainable export industry: a case study of the boundary waters wilderness. Ecol. Econ. **146**, 60–68 (2018)
12. Im, J.Y., Hancer, M.: Shaping travelers' attitude toward travel mobile applications. J. Hosp. Tour. Technol. **5**(2), 177–193 (2014)
13. Kaliappan, S.R., Ahmad, S.A., Ismail, N.W.: Service export and economic growth in the selected developing Asian countries. Int. J. Econ. Manag. **11**(2), 393–418 (2017)
14. Kotorri, M., Krasniqi, B.A.: Managerial characteristics and export performance – empirical evidence from Kosovo. South East Eur. J. Econ. Bus. **13**(2), 32–48 (2018)
15. Krammer, S.M., Strange, R., Lashitew, A.: The export performance of emerging economy firms: the influence of firm capabilities and institutional environments. Int. Bus. Rev. **27**(1), 218–230 (2018)
16. Lopes, A.P.F., Muñoz, M.M., Alarcón-Urbistondo, P.: Regional tourism competitiveness using the PROMETHEE approach. Ann. Tour. Res. **73**, 1–13 (2018)
17. Mahmoodi, M., Mahmoodi, E.: Foreign direct investment, exports and economic growth: evidence from two panels of developing countries. Econ. Res. - Ekonomska Istraživanja **29**(1), 938–949 (2016)
18. Mazaraki, A., Boiko, M., Bosovska, M., Vedmid, N., Okhrimenko, A.: Formation of the national tourism system of Ukraine. Prob. Perspect. Manag. **16**(1), 68–84 (2018)
19. Milman, A., Okumus, F., Dickson, D.: The contribution of theme parks and attractions to the social and economic sustainability of destinations. Worldw. Hosp. Tour. Themes **2**(3), 338–345 (2010)
20. Moreno-Rojas, J., González-Rodríguez, M.R., Martín-Samper, R.C.: Determinants of the effective tax rate in the tourism sector. A dynamic panel data model. Tour. Manag. Stud. **13**(3), 31–38 (2017)
21. Okumus, F., Wong, K.: A critical review and evaluation of teaching methods of strategic management in tourism and hospitality schools. J. Hosp. Tour. Educ. **16**(2), 22–33 (2004)
22. Paraskevaidis, P., Andriotis, K.: Altruism in tourism: social exchange theory vs altruistic surplus phenomenon in host volunteering. Ann. Tour. Res. **62**, 26–37 (2017)
23. Pavlov, D., Sheresheva, M., Perello, M.: The intergenerational small family enterprises as strategic entities for the future of the european civilization – a point of view. J. Entrepreneurship Innov. **9**, 26–38 (2017)
24. Russian Export Center announced the creation of the ranking of inbound tourism countries. https://www.exportcenter.ru/press_center/news/rets-anonsiroval-sozdanie-renkinga-stran-vezdnogo-turizma/. Accessed 07 Mar 2019
25. Sheresheva, M., Kopiski, J.: The main trends, challenges and success factors in the Russian hospitality and tourism market. Worldw. Hosp. Tour. Themes **8**(3), 260–272 (2016)

26. Sheresheva, M.: Russian hospitality and tourism: what needs to be addressed? Worldw. Hosp. Tour. Themes **8**(3), 380–396 (2016)
27. Sheresheva, M.Y.: The Russian hospitality and tourism market: what factors affect diversity and new destination development? Worldw. Hosp. Tour. Themes **10**(4), 510–522 (2018)
28. Sheresheva, M.Y.: The Russian tourism and hospitality market: new challenges and destinations. Worldw. Hosp. Tour. Themes **10**(4), 400–411 (2018)
29. Stylidis, D., Shani, A., Belhassen, Y.: Testing an integrated destination image model across residents and tourists. Tour. Manag. **58**, 184–195 (2017)
30. Stylidis, D.: Residents' place image: a cluster analysis and its links to place attachment and support for tourism. J. Sustain. Tour. **26**(6), 1007–1026 (2018)
31. Tourism is Export. http://www.tourismandmore.com/tidbits/tourism-is-export/. Accessed 07 Mar 2019
32. UNWTO Annual Report 2015. http://cf.cdn.unwto.org/sites/all/files/pdf/annual_report_2015_lr.pdf. Accessed 07 Mar 2019
33. UNWTO Annual Report 2016. https://www.e-unwto.org/doi/pdf/10.18111/9789284418725. Accessed 07 Mar 2019
34. UNWTO Annual Report 2017. https://www.e-unwto.org/doi/pdf/10.18111/9789284419807. Accessed 07 Mar 2019
35. UNWTO Tourism Highlights 2018. https://www.e-unwto.org/doi/pdf/10.18111/9789284419876. Accessed 07 Mar 2019
36. White, E., Bowker, J.M., Askew, A.E., Langner, L.L., Arnold, J.R., English, D.B.: Federal outdoor recreation trends: effects on economic opportunities. US Department of Agriculture, Forest Service, Pacific Northwest Station, Portland (2016)
37. Zengeni, N., Zengeni, D.M.F.: The impact of current visa regime policy on tourism recovery and development in Zimbabwe. Int. J. Dev. Sustain. **1**(3), 1008–1025 (2012)

Management of Organizational Knowledge as a Basis for the Competitiveness of Enterprises in the Digital Economy

Zhanna Mingaleva$^{(\boxtimes)}$ ⓘ, Ludmila Deputatova ⓘ,
and Yurii Starkov ⓘ

Perm National Research Polytechnic University, 614000 Perm, Russia
mingal1@psu.ru

Abstract. The development of a system of industrial and organizational relations based on the use of digital technologies is an important task of modern
enterprises. Management of organizational knowledge of the company is aimed
at the creation and introduction of new products sold in the digital economy of
the country. The article presents a scheme for the transformation of knowledge
with the establishment of stages of the circuit from creating an idea to its
commercialization in the market. The process of transforming organizational
knowledge involves changing the type of knowledge along a chain from personal to codified and competence-based to materialized. The author's scheme
substantiates the key stages necessary for the most effective management of the
turnover of organizational knowledge and the creation of innovations in the
enterprise. The knowledge transformation scheme was the basis for the development of an organizational knowledge management algorithm. The presented
knowledge management algorithm includes the creation, evaluation, adaptation,
testing in practice of new/improved knowledge. The introduction of the HADI-
cycle at the stage of testing a new/improved knowledge in the practice of the
enterprise improves the efficiency of the preparation of the finished intellectual
product. Evaluation of the effectiveness of innovative projects themselves as a
result of the creation of new knowledge is not the aim of the study.

Keywords: Digital economy · Competitiveness of enterprises ·
Organizational knowledge · HADI-cycle

1 Introduction

The digital economy is digital-based manufacturing. The program "Strategies for the
development of the information society in the RF for 2017–2030" provides for a set of
measures to create conditions for the development of digital society in our country [1].
The program provides for an increase in the quality of life of citizens and an increase in
their well-being through increasing the availability and quality of goods and services
produced using modern digital technologies. The Program also sets targets for raising
the level of awareness and digital literacy of the Russian population, improving the
availability and quality of public services for citizens, and ensuring security both inside

© Springer Nature Switzerland AG 2020
T. Antipova (Ed.): ICIS 2019, LNNS 78, pp. 203–212, 2020.
https://doi.org/10.1007/978-3-030-22493-6_18

the country and abroad. The program defines goals and objectives within the framework of 5 basic directions of development of the digital economy in the Russia up to 2024.

The basic areas include (1) regulation, (2) human resources and education, (3) the formation of technical reserves and research competencies, (4) information infrastructure, (5) information security. Such areas as "personnel and education" and "the formation of research competencies" are designed to solve the problem of creating and implementing new organizational knowledge and its management.

The report of the World Economic Forum on Global Competitiveness 2016–2017 emphasizes the special importance of investing in innovation along with the development of infrastructure, skills and efficient markets. In the international ranking, the Russian Federation ranks 43rd, significantly lagging behind many of the most competitive economies in the world. The observed technological lag of Russian industrial enterprises from companies of leading countries has lasted for many years [2]. This lag is due to the insufficient use of modern technologies - the share of organizations introducing technological innovations in Russia does not exceed 10%, and in developed countries more than 50%.

The purpose of the article is to develop a mechanism to accelerate the introduction of innovations in Russian enterprises. Since the greatest difficulties in financing and introducing innovations are experienced by small and medium-sized enterprises, the area of analysis in the article is limited to small and medium-sized enterprises [3]. Large enterprises tend to carry out innovative activities on an ongoing basis. Also, the area of research in the article is limited to the creation and theoretical proof of the effectiveness of the organizational mechanism to stimulate the creation of new knowledge as the basis for the innovation process in enterprises. Evaluation of the effectiveness of innovative projects themselves is not included in the framework of the study.

2   Theory

The competitiveness of enterprises in modern conditions is determined by the state of its human and intellectual capital. The researchers emphasize the importance of the question of managing the processes of innovation and the generation of ideas, the formation of organizational knowledge a long time ago [4]. In the most competitive, complex and turbulent environment, knowledge is the most important resource for the development of companies and countries. In such conditions, organizations should use their knowledge to increase their competitiveness [5].

Innovation acts as a tool to strengthen competitiveness. Bhushan identifies two types of innovations: stratified innovations and consistent innovations. These two types of innovation are the forerunners of each other, thereby spawning a stimulated innovation cycle [6].

The results of researchers Nanda, Rhodes-Kropf show that venture capital investors invest in more risky and innovative start-ups in hot markets [7].

According to studies conducted by Adner, Kapoor. The success of an innovative company often depends on the efforts of other innovators in their environment [8].

Scientists propose to improve the efficiency of vertical integration as a strategy for managing the interdependence of ecosystems throughout the entire life cycle of a technology. Based on the obtained empirical data, this contributes to increased competitiveness.

The theoretical approach to the analysis and evaluation of the characteristics of the formation and development of organizational knowledge developed by Hall, Andriani, includes the division of innovation into radical and additional, the concept of implicit and explicit knowledge, as well as highlighting the five basic processes of knowledge management in an organization: externalization, distribution, internalization, social-ization and intermittent learning [9]. Hall and Andriani believe that in the process of managing innovation and organizational knowledge, each participating party must determine the characteristics that they believe should be a successful innovation.

This involves identifying gaps in knowledge that need to be addressed to achieve development goals. Such gaps in knowledge "constitute the units of analysis. For each unit of analysis/knowledge gap, the size of the gap, and the nature of the required knowledge are estimated subjectively by each project team member. This allows both the identification of units which have high risk and the nature of the knowledge transformation processes, which need to be managed" [9, p. 145].

In turn, to successfully implement the process of transforming knowledge in accordance with modern business conditions and competition, a special management mechanism is needed, which differs from different mechanisms for managing business processes and production processes, as well as from well-known innovation manage-ment mechanisms, project management mechanisms, and knowledge, including through employee training, strategies of organizational behavior, organizational cul-ture, etc.

Reed, Card confirms that among many tools and methods of knowledge manage-ment and innovative projects, the plan-do-study-act (PDSA) cycle is one of the few that focuses on the essence of change, on translating ideas and intentions into action. PDSA provides a structured experimental approach to testing changes [10]. As a result, the PDSA cycle and the concept of iterative change tests are central to many quality management approaches.

Numerous and diverse models of quality management used by various organiza-tions and firms are based on the work of Deming. Deming's PDSA Cycle ("Plan-Do-Study-Act") is a cyclically iterative decision-making process used in quality manage-ment [11]. Also at present, The Model for Improvement [12, 13] is widely used in production management practice as a basic framework for the improvement of the innovation process. This model allows improvement of any projects «(e.g. introduction of a new product line or service for a major organization)» [13, p. 8, 14]. Raudeliu-niene proposed a system for assessing the capacity of an organization and the Knowledge management process model [15].

However, in modern conditions of digitization of the economy and production processes, these models require adjustment and adaptation to new requirements. As noted by Huarng, Mas-Tur, Calabuig, information technology management facilitates knowledge creation [5].

3   Model

Based on the analysis of scientific and empirical approaches to the need to develop a mechanism for managing organizational knowledge in order to increase the competitiveness of enterprises, we have proposed an author's approach to identifying the main types of organizational knowledge that affect the implementation of innovation activities.

These are: personal, codified, competence and materialized knowledge. Personal knowledge is the knowledge of a person, accumulated as a result of the formation of a creative personality, the creation of human capital. Codified knowledge - knowledge accumulated in knowledge bases and enterprise databases, as a result of coding and distribution of personal knowledge. Competence knowledge - the knowledge of employees realized in the process of intellectual labor, as a result of the development and maintenance of social capital. Materialized knowledge - embodied new knowledge in products and services.

Stages of knowledge transformation provide a cycle of new knowledge in the company, which was named by us as the HADI cycle [16].

The HADI cycle is a cyclically iterative process of testing ideas that influence the improvement of key indicators for the implementation of innovative projects, the results of start-ups, enterprises.

Four letters mean four words - "Hypothesis" - "Action" - "Data" - "Insights", and the mechanism itself is aimed at testing the hypothesis by means of action followed by collecting analytical information and formulating relevant conclusions. The HADI (Hypothesis-Actions-Data-Acquisition-Understanding) cycle allows you to test created intellectual products in order to determine their effectiveness, feasibility of introduction into the production of an enterprise, commercial viability, patentability. The formulated hypothesis triggers a mechanism for testing an intellectual product according to the main indicators of applicability and commercial significance [17].

The use of the HADI cycle to test a new knowledge and its transformation into an innovative activity of a company allows organizing a mechanism for transforming the whole complex of knowledge to increase the likelihood of innovation [18], for accelerating the process from idea to the finished result [19], for organizing a continuous and sustainable innovation process [16], improvements in patent applications [20] etc.

The knowledge transformation scheme in the framework of the HADI cycle is presented in Fig. 1.

The algorithm for managing organizational knowledge from the stage of generating ideas to introducing innovations is presented in Fig. 2. An important component of the developed algorithm is the previously presented process of transforming organizational knowledge along a chain from personal to codified and competence-based to materialize. This corresponds to the process of transformation of embedded knowledge to embodied knowledge [21].

The presented knowledge management algorithm includes the creation, evaluation, adaptation, testing in practice of new/improved knowledge. The introduction of the HADI-cycle at the stage of testing a new/improved knowledge in the practice of the enterprise improves the efficiency of the preparation of the finished intellectual product.

The knowledge base is a set of files (HTML, MS Office), where each file has a search pattern when using the neural network module [22].

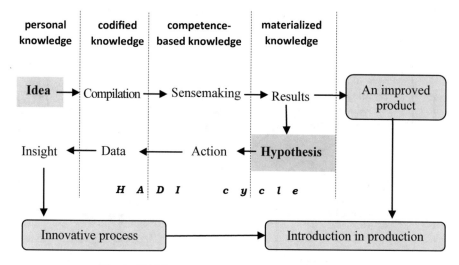

Fig. 1. HADI cycle knowledge transformation scheme.

The corporate knowledge base for each professional task contains group and individual theoretical, methodological, methodical, situational, reference, orienting knowledge and differs from the existing one by structuring knowledge on professional tasks of various management levels [23]. This allows accumulating individual and group knowledge at strategic, tactical and operational levels management [24].

Creating a system of knowledge transformation adequate to the requirements of the digital economy is impossible without a corresponding adjustment of the personnel management system [25]. The relationship of the personnel management system with the system of knowledge transformation in the enterprise is shown in Fig. 3.

The personnel management system creates the conditions for the creation of innovations in the enterprise, thereby forming intellectual capital [26]. Knowledge management system is necessary as a condition for the implementation of innovation in the enterprise [4].

For the successful development of the enterprise and maintaining its competitiveness in the market, it is necessary to create a special organizational culture that perceives innovation as an integral element of production and employment [27]. In this regard, it is important to accept the core values of the enterprise's activities among its employees. The main corporate values in modern conditions include: development, team, power, cooperation, remuneration [28]. The company's personnel management and knowledge management systems are aimed at the formation and development of these values in the company (Table 1).

Thus, it is necessary to clearly separate the two cycles: the cycle of creating innovations and the cycle of introducing innovations, and develop personnel management tools in accordance with the features of both of these cycles. This will allow organizing the correct impact of human resource management on organizational performance [29]. It can also be the basis for the introduction of modern methods and models of knowledge management [30].

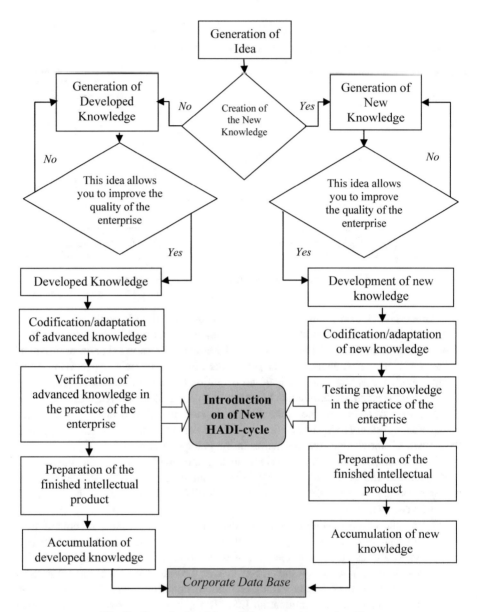

Fig. 2. Organizational knowledge management algorithm

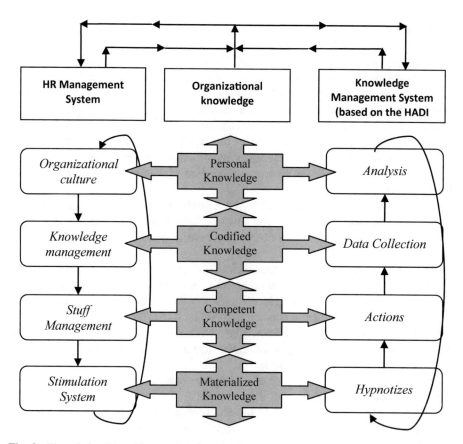

Fig. 3. The relationship of the personnel management system with the knowledge management system.

Table 1. Values in enterprise management systems

Values	Enterprise Management Systems	
	HR Management System	Knowledge management system (based on HADI cycle)
Development	Human Resource Management	Participation in project groups
Team	Corporate culture	Teamwork for project implementation
Power	Human Resource Management	Introduction of own rational proposal (innovation)
Cooperation	Human Resource Management	Freedom to implement ideas
Award	Stimulation System	Free distribution of monetary reward according to merit in the project team

4   Conclusion

The development of the digital economy depends on the quality of the intellectual capital of the staff of modern enterprises and on their willingness to implement worthy ideas. The implementation of the values of development, maintenance and preservation of intellectual potential is one of the main factors for improving the competitiveness of enterprises. The competitiveness of the economy in general and the company in particular is determined by the effectiveness of knowledge management and personnel management. The interaction of these key management systems allows you to create the conditions for the creation and implementation of innovations necessary for the digital economy.

The goal of the study was to develop and theoretically substantiate the effectiveness of the use of the mechanism for accelerating the introduction of innovations at Russian enterprises based on the process of managing new knowledge. The HADI-cycle provides the solution to this problem.

The HADI-cycle in the organizational knowledge management system ensures the introduction of innovations, approbation of a new idea, and rapid acquisition of knowledge about the degree of efficiency in the use of an idea in an enterprise's activity. Adaptation, codification and accumulation of acquired knowledge ensures the transformation of organizational knowledge. Evaluation of each stage of the HADI cycle and return to previous stages with a negative assessment ensure the achievement of the task of introducing innovations.

All new or improved knowledge created in the organization should be accumulated in the corporate knowledge base of the company.

Effective management of organizational knowledge of the company is aimed at creating and implementing innovative products, innovative processes that meet the requirements of the digital economy. A high level of innovation is the basis of competitiveness and growth of the company's market parameters.

The further research will focus on studying the specific interrelations of various aspects of the development and accumulation of organizational knowledge of enterprises with various types of innovative abilities of staff. The study will also continue in the direction of studying the mechanisms and extent of the influence of the HADI cycle on the management of organizational knowledge. Another direction in the development of research will be to study the influence of the mass distribution of digital technologies in management on the firm's competitiveness, on the staff's ability to search for innovative, innovative problem solutions, on the development of the enterprise's intellectual capital. Finally, recognizing that the links between organizational knowledge and innovative activity of firms increase as the digital economy develops, in the future we intend to identify and calculate quantitative indicators of this dependence.

Acknowledgment. The work is carried out based on the task on fulfilment of government contractual work in the field of scientific activities as a part of base portion of the state task of the Ministry of Education and Science of the Russian Federation to Perm National Research Polytechnic University (topic # 26.6884.2017/8.9 "Sustainable development of urban areas and the improvement of the human environment").

References

1. Presidential Decree of the Russian Federation of 09.05.2017 N 203 "On the Strategy for the Information Society Development in the Russian Federation for 2017–2030". https://bazanpa.ru/prezident-rf-ukaz-n203-ot09052017-h2985187. Accessed 02 Apr 2019
2. Mingaleva, Z., Mirskikh, I.: On innovation and knowledge economy in Russia. World Acad. Sci. Eng. Technol. **66**, 1032–1041 (2010)
3. Mingaleva, Z., Mirskikh, I.: Small innovative enterprise: the problems of protection of commercial confidential information and know-how. Middle East J. Sci. Res. **13** (SPLISSUE), 97–101 (2013)
4. Daneels, E.: The dynamics of product innovation and firm competencies. Strateg. Manag. J. **23**, 1095–2021 (2002)
5. Huarng, K.-H., Mas-Tur, A., Calabuig, M.F.: Innovation, knowledge, judgment, and decision-making as virtuous cycles. J. Bus. Res. **88**, 278–281 (2018)
6. Bhushan, B.: Stimulated innovation cycle to serve the poor: a case of Mann DeshiMahila group XLRI, Jamshedpur, India. In: Social Entrepreneurship and Sustainable Business Models: The Case of India, pp. 177–205, May 2018
7. Nanda, R., Rhodes-Kropf, M.: Investment cycles and startup innovation. J. Financ. Econ. **110**(2), 403–418 (2013)
8. Adner, R., Kapoor, R.: Value creation in innovation ecosystems: how the structure of technological interdependence affects firm performance in new technology generations. Strateg. Manag. J. **31**(3), 306–333 (2010)
9. Hall, R., Andriani, P.: Managing knowledge associated with innovation. J. Bus. Res. **56**(2), 145–152 (2003)
10. Reed, J.E., Card, A.J.: The problem with Plan-Do-Study-Act cycles. BMJ Qual. Saf. **25**(3), 147–152 (2016)
11. Deming, W.E.: The New Economics. Massachusetts Institute of Technology Press (1993)
12. Mayer, R., Davis, J., Schoorman, F.: An integrative model of organizational trust. Acad. Manag. Rev. **20**(3), 709–734 (1995)
13. Moen, R., Norman, C.: The History of the PDCA Cycle. https://deming.org/uploads/paper/PDSA_History_Ron_Moen.pdf. Accessed 02 Feb 2019
14. Weekes, L., Lawson, T., Hill, M.: How to start a quality improvement project. BJA Educ. **18**(4), 122–127 (2018)
15. Raudeliuniene, J., Davidaviciene, V., Jakubavicius, A.: Knowledge management process model. Entrepreneurship Sustain. Issues **5**(3), 542–554 (2018)
16. Mingaleva, Z., Deputatova, L.: Stimulating of entrepreneurship through the use of HADI cycle technology. In: Insights and Potential Sources of New Entrepreneurial Growth, pp. 72–85. Filodiritto Publisher, Bologna (2017)
17. Costa, V., Monteiro, S.: Key knowledge management processes for innovation: a systematic literature review. VINE J. Inf. Knowl. Manage. Syst. **46**(3), 386–410 (2016)
18. Østergaard, C.R., Timmermans, B., Kristinsson, K.: Does a different view create something new? The effect of employee diversity on innovation. Res. Policy **40**(3), 500–509 (2011)
19. Cooper, R.G.: Winning at New Products: Accelerating the Process from Idea to Launch. Perseus Books, Cambridge (2001)
20. Markus, A., Kongsted, H.C.: It all starts with education: R&D worker hiring, educational background and firm exploration. In: Academy of Management Proceedings, vol. 1, p. 14296 (2013)
21. Madhavan, R., Grover, R.: From embedded knowledge to embodied knowledge: new product development as knowledge management. J. Mark. **62**(4), 1–12 (1998)

22. Tuzovsky, A.F., Chirikov, S.V., Yampolsky, V.Z.: Knowledge management systems (methods and technologies)/Under total. ed. V.Z. Yampolsky. Publishing house NTL, Tomsk (2005)
23. Pogorelova, E.V.: Theoretical and methodological bases of knowledge management in an organization, dissertation for the degree of Doctor of Economic Sciences. Samara State University of Economics, Samara (2011)
24. Brown, J.S., Duguid, P.: Organizational learning and communities of practice: toward a unified view of working, learning and innovation. Organ. Sci. 2, 40–57 (1991)
25. Danilina, E.I., Mingaleva, Z.A., Malikova, Y.I.: Strategic personnel management within innovational development of companies. J. Adv. Res. Law Econ. VII 5(19), 1004–1013 (2016)
26. Stewart, T.A.: Intellectual Capital. Doubleday-Currency, New York (1997)
27. Davenport, T.H., Prusak, L.: Working Knowledge: How Organizations Manage What They Know. Harvard Business School Press, Boston (1998)
28. Mingaleva, Z., Deputatova, L., Starkov, Y.: Values and norms in the modern organization as the basis for innovative development. Int. J. Appl. Bus. Econ. Res. 14(9), 5799–5808 (2016)
29. Becker, B., Gerhart, B.: The impact of human resource management on organizational performance: progress and prospects. Acad. Manag. J. 39(4), 779–801 (1996)
30. Garcia-Fernandez, M.: How to measure knowledge management: dimensions and model. VINE 45(1), 107–125 (2015)

The Use of Digital Technologies for the Modernization of the Management System of Organizations

Alexander Bashminov[1,2] and Zhanna Mingaleva[1,3]($\boxtimes$)

[1] Perm National Research Polytechnic University, Perm 614000, Russia
mingall@psu.ru
[2] Perm Basketball Development Foundation "PARMA",
13, Lebedeva St., Perm 614107, Russia
[3] Perm State Agrarian and Technological University named after ak.
D.N. Pryanishnikov, Perm 614000, Russia

Abstract. The digitalization covers all units and subsystems of the organization's management mechanism. However, the importance, necessity, expediency and possibility of applying digital technologies for different elements of the organization's management system differ. The purpose of the work is to identify the possibility and expediency of applying digital technologies for different elements of the organization's management system. The methods of structural and hierarchical analysis are used to select blocks and individual elements of the control system. In order to improve the accuracy of the study, 3 management control units were identified. The main elements of the control system are defined within the framework of each control unit. Also, four behavioral information subsystems existing in Russian organizations are highlighted. The result of the study is an assessment of the possibility and expediency of applying digital technologies for different elements of an organization's management system within different types of information and behavioral subsystems. The conclusion from this study is that in some cases the use of digital technologies is not necessary and possible.

Keywords: Digital technologies · Management control units ·
Organization's management mechanism · Behavioral information subsystems

1 Introduction

The successful functioning and development of the organization in modern conditions is possible if two key management requirements are met: (1) the creation and maintenance of sustainable competitive advantages and (2) the availability of an effective management system. The specified requirements are applicable both to commercial firms and non-profit organizations. Any organization should quickly and adequately respond to changes in the situation in society and the market, make appropriate changes in their activities.

But the technology, methods and management tools are more stable parts than the control system as a whole. As a result, the depth and speed of changes in different

© Springer Nature Switzerland AG 2020
T. Antipova (Ed.): ICIS 2019, LNNS 78, pp. 213–220, 2020.
https://doi.org/10.1007/978-3-030-22493-6_19

subsystems of the organization will be different [1]. This is especially evident in the process of introducing digital technologies into different elements of the control system [2].

The intercalation of digital technology makes major changes in the structure and mechanism of management of the organization. But such changes cause corresponding changes in the corporate culture of the organization and managers and employees may not accept them [3, 4]. As a result, instead of positive changes and improving the efficiency of the organization, the digitization of individual elements may lead to a deterioration of the situation. Therefore, to ensure the main goal of the organization's management - ensuring its competitiveness and sustainable development, it is necessary that the changes made to the structure and the mechanism of the organization's management are consistent with the goals and objectives of management as a whole and for each individual element activity [5].

An important and still controversial issue in the theory of management is the requirement to abandon the principle of increasing the functional qualities of new management bodies in comparison with the previous state and the principle of increasing their effectiveness [6]. This is justified by the fact that the formation of a new management system of an organization should be focused on the content of new tasks that arise before the organization [7]. Thus, the real need for the development of the management system and the direction of this development are determined by the emergence of new tasks facing the organization, and not by the imperfection of certain aspects of management or by the lack of the possibility of improving management effectiveness.

Therefore, the purpose of the research in this article is to study the possibility of digitization of the main blocks and elements of the organization management system.

The methods of research are the structural and hierarchical analysis. The authors use a qualitative approach to explore information support system in the organization's activities.

2   Theory and Method

Modern development assumes digitization of management systems by organizations. But so far the majority of managerial approaches to solving the problems of the organization's development are focused on taking into account past experience in solving similar problems. This is explained by the fact that successful past experience is quickly transformed into stereotypes of assessing the situation and solving problems and, accordingly, in stereotypical decisions and behavior patterns. And if in the recent past such an approach has yielded positive results, in the current conditions of avalanche-like growth of information, a sharp increase in changes in the external environment and their increasing unpredictability, this approach ceases to justify itself [8–10]. "Some of the benefits business organization seek to achieve through information systems include: better safety, competitive advantage, fewer errors, greater accuracy, higher quality products, improved communications, increased efficiency and productivity, more efficient administration, superior financial and managerial decision making" [11, p. 101].

Based on the study of the management system of sports organizations, we previously identified a system of information support for the organization and elements of information content in modern conditions [12]. The universal model of information support of the organization was built on the basis of these results and using modern results of foreign studies on the construction of integrated management systems [13]. An adapted model is presented in Fig. 1. The universal model of information support of the organization was built on the basis of these results and used for the present study.

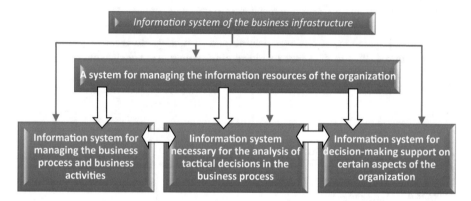

Fig. 1. The system of information support for an organization. *Source: adapted by authors from* [6].

Each of the elements, present in the Fig. 1, has a specific content, ensuring the performance of management functions at the levels of management of the organization and the features of the transfer and presentation of information.

In general, all these elements provide an increase in the economic performance of the organization, improve the efficiency of resource use and management [14, 15]. At the same time, the system of information support of business activities is characterized by a number of specific features. This determines the adaptation of common approaches to building networking in such organizations [16, 17].

The model is based on compliance with the following methodological principles:

(1) the need for a clear and complete transfer of all available information about the changes and possible changes in the external and internal environment to the whole team, each structural unit;

(2) the need for those employees of the organization who are responsible for making operational and strategic decisions to have sufficient individual skills to respond to rapid changes in the situation, to develop and implement programs to respond to changes;

(3) the need for a clear management program in the face of change;

(4) the need for a clear formulation of key management tasks, the definition and creation of conditions under which these tasks can be solved;

(5) the development and implementation in the management system of acceptable standards for each organization to achieve success.

The successful application of all these principles depends on the type of information-behavioral subsystem existing in the organization. In the course of the analysis, 4 types of information-behavioral subsystems existing in Russian organizations were identified. The criteria of centralization of control and destructive behavior underlie the classification of types of information-behavioral subsystems. The following Table 1 gives the main differences between the types. Also Table 1 indicates the possibilities of sustainable long-term existence of the organization when implementing different types of information-behavioral subsystems.

As can be seen from Table 1, a resistance to conflicts, the presence of interest in the final results of the functioning of the management system, the interest in the development of all members of the organization and in raising the level of joint activities characteristic of 3rd and 4th types of information-behavioral subsystems ensure a sustainable and long-term existence of the organization.

Table 1. Comparative analysis of types of information-behavioral subsystems of organizations

The type of information-behavioral subsystem	Basic organization characteristics		
	Feature of the management structure	*Presence of conditions for destructive behavior*	*Implications for the existence of the organization*
1st type – centralized	The formal structure is explicitly expressed	The presence of conflict situations, unwillingness to help colleagues, indifferent attitude to the final results of the organization	The high risk of liquidation of the organization, there are no incentives for development
2nd type – transition	The formal structure is expressed	The presence of conflict situations, the desire of individual members of the team to obtain changes primarily through other members or through organizational restructuring	The high risk of liquidation of the organization, the incentives for development are limited to the interests of leaders
3rd type – hybrid	Formal structure with elements of decentralization	The presence of interest in the final results of the functioning of the management system, development of all members of the organization and increase of the level of joint activity	All team members are interested in long-term work of the organization
4th type – decentralized	Flexible management structure aimed at self-development	High interest in the final results, in the development of all members of the organization and in raising the level of joint activity	High incentives for development and improvement, the threat of liquidation of the organization is minimal

Source: own processing

Modern management conditions that characterize the increasing volume and speed of information dissemination, as well as the accelerating degree of obsolescence, no longer allow all decisions to be carried out strictly along the command chain within a strictly centralized management structure (1st and 2nd types). The essence of the new approach to management is to transfer the authority to define goals, action programs, the ability to make operational decisions from senior management to the lowest possible management level or directly to the performers themselves. With the digitalization of the management system, managers and executors can react to the occurrence of a problem quickly, as well as make better and more effective decisions. Therefore, the organization, which is still used 1st or 2nd types, in order to achieve the long-term normal business activity should make the transition to the 3rd or 4th types. Modern digital technologies can provide a quick and efficient transition to these types of information-behavioral subsystems.

3 Results and Discussions

The application of a specific model of the information-behavioral subsystem of the organization should correspond to the chosen approach to solving the problems of improving the management system in general and the business strategy of the enterprise development, taking into account the issues of ensuring its economic security [7–9].

We have identified separate control units in order to identify the possibilities of using digital technologies for different types of subsystems and for individual elements of subsystems. For this study, we have identified 3 control units:

1. organizational and management unit;
2. personnel management unit;
3. information and management unit.

Further, in each of these control units, we have identified individual elements. The availability of digital technology has been evaluated with respect to these specific elements. The list of elements in the control units is given in Fig. 2.

The next step of the analysis included a description of each control from 3 identified control units with respect to 4 types of information-behavioral subsystems.

Fig. 2. Basic elements of control units. *Source: own processing*

In this article we will focus on the results of the analysis of the possibility and necessary (appropriate) of digitization of the basic elements of an organizational and management unit. The characteristics and features of the manifestation of each element of the organizational and management unit in different types of information-behavioral subsystems are given in Table 2.

Table 2. The features of elements of the organizational and management unit with 4 types of information-behavioral subsystems of organizations

Type of information-behavioral subsystem	The element of the organizational and management unit		
	Sources of initiation of actions and activities in general	*Presence of formal performance control*	*Presence of informal relations*
1st type – centralized	Orders	It is necessary and highly developed, it is an important management function	Associated with the interests outside the organization
2nd type – transition	Orders	It is necessary and highly developed	Extensive practice of closed discussions and discussions in a narrow circle of elected persons
3rd type – hybrid	The emergence of a creative attitude to work, the emergence of amateur beginnings	Formal control loses its leading importance, orientation toward obtaining greater results due to improvement of internal relations	The desire to solve conflict situations, pre-frugality in relations, openness in speech, tendencies towards cooperation
4th type – decentralized	Creative attitude to work, self-organization, self-tuning	It is minimized	The desire to solve conflict situations, the willingness to deal with relationships, the openness of sayings, the inclination to cooperate

Source: own processing

Assessment of the possibility and feasibility of applying digital technologies for the modernization of each selected block element to improve the organization's activities is given in Table 3. It is important to remember that the digitization of each organization's management system element should not degrade the results of organization's core activities. Therefore, the possibility and feasibility of applying digital technology should be assessed very carefully.

Table 3 confirms the previous claims that the digitization of individual elements may lead to a deterioration of the situation instead of positive changes and improving the efficiency of the organization. This is true of such elements as "The presence of formal control of execution" in 4th type of information-behavioral subsystems (decentralized) and "The presence of informal relations" in 2nd type of information-behavioral subsystems (transition).

Table 3. Evaluation of the possibility and feasibility of applying digital technologies for the modernization of the organizational and management unit of organizations

Type of information-behavioral subsystem	The element of the organizational and management unit		
	Sources of initiation of actions and activities in general	*Presence of formal performance control*	*Presence of informal relations*
1st type – centralized	Possible and appropriate	Possible and very important	It is not possible
2nd type – transition	Possible and appropriate	Possible and very important	It is not necessary
3rd type – hybrid	Possible and necessary	Possible but not important	Possible and appropriate
4th type – decentralized	Possible and necessary	It is not necessary	Possible and appropriate

Source: own processing

The analysis shows that the transformation of the management system of the organization is a complex and not unambiguous in the theoretical plan process, which requires constant correlation of the actions taken to reform management with the tasks facing management, as well as the opportunities that the organization as a whole and the system that exists in it management.

4 Conclusions

In the process of information system creation for the management of the organization activities, it is also necessary to take into account the requirements for the effectiveness of such system and the process of constant improvement and development of management systems. The study showed that the use of digital technology should be strictly supported by the entire management system of the organization. The results of the study showed the possibility of changing the internal environment of the organization as a result of the application of digital technologies.

The result of the study is the construction of a classification of information-behavioral subsystems existing in Russian organizations, and an assessment of the possibility and feasibility of using digital technologies for different elements of the organization's management system within these types.

Further research will be continued in the direction of a detailed analysis of the possibility of digital technologies for the two other dedicated control units: (1) the personnel management unit and (2) the information management unit. Comparative analysis will also be carried out in relation to 5 types of information-behavioral subsystems.

References

1. Bonazzi, R., Hussami, L., Pigneur, Y.: Compliance management is becoming a major issue in IS design. In: D'Atri, A., Saccà, D. (eds.) Information Systems: People, Organizations, Institutions, and Technologies, pp. 391–398. Springer, Heidelberg (2010)
2. Schuh, G., Bernardy, A., Zeller, V., Stich, V.: New requirement analysis approach for cyber-physical systems in an intralogistics use case. In: IFIP Advances in Information and Communication Technology (IFIP AICT) 506, pp. 149–156 (2017)
3. Barney, J.: Organizational culture: can it be a source of sustained competitive advantage? Acad. Manag. Rev. **11**(3), 656–665 (1986)
4. Mingaleva, Z., Deputatova, L., Starkov, Y.: Values and norms in the modern organization as the basis for innovative development. Int. J. Appl. Bus. Econ. Res. (IJABER) **14**(10), 124–133 (2016)
5. Iden, J.: Investigating process management in firms with quality systems: a multi-case study. Bus. Process Manag. J. **18**(1), 104–121 (2012)
6. Bashminov, A.V.: Creation of an effective system for managing organizations on the basis of a change management approach. Econ. Entrepreneurship **12**(4), 572–574 (2013)
7. Nayan, N.M., Zaman, H.B.: Information system development model: theories analysis and guidelines. In: International Visual Informatics Conference IVIC 2009: Visual Informatics: Bridging Research and Practice, pp. 894–904 (2009)
8. Anicic, D., Rudolph, S., Fodor, P., Stojanovic, N.: Real-time complex event recognition and reasoning – a logic programming approach. Applied Artificial Intelligence **26**(Special Issue on Event Recognition), 6–57 (2012)
9. Luckham, D.C.: Event Processing for Business. Wiley, Hoboken (2012)
10. Belz, R., Mertens, P.: Combining knowledge-based systems and simulation to solve rescheduling problems. Decis. Support Syst. **17**, 141–157 (1996)
11. Rosca, D., Banica, L., Sirbu, M.: Building successful information systems – a key for successful organization. Ann. "Dunarea de Jos" Univ. Galati Fascicle I: Econ. Appl. Inform. **2**, 101–108 (2010)
12. Bashminov, A.V.: Development of information system of management decisions in sports organizations. In: Conference Proceedings – Recent Advances in Information, Tourism, Economics, Management and Agriculture – ITEMA, pp. 758–766. Budapest, Hungary (2017)
13. Jørgensen, T.H., Remmen, A., Mellado, M.D.: Integrated management systems – three different levels of integration. J. Clean. Prod. **14**(8), 713–722 (2006)
14. Barata J., Da Cunha, P.R.: Towards a business process quality culture: from high-level guidelines to grassroots actions. In: Proceedings of the 23rd International Conference on Information Systems Development, pp. 6–13. Varaždin, Croatia (2014)
15. Perez-Arostegui, M.N., Benitez-Amado, J., Tamayo-Torres, J.: Information technology enabled quality performance: an exploratory study. Ind. Manag. Data Syst. **112**(3), 502–518 (2012)
16. Mingaleva, Z., Bykova, E., Plotnikova, E.: Potential of the network concept for an assessment of organizational structure. Procedia Soc. Behav. Sci. **81**, 126–130 (2013)
17. Stich, V., Kurz, M., Optehostert, F.: Framework conditions for forming collaborative networks on smart service platforms. In: Risks and Resilience of Collaborative Networks. 16th IFIP WG 5.5 Working Conference on Virtual Enterprises, PRO-VE, pp. 193–200 (2015)

New Method for Digital Economy User's Protection

Valery Konyavsky[1](✉) ⓘ and Gennady Ross[2](✉)

[1] Plekhanov Russian University of Economics, Moscow, Russia
Konyavskiy@gospochta.ru
[2] Financial University, Moscow, Russia
ross-49@mail.ru

Abstract. This paper analyses the existing approaches to the identification/authentication of users in the digital economy systems, and consideres options for functioning in trusted/untrusted environments. There is shown that currently known biometric features, which are effectively used in forensic science, do not allow the digital economy to guarantee protection against the influence of malicious software. As result of this study a new approach have been determined to the specific security requirements for computer systems in the digital economy conditions.

This approach will allow to establish the authenticity of the data source and increase the reliability of identification, and the combination of several biometric modalities, supplemented by an analysis of at least one of the possible physiological (reflex) reactions, will significantly increase the accuracy of biometric identification and provide a solution to the vital problem. As the result of this research, a Patent was obtained for a new "Interactive method of biometric user authentication".

Keywords: Identification · Authentication · Digital citizen · Digital economy ·
Trusted environment · Untrusted environment ·
Static and dynamic behavioral signs · Neural networks

1 Introduction

One of the main distinguishing features of the digital economy is the presentation of services in the form of a single business process, within which information processes from fragmented ones become holistic and complete. In accordance with the concept of the digital economy, a new concept of "digital citizen" is introduced - this is a person who must have extensive skills, knowledge and access to using the Internet through computers, mobile phones and other network devices to interact with various organizations.

Economy digitization is impossible without a reliable system of integrated security of developed computer systems. For targeted execution of operations, at least a confirmation of the identity of a "digital citizen" (user) is required. At the same time, the existing mechanisms for identification and authentication of the user of the network services of the digital economy are hopelessly outdated. The use of biometric remote

© Springer Nature Switzerland AG 2020
T. Antipova (Ed.): ICIS 2019, LNNS 78, pp. 221–230, 2020.
https://doi.org/10.1007/978-3-030-22493-6_20

identification systems stimulates the development of entire areas, for example, the banking market, retail, public services, ensuring a high level of security.

Mobile devices are the key to many areas of modern life, and access to their data is the goal of cyber fraudsters. That is why methods of user identification for access to computer equipment's (CE) (smartphones, tablets, etc.) and their data are rapidly developing. But the identification tools placed on the mobile CE do not provide their power of attorney, and the software may contain malicious components that can discredit any locally performed control operations. Mobile devices are the key to many areas of modern life, and access to their data is the goal of cyber fraudsters. That is why methods of user identification for access to CE (smartphones, tablets, etc.) and their data are rapidly developing. But the identification tools placed on the mobile CE do not provide their power of attorney, and the software may contain malicious components that can discredit any locally performed control operations.

CE differ in the degree of attorney's power. If you can achieve sufficiently high security in a corporate (closed) environment, then there is no trusted route in an open network environment between the interacting parties of identification/authentication. If the power of the data center (DC) can be provided, then the power of the client terminal can neither be installed nor maintained, which means that in general the data transmitted by the subject may not coincide with the data obtained and used for the control procedures. The uncertainty of client tools is a fundamental difference, and identification/authentication tools should take this into account. However, the creation of such means of identification/authentication in the traditional way is impossible, since any implemented control mechanism will always be "easy prey" for a variety of malware attacks. In the conditions of untrusted client terminals, the use of cryptographic security tools on them is completely meaningless, since it is impossible to properly keep secret cryptographic keys.

Taking into account the specific security requirements of computer systems of the digital economy, the most promising development of an identification/authentication system, in our opinion, is possible in two directions:

1. To impose all measures of protection on the data processing center (DPC), resigned to the untrusted mobile devices of users. Some experts believe such an approach is sufficient from the point of view of risks, but this is not the case - the offender can always find a gap in protection sufficient for the penetration of the malicious software. Reducing risks to an acceptable level can be provided only with substantial financial investments both in the development of an information protection system (IPS) in the data center, and in insurance of information risks associated with deficiencies in software and hardware protection.
2. Development of a distributed "immune" IPS, ensuring an acceptable level of client transaction security even with untrusted equipment. The basis of this trend can be systems for detecting malicious activity, built on the principle of multi-agent systems [1], as well as systems for multimodal biometric identification using the dynamics of human reflex reactions. In fact, we have to "replace" the complexity of cryptographic transformations by the complexity of a person's reflex reaction (there is no human model—the response to a stimulus is qualitatively understandable, but it is impossible to quantify them).

Analysis of the concepts of identification in forensic science and the digital economy. It is possible to treat identification as identification in forensic science and in the digital economy. However, this concept for each of the areas of activity has its own specifics. We introduce the concept of the null identification hypothesis, the essence of which is that "The identifiable object is the one for whom he claims to be (for whom the subject takes it)".

Forensic science deals with people (suspects or criminals) who are not interested in correct identification, i.e. in the proof of the fact of the commission of illegal actions (access) of the object. Active opposition here, as a rule, is either absent or aimed at identifying a violation - to prove that it was not, did not participate, did not violate. The identification tools used for this are trusted. They are specially developed, protected by certified means, undergo regulatory control procedures and so on. The purpose of counteraction (inaction) in forensic science is to reject the null hypothesis while it is true. Opposition (from the subject, or accomplices, or difficulties associated with lack of data) is aimed at achieving the error "false positive", called the error of the first kind.

In the digital economy - the object of identification is a lively and respectable user, whose needs are aimed at gaining access to certain resources. He is ready to cooperate, performs the necessary actions in order to get the service he needs after successful identification, that is, he is interested in correct identification. Active hackers in the digital economy can be carried out by a hacker (attacker), seeking for their own sake a false identification when using mobile technical tools that are not protected from malicious software. Under these conditions, the attacker's goal is to counter the user's normal access to the necessary resources, impersonating him, in order to force the subject to accept the null hypothesis while it is false. Counteraction (from hackers) is aimed at achieving the error "false negative" - errors of the second kind.

The simplest and most commonly used identifier in smartphones are passwords presented in graphic or digital form. Their complexity can be increased by additional financial means, but this will make them inconvenient to use, security will not increase, since they can be intercepted by malicious software. In this case, it is advisable to use biometric signs, inalienable from the person.

2   State-of-Arts for Biometric Authentications' Methods

A fundamentally important advantage of using biometric features is their inseparability from the identity of the owner and the elimination of intermediate information carriers (magnetic cards, smart cards, tokens, diskettes, etc.). The existing methods can be divided into two groups: with static features and dynamic features.

Static signs. All static features can be divided into two types: immutable and slowly immutable. Examples of immutable attributes are fingerprints, iris of the eye, palm geometry, face geometry, DNA code, with some degree of conventionality - RFID chips stitched in. Examples of slowly unchangeable signs that undergo changes in the process of human aging are: retina of the eye, drawing of the palm and finger of the vascular bed, data on the physiological processes of the body taken by biometric chips and others.

In the event that user identification is carried out under the conditions of trusted hardware, the immutability of static features allows them to be used to prove the presence of an object at the data collection point. If untrusted devices are used, the immutability of the attribute allows the attacker to replace the data.

The successful experience of using static biometrics at the moment is connected only with forensic identification. It is assumed that in the databases no one will substitute the prints, the citizen will not put on a glove during registration and will not give it to the attacker afterwards, and the identification tools used by the police will be trusted.

Analysis of static methods of biometric authentication shows that due to the simplicity and static nature of the used modalities (papillary pattern, iris and retina, vascular bed, etc.) are easily reproduced and modeled, which not only does not reduce the risk of erroneous identification, but also allows you to directly influence her results. Traditional (invariant) biometric modalities do not and cannot provide a sufficient level of power of attorney for identification on an untrusted device.

Thus, static biometric features can effectively be applied in forensic science due to their invariance to external factors, in full or in part, based on the assumption of the power of attorney of technical processing means. For the digital economy, this assumption is wrong, and that is why it is necessary to change the approach to biometric characteristics as invariants.

Dynamic signs. Dynamic features of biometric authentication are based on the behavioral (dynamic) characteristics of a person, that is, they are built on features characteristic of subconscious movements in the process of reproducing an action. These authentication features include:

- Voice. Its characteristics are associated with the combination of both soft and hard tissues of a person and can characterize the peculiarities of his emotional state, but this does not exclude the possibility of voice modeling. Such models are well developed and for the digital economy this authentication method is quite risky.
- Facial expression. This method allows you to define the "vital" (vitality) of the object, i.e. establish that it is not a photograph that is presented, but the data of the person present This is achieved by analyzing the execution of simple commands, for example, "look left, right, smile." However, there are well-designed models that allow an attacker to fake a reaction to a team.
- Handwriting. The main feature of this method is that for identification it is necessary for the user to do a certain set of actions, characteristic only for him, but the features of the underscore are easily modeled, which allows an attacker to use VDP for untrusted devices.
- Implantable chips. These chips can analyze the pulse, the composition of the gastric juice and other biometric data. The application of this procedure will allow access to the smartphone, but will not provide the power of attorney of the smartphone.

Biometric features that can be used to construct authentication procedures can be characterized according to the principle of time variability and variability under the influence of external stimuli. If the time axis is directed from the bottom up, then at the very top will be DNA, as an unchangeable sign, below will be almost unchanging papillary pattern, retina and iris, slowly changing venous pattern, and so on, up to saccadic movements, recorded even in sleeping and blind. Those typical features are shown on Fig. 1 below.

DNA	-
Papillary pattern	-
Retina and iris	-
Venous drawing (finger, palm)	-
The structure of solid tissue	-
Face shape, palm	-
Vote	*
Behavioral signs:	
- gait	*
- handwriting	*
- handwriting keyboard	*
Kinesiological characteristics	*
Pulse wave	*
Eye movements:	
- saccadic	*
- while tracking the stimulus	+
- when searching for an object	+

Fig. 1. Dependence of biometric features on time and external stimuli.

On Fig. 1 signs "–" mean independent of external stimuli; "*" signs mean the dependence of which on external stimuli is manifested, but slowly, in the on-line mode; and the "+" sign depending in the in-line mode, which manifest themselves quickly. Signs marked as "–" can be used in forensic science, "+" sign - in digital economy systems, "*" sign are usually used as additional modalities used to clarify the authentication solution.

In connection with the above, the task of creating a new method and algorithm for identifying a user according to his individual biometric characteristics becomes relevant, allowing to perform trusted authentication on untrusted devices.

3 New Method of Interactive Identification Using External Stimuli

Identification by dynamic indicators such as "stimulus-response". To eliminate the vulnerabilities associated with the simplicity of the substitution of measurements on untrusted devices, it is necessary to move from static and dynamic behavioral indicators to dynamic indicators of the "stimulus-response" type with complex communication dynamics—to interactive identification. The dynamic link, which is extremely difficult to model, is the human nervous and vegetative systems and the associated features of the physiology of movements. In particular, involuntary reactions to external stimuli (for example, audio and video stimuli) are individual.

The possible reaction of the user to stimuli can be recorded by the sensors of the client device and processed using artificial intelligence methods, for example, artificial neural networks [2]. Such a procedure will determine the source of data streams and increase the reliability of identification, and the combination of several biometric modalities, supplemented by an analysis of at least one of the possible physiological (reflex) reactions, will significantly increase the accuracy of biometric identification and provide a solution to the problem of vitality.

On an untrusted client terminal, it is not difficult to fake a voice, a face, change fingerprints and a vascular bed pattern, imitate eye movement, but using an additional biometric feature "reaction to stimulus" consistent with them will make it impossible to fake data streams. This is due to the fact that in order to reconcile fake data, a model of reactions of a specific person is needed, which is impossible due to the high complexity of such a model.

Thus, a contradiction arises between the need of society for the trusted identification of users using untrusted terminals and the lack of reasonable approaches to solving this problem. The possibility of resolving this contradiction, in our opinion, is associated with the use of human reflex responses to external stimuli, which makes it possible to consider research in this area as relevant.

As external stimuli, you can use the capabilities of a smartphone - sound, color and light, vibration, text display (including with random changes in the number of spaces).

For movement on the development of this approach it is necessary:

- check the method of objective measurements of the hypothesis about the effect of various external stimuli on reflex reactions;
- create algorithms for collecting primary data on user devices (smartphones)
- to develop an artificial neural network for analyzing the data taken by the smartphone, or to use other methods of analysis adequate to the task.

Known works in which eye movement was considered as a biometric feature [3, 4]. In fact, the results obtained in them can be regarded as confirmation of the hypothesis that there is biometric information in the movement of the eyes (indirectly in movements associated with reflex responses). Based on these results, you need to reproduce them, study the required resolution, and then design identification algorithms based on random stimuli generated remotely and remote processing and decision making. At the same time, the client terminal will be the only means of displaying and retrieving the primary data, which can in no way affect the security of identification. In a number of studies, the removal of indicators was carried out using an infrared camera and special calibrated devices that fix the position of the user's head, but this approach is inapplicable for widespread use.

Thus, it is necessary to study the possibilities of removing dynamic indicators of eye movement in response to external stimuli with the help of mass user devices: home video cameras, embedded cameras of tablet devices. The key tasks are to study the possibilities of taking these indicators using RGB-cameras (optical range) with a low frame rate (30–60 fps), and to study the possibilities of geometric calibration of the "man + camera" system just before the measurements or during the measurements.

Learning identification algorithm. One of the variants of the identification algorithm can be neural networks. Artificial neural networks have gained tremendous distribution in various fields due to their ability to learn from data and model a function of almost any degree of complexity. However, neural network algorithms are characterized by a huge number of parameters and long learning. Great progress in the use of neural networks has occurred in relatively recent times after the accumulation of large data collections and the growth of computing power.

When solving problems of biometric identification and authentication, neural networks are also used. In particular, when the image is a biometric identifier, convolutional neural networks are used [5]. At the same time, the Siamese network architecture is traditionally used [6]. With this architecture, the network is applied to the incoming image and the reference image, and builds a compact vector representation of the images. The decision on the conformity of the biometric sample and the standard is made either on the basis of metric similarity of the description vectors, or using another neural network that accepts two received descriptions [6, 7]. The convenience of this architecture is that the feature representations of all objects from the database can be calculated in advance and stored in the form of their compact representations, and only one input can be processed. In the case of dynamic biometric data, the overall network architecture remains the same.

Conventional fully connected networks can be used to process reaction sequences, if a low-level indicative representation of the data is constructed. However, if there is enough training data, you can try to eliminate the "feature engineering" step and feed the raw data to the network input. In this case, recurrent neural networks, in particular LSTM [8] and GRU [9], are traditionally effectively used for sequence processing. In this case, more data will be required, as the network will contain much more trained parameters.

In the case of the study of dynamic reactions, it is required to assign a "stimulus-response" to the person for whom such reactions are characteristic. It should be noted that the task is quite complicated, since it is assumed that the incentives will not be repeated. However, the ability of neural networks to build generalizations should allow identifying common elements in stimuli and assessing the degree of similarity of reactions in similar situations, based on this information and the training sample of the network implicitly reflected in the parameters, determine for which person such reactions are characteristic. On the other hand, unlike the task of modeling the nervous system, in this case the task of predicting a person's response to a stimulus is not worth it. Determining the similarity of reactions is an order of magnitude simpler, descriptive task, whereas the previous one is generative.

Technologically, to create an identifying neural network, you need to collect a collection of data for training. The data collection should contain stimulus-response pairs. And for each person there should be several pairs. The key task for the marked data is to train the network to build representations for reflex reactions. The presentation should be such as to reflect the reactions of the same person at close points, and of different people at distant points. This can be achieved, for example, by using the triplet loss function [7]. After learning the neural network, adding or deleting a new entry into the base of biometric identifiers will not require retraining the neural network, but will only require the calculation of the representation.

The complexity and number of parameters of the neural network determines the security of this approach. Since the potential attacker does not know which parameters of the reactions are analyzed by the neural network, it is extremely difficult to fake the reaction itself. On the other hand, if an attacker suddenly gets a set of data for which the neural network has been trained, this will not lead to the fact that he can reproduce it. Even if a network of a similar architecture is built and trained, the resulting presentation will be different from the existing one. The result of the network learning, the characteristics of the output and low-level neurons are of a stochastic nature, determined by random initial parameters and the order of data arrival. For example, rearranging the output neurons in places will not change the accuracy of the network, but it will fundamentally change the representation of the reaction, that is, the identifier of the person who will be compared on the server with identifiers from the database.

Thus, the technology of neutron networks is fundamentally different from traditional identification methods. Its ability is that it does not require a preliminary study of the nature of the data, manual determination of model parameters (choice of key features, their relationships, etc.). The neutron network extracts the model parameters automatically in the best way possible during the learning process. Another distinctive feature of this technology is the possibility of implementing parallel computing.

The basis of the new approach is the hypothesis that the dependence of human reactions to external stimuli substantially depends on the cognitive and kinesiological characteristics of a person, is dynamic in nature and is reflected in measurements to an extent sufficient for analysis.

The principal features of the stimulus-response system are:

- the presence of the human nervous system as a link between the stimulus and the reaction. If there are already quite a lot of approaches to the simulation of intelligence, then there is practically no work to simulate the nervous system. The nervous systems of people are extremely different among themselves and difficult to model. Unlike intellectual tasks, the nervous system does not have a "right answer" to teach a machine.
- random, non-repetitive incentives. Such an approach would fundamentally exclude the possibility of reproducing previously recorded user reactions, that is, forgery of primary data.
- processing of the stimulus-response pair can be done on a remote trusted device.

Wherein:

Untrusted client terminal does not affect the results, since the stimulus generation and reaction analysis is performed on the alien trusted resources, and distortion of the reaction will not allow the attacker to obtain the desired result for him.

It does not make sense to intercept the stimulus, since knowing the stimulus, it is impossible to generate a reaction due to the absence of a human model.

It is impossible to extract the parameters of the neural network by testing it under the given conditions, and the identification acts constantly specify the parameters of the neural network, and therefore even total observation will not allow the network to fully reproduce.

It is easy to see that the main feature that ensures the security of identification on an untrusted device is interactivity - neither the client terminal, nor the center itself will not perform identification. The procedure is essentially interactive, which allows the generation of the stimulus and the decision making to be assigned to a trusted center, and the removal of information is carried out on a personal device belonging to the client [10].

Figure 2 illustrates the interactivity of a trusted identification process using an untrusted smartphone.

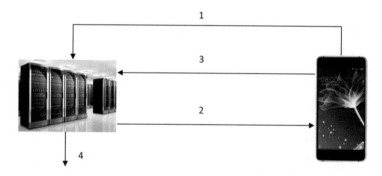

Fig. 2. Trusted identification process.

Shown on Fig. 2, trusted identification process consists of following stages:

1. The user initiates the process by sending a request to a trusted identification center.
2. The center generates a random stimulus and sends it to the user's smartphone.
3. The user performs the task of tracking the stimulus (or searching for an object), and data on eye movement are sent to the Center.
4. After data processing and comparison with the stimulus, the Center makes a decision on identification.

Sure, a complete security analysis has yet to be performed. It may be worth looking for an alternative neural network decision-making mechanism.

4   Conclusion

As the result of this research, the Patent [10] was obtained for a new "Interactive method of biometric user authentication". The features of this patented method are as follows:

- registration of one or several indicators of the user's heart activity (heart rate and/or pulse wave);
- registration of the user's eye pupil motility in response to one of the following pathogens (brightness, color of the screen, frequency or loudness of the sound of the electroacoustic converter of computer means;
- the current changes in the diameters of the pupils in relation to the initial ones: at a given periodicity, display the user's face so that on each display it recognizes the pupils and calculates - for each individual eye;
- a detailed examination of the object in the user's field of view, fixing the position of the eyes - pictures with fine details.

Thus, the conducted studies suggest that a new method of identification/authentication of a digital citizen has been found, which provides a reliable guarantee against making an unreasonable decision.

References

1. Liechtenstein, V.E., Konyavsky, V.A., Ross, G.V., Los', V.P.: Multi-agent systems: self-organization and development, p. 264. Finance and Statistics, Moscow (2018)
2. LeCun, Y., Bengio, Y., Hinton, G.: Deep learning. Nature **521**, 436–444 (2015)
3. Bednarik, R., Kinnunen, T., Mihaila, A., Fränti, P.: Eye-movements as a biometric. In: Kalviainen, H., et al. (eds.) Scandinavian Conference on Image Analysis, pp. 780–789. Springer, Berlin, Heidelberg (2005)
4. Bargary, G., et al.: Individual differences in human eye movements: an oculomotor signature? Vision. Res. **141**, 157–169 (2017)
5. Krizhevsky, A., Sutskever, I., Hinton, G.: ImageNet classification with deep convolutional neural networks. In Proceedings: Advances in Neural Information Processing Systems, vol. 25, pp. 1090–1098 (2012)
6. Taigman, Y., Yang, M., Ranzato, M.A., Wolf, L.: DeepFace: closing the gap to human-level performance in face verification. In: Conference on Computer Vision and Pattern Recognition (CVPR), Columbus, Ohio, USA (2014)
7. Schroff, F., Kalenichenko, D., Philbin, J.W.: Facenet: a unified embedding for face recognition and clustering. In: IEEE Conference on Computer Vision and Pattern Recognition (CVPR), Boston, MA, USA (2015)
8. Hochreiter, S., Schmidhuber, J.: Long short-term memory. Neural Comput. **9**, 1735–1780 (1997)
9. Chung, J., Gulcehre, C., Cho, K., Bengio, Y.: Empirical evaluation of gated recurrent neural networks on sequence modeling. In: NIPS 2014 Workshop on Deep Learning, Montréal, Montréal Canada (2014)
10. Konyavsky, V.A.: Interactive biometric user authentication method. The patent for the invention No. 2670648, 10.24.2018, bull. No. 30

Use of Information Technologies for Managing Executive Compensations in Network Companies

Zhanna Mingaleva[1,2](✉) ⓘ, Anna Oborina[1,3] ⓘ,
and Irena Esaulova[1] ⓘ

[1] Perm National Research Polytechnic University,
Perm 614990, Russian Federation
mingall@pstu.ru
[2] Perm State University, Perm 614990, Russian Federation
[3] RosMetTrade Company, 2, Plekhanov St., Perm 614065, Russia

Abstract. The article reveals the model of executive compensations in Russian company operating in the industry of ferrous and non-ferrous metals scrap and the ways of digitization and automation of the business process of calculating material remuneration. The purpose of the article is to suggest for Russian companies an appropriate model of executive compensations with possibility of digitization by means of information technologies. The ways of digitization are discussed. Grouping and pyramid of key performance indicators of managers in the company are proposed in the article. The examples of mapping key indicators for managers of different departments of the company are given in the article and the method of their calculation for a specific company is given. The methodic recommendations for top managers motivation are adopted to specific features appropriate to Russian industry of ferrous and non-ferrous metals scrap.

Keywords: Executive compensations · Key performance indicators ·
Information technologies · Digitization

1 Introduction

The problem of top managers motivation onto implementing current and strategic goals are widely recognized among theoretic and field researchers [1–4] in area of management of an organization. At this moment there are many applied approaches to top managers motivation, but none of them is universal.

The complexity or impossibility of applying the proposed approaches to the determination of top managers' material remuneration is a disadvantage of some systems. For example, you cannot use executive compensation techniques based on the company's stock price if the company does not have a listing. The high cost of using other reward systems is a disadvantage, which largely negates the value of these systems and their effectiveness.

Each motivation's approach should be adopted to operating activities of existing company [5].

T. Antipova (Ed.): ICIS 2019, LNNS 78, pp. 231–241, 2020.
https://doi.org/10.1007/978-3-030-22493-6_21

Accordingly, there are not the companies that have the opportunity to adopt, implement and maintain "pure" standard motivation system. And the more important is the task of digitization of the business process of top managers' material remuneration. This is especially important in large or widely diversified companies.

RosMetTrade company is a leading scrap metal trader in Perm region situated at central part of Russian Federation. More than 200 employees are at work in the company, the number of a managers at the different levels of organizational hierarchy exceeds 30 persons. Its operational activities of RosMetTrade company have a set of features. The most important characteristic of the company is the network of scrap metal processing and storage workshops operating in regions that are far removed from steel-making plants consuming scrap metal in cast iron production. RosMetTrade company is the network company. As a consequence, there is strong need to develop the model of executive compensations appropriate for Russian company operating in the industry of ferrous and non-ferrous metals scrap. This remuneration model should have scientific evidence, be fit to any company operating in mentioned industry and take into account the peculiarities of its organizational structure and network nature [6]. Consequently, the purpose of the article is to suggest for Russian companies an appropriate model of executive compensations with possibility of digitization by means of information technologies in the network company.

2 Review of Scientific Literature and the Methodology of Current Research

In foreign scientific literature there is a number of the works devoted to top manager's motivation in a company. At a turn of the 20–21st centuries such aspects of a top managers remuneration system were most actively discussed: level and pay structure of the CEO [7, 8]; a ratio between the salary of the CEO and performance of a firm (the so-called theory "sensitivity for payment work") [9–11]. Much attention was paid to the analysis of dismissal of directors as sequence of bad firm's performance [12, 13], studying of systems of motivation of directors for performance measured on the relation to the market or industry, calculation of annual rewards plans, share options for heads and a technique of assessment of options, etc. [2, 14, 15].

Also, the international differences in CEOs compensations [16–18], process of corporate governance and top executive compensation structure [19], communication between remuneration and firm's past performance [20], communication between remuneration and firm's economic future [21, 22] and many other questions were studied. For example, studies of Fey, C.F. and Furu, P. are devoted to additional bonuses for top managers on the basis of multinational corporation's performance and knowledge sharing between its divisions [23]. Studying of foreign scientific literature allowed to summarize the key aspects of the approaches to top managers motivation in the foreign companies [24–29]:

1. *Remuneration is paying only for the real results.* The top manager has to know what results from him (in long-term and short terms) are expected and what will be remuneration for him – in the form of an hourly wage rate or in advance specified

amount for each executed goal (task). The different indicators will be taken under such results in different countries. US firms used stock-based performance measures such as total shareholders return (TSR) [30, 1142]. In Japan the executive compensation contracts are as follows, in order of frequency net income or income before taxes (46%); operating income (43%); ordinary income (36%); and sales (35%) [30, 1142].

2. *Compliance between compensations to be paid and a scale and a complexity of the work performed by top manager.* Remuneration is paid not for a position within the company, and for the work performed by top manager.

3. *The operational and strategic objectives are accepted as the criterions for top managers remuneration.* The top managers have to be motivated to implementation of specified objectives and tasks.

4. *Balanced scorecards.* When remuneration is paid to top manager on a basis of key performance indicators, he will be able to see any result from the work performed by him and thus could impact on amount of expected payment.

5. *The amount of remuneration of top managers* depends from (a) complexity and risk level of performed work and (b) their personal values, interests and corporate style of behavior.

In the foreign practice of remuneration, there are options schemes which give to top managers the chance in the future to take shares of the company managed by them at the current price which is in advance stipulated. So, the options stimulate top managers to achieve superior performance of a company during long time that influences its share price and, therefore, increases the expected remuneration.

In Russian scientific literature, the problem of top managers motivation is well recognized, the best foreign and national practices of motivation are investigated. However, it is necessary to note a lack of universal methodical developments in this field. Usually, all approaches offered in the Russian researchers come down to balanced scorecard using [see, e.g., 31–34]. It can be explained with the fact that in the Russian companies the most factors which exist in foreign companies and define their motivation systems are absent. For example, Russian managers have managed a specified company during short period: usually, they come "from the outside", and did not have a long career. In many cases managerial experience of the Russian managers does not comply to industry in that a company is operating at the moment. Further, the most of Russian companies are not joint-stock and are not managed by boards, so, there are no committees on compensations. These and many other circumstances lead to the fact that the accumulated international experience on top managers motivation can give a little benefit to the Russian companies. Therefore, at the basis of the motivation systems which are the most applicable in Russian companies lie various cards of key performance indicators. There are not big companies in Russian company operating in the industry of ferrous and non-ferrous metals scrap. According Ruslom agency in this industry 60 thousand employees are engaged in 5,000 enterprises [35]. Therefore, no one of scrap metal market's players is able to develop and support advanced motivation systems both complicated and expensive.

The authors take as a basis of current research the creation of balanced scorecard for calculating the top managers remuneration. The methodology of the study includes the results of modern work on assessing the complexity of the work of senior managers, which is measured with proxy variables associated with operating activities [5].

The object is a company operating in Russian industry of ferrous and non-ferrous metals. The applied side of research is digitization of top managers motivation model by means of modern information technologies.

3   Research and Results

3.1   Principles of Top Managers Motivation

The creation of balanced scorecard for calculating the top managers remuneration at RosMetTrade company presented in Table 1.

The top managers motivation at RosMetTrade company is based upon the following:

(1)   each manager, division or team is having the set of key performance indicators in which achievement of short-term and long-term goals (tasks) is shown. The amount of remuneration directly coordinates with the key performance indicators;

(2)   the set of key performance indicators of any manager has to be fully reflecting his personal deposit into operating activities of the company;

(3)   the remuneration paid for short-term and long-term goals achieved has to compensate all efforts made by top manager;

(4)   all heads who participated in it have to receive remuneration for accomplishment of short-term and long-term goals (Fig. 1).

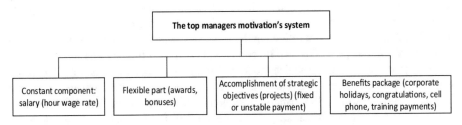

Fig. 1.   The top managers motivation's system at RosMetTrade company

The top managers motivation's system at RosMetTrade company consists of four parts:

(1)   constant component: salary (hour wage rate);

(2)   flexible part: accomplishment of planned values according the set of key performance indicators;

(3)   accomplishment of strategic objectives (projects): the fixed or unstable payment;

(4)   benefits package: corporate holidays, congratulations, cell phone, training payment, etc.

The first three components form monetary encouragement in the following structure: constant component of 30–40%, flexible part of 40–50%, accomplishment of strategic objectives: 10–20%.

3.2   Key Performance Indicators

Figure 2 presents the core perspectives of key performance indicators that must be included in the system of calculating of top manager's remuneration and financial motivation at RosMetTrade company.

The first group, *financial perspective*, assesses the accomplishment of OPEX and CAPEX budgets of entire company, its separate activities and divisions. Among them are such indicators as revenue, the average sales price, level of a marginal profit, production costs, the average salary of personnel, etc.

The second group of indicators, *production perspective*, reflect outcomes of operational activities: the volume of scrap metal shipment to consumers, the volume of scrap metal refined, the scrap metal at storage places, production costs, loss rate at sales to steel-making plants, productivity, average number of employees, etc.

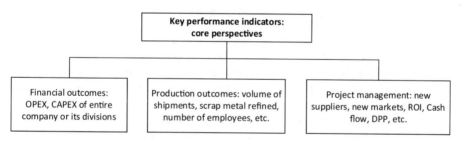

Fig. 2. Key performance indicators at RosMetTrade company

The third group of indicators treats *project management*. The RosMetTrade company participates in the tenders held on stock exchanges executed by such companies as PAO "LUKOIL", PAO "Rosneft", PAO "RZhD". Among them are such indicators as the volume of the tenders won for the period, the average price of scrap metal purchases, return on sales by separate lots, return on investment, etc.

Features of the company's operating activities determine the importance and necessity of including all these indicators in the system of material remuneration of top managers in terms of firm performance.

3.3   The Sets of Key Performance Indicators for Top Managers

Figure 3 shows the hierarchy of key performance indicators at RosMetTrade company. It should be noted that key indicators on the bottom levels do not enter in set of indicators on higher level. Thus, the personal zone of responsibility for heads on each level of organizational hierarchy remains invariable.

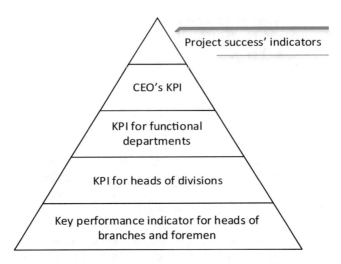

Fig. 3. The hierarchy of key performance indicators at RosMetTrade company

At the first level there is set of key performance indicators for heads of scrap metal processing and storage workshops which are located at various regions, for example, at the Perm region, the Komi Republic, Western Siberia, etc.

Tables 1–3 present the forms applicable to calculation of top manager's remuneration at different levels of organizational hierarchy at RosMetTrade company. The digital technologies of collecting and data handling, measure calculation, expeditious representation of results are used for filling this forms.

Table 1 shows the set of key performance indicators for the first level. Priority of the first level is the efficiency of operating activities which in the industry of scrap of ferrous and non-ferrous metals depends on the sales amount of scrap metal, production costs, fund of working hours and level of business expenses.

Table 1. The calculation form for managers at branch's level

Key performance indicator	Planned value	Actual value	Plan accomplished, %	Bonus size, rub	Bonus rate, %	Amount to be paid, rubs
Scrap metal sales, tons	x	x	x	x	X	x
Production costs, mln rubs	x	x	x	x	X	x
Fund of working hours per month, hours	x	x	x	x	X	x
Business expenses, mln rubs	x	x	x	x	X	x

Accomplishment of the plan is evaluated as a ratio of the actual and planned indicators. At accomplishment of the plan for each KPI the award in the fixed size is charged. The amount of payment for each indicator is defined by awarding coefficient which is defined, for example, so: if achievement of plan according KPI is more 95 an award will be paid, if actual value is lower than 80% of what planned, the award on this indicator will be not paid. All bonus payments are adjusted on a ratio of the number of days which are actually a manager worked to the number of the working days in current period (month, quarter).

At the second level – key performance indicators for heads of divisions (see *Table 2*). The cost efficiency is a priority of operational activities. Marginal income is the main incentive for division's leader.

Table 2. The calculation form for top managers at division's level

Key performance indicator	Planned value	Actual value	Plan accomplished, %	Bonus size, rub	Bonus rate, %	Amount to be paid, rubs
Scrap metal sales, mln rubs	x	x	x	x	X	x
Non-production costs, mln rubs	x	x	x	x	X	x
Productivity rate	x	x	x	x	X	x
Production costs, rub per a ton	x	x	x	x	X	x

Heads of functional divisions at office also have individual sets of KPI. As an example, here is a set of key performance indicators for a head of logistics department (see *Table 3*). The main incentive for him is the cost efficiency of motor transport.

Table 3. The calculation form for a head of logistics department

Key performance indicator	Planned value	Actual value	Plan accomplished, %	Bonus size, rub	Bonus rate, %	Amount to be paid, rubs
Expenses on all types of motor transport, per ton of scrap metal's shipment, rub	x	x	x	x	X	x
Specific prime cost of one machine-hour (ton-kilometer), rub	x	x	x	x	X	x
Share of expenses on motor transport in constant expenses, %	x	x	x	x	X	x
Fund of working hours in a month, hours	x	x	x	x	X	x

At the top level there are key performance indicators for the CEO and for the entire company. As a rule, these two levels in many respects coincide. The priority is the market value of a business. Examples of KPI are rate of annual increase in market value of a business, earnings per share, return on the capital invested, return on equity and others.

4 Discussion

4.1 Structure of Business Process "Calculation of Remuneration of a Manager"

The complexity of some approaches to top managers' material remuneration determines their high cost that hamper their efficacy. Another approaches to motivation have to be realized through complicated software application that significantly increases their cost and makes using unsuitable for small company.

Now consider the structure of business process "Calculation of remuneration of a manager" (see Fig. 4). The main stages are determination of planned values on the KPI set. It could be carried out by assessing the accomplishment of operation plans and budgets. In a process of operating activities, the top manager and accounting department keep track of the current results.

The head of department takes measures for accomplishment of the operation plan, accounting department make the plan-fact analysis according results of operational activities at the reporting period. According results of operational activities in the reporting period the set of key performance indicators for manager are filled and calculation of the lamenting remuneration is performed.

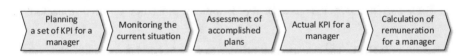

Fig. 4. Structure of business process "Calculation of remuneration of a manager" at RosMetTrade company

Heads of divisions and heads of branches get awards for long-term goals or special tasks which accomplishment is beyond planning period (month, quarter). Transporting of scrap of ferrous and non-ferrous metals from divisions of PAO "LUKOIL" located at the southern coast of the White Sea in extremely difficult geological and climatic conditions can be an example of accomplishment of a specific task. The separate action plan was developed for this task, comprehensive preparation was performed (motor transport, temporary points of storage, the vacation spot, crew, logistics and so forth). According outcomes of such task's accomplishment the managers, including foremen, earned a reward corresponding with results received.

4.2   Decisions Concerning Digitization of Business Process "Calculation of Remuneration of a Manager"

Digitization and automation of business process "Calculation of remuneration of a head" are possible at all its stages. At RosMetTrade company the main applied information systems used for digitization and business process automation are "Bitrix" and "1C: Enterprise". In the company the following decisions on digitization of calculation of material remuneration of a head are realized:

1. **Determination of the KPI's planned values.** Calculation of plans and budgets is performed by means of MS Office that saves the company from expensive decisions.
2. **Monitoring the current results and assessment of accomplishment of long-term goals** and specific tasks are realized in "Bitrix" software application. Also, by means of this application accounting of working hours is kept.
3. **The assessment of accomplishment of operation plans and budgets** is implemented by means of "1C: Enterprise" software application: forming of management reports about results of operational activities at the reporting period.
4. **Counting actual values of KPI for managers** is performed automatic by means of "1C: Enterprise" software application according accounting transactions for reporting period.
5. **Calculation of remuneration for managers** is automated partially by means of MS Office and "Bitrix" that allows even to novice users to use calculation results.

The business process automation will be deepening while RosMetTrade company's operating activities is expanding. Appropriate information technologies will be used for this purpose.

5   Conclusion

In the course of current research, the opportunity of developing and further digitization of the top managers motivation's system at RosMetTrade company was examined. Modern digital technologies purposed to digitalize the managerial business processes are able to increase cost and complexity of operations. Nevertheless, their use for top managers motivation's system could bring superior financial outcomes for a company.

In article it is shown that use of digital technologies does the mechanism of collecting of primary data accurate and consecutive, facilitates making decision on a form and methods stimulation of top managers. Also, the author reveals the use of KPI's sets for various levels of organizational hierarchy at RosMetTrade company. The method of calculation of an award for accomplishment of key indicators of activity is provided. The article presents the approach and the technique of digitization and automation of business process "Calculation of remuneration of a manager" which can be used in other companies operating in Russian industry of scrap of ferrous and non-ferrous metals.

Further researches will be conducted regarding use of cloud computing, network analytics, architectural approach and interfaces on the standards basis. Another aspects digitization and automation of business processes in a company will be investigated.

Acknowledgment. The work is carried out based on the task on fulfilment of government contractual work in the field of scientific activities as a part of base portion of the state task of the Ministry of Education and Science of the Russian Federation to Perm National Research Polytechnic University (topic # 26.6884.2017/8.9 "Sustainable development of urban areas and the improvement of the human environment").

References

1. Anderson, R.C., Bizjak, J.M.: An empirical examination of the role of the CEO and the compensation committee in structuring executive pay. J. Bank. Finan. **27**(7), 1323–1348 (2003)
2. Barontini, R., Bozzi, S.: Board compensation and ownership structure: empirical evidence for Italian listed companies. J. Manag. Gov. **15**(1), 59–89 (2011)
3. Coughlan, A.T., Sehmidt, R.M.: Executive compensation, management turnover and firm performance: an empirical investigation. J. Acc. Econ. **7**, 43–66 (1985)
4. Murphy, K.: Executive compensation. In: Ashenfelter, O., Card, D. (eds.) Handbook of Labor Economics, vol. 3, pp. 2486–2557. Elsevier, Amsterdam (1999)
5. Lin, Y.-L., Chen, Y.-J.: Direct and indirect effects of job complexity of senior managers on their compensation and operating performances. Adv. Intell. Syst. Comput. **773**, 812–821 (2019)
6. Mingaleva, Z., Bykova, E., Plotnikova, E.: Potential of the network concept for an assessment of organizational structure. Procedia-Soc. Behav. Sci. **81**, 126–130 (2013)
7. Jensen, M.C., Murphy, K.J.: CEO incentives: it's not how much, but how. Harvard Bus. Rev. **68**(3), 138–153 (1990)
8. Farid, M., Conte, V., Lazarus, H.: Toward a general model for executive compensation. J. Manag. Dev. **30**(1), 61–74 (2011)
9. Bąk, P., Michalak, A.: The problem of managers' remuneration in state-owned enterprises in the context of corporate governance. Gospodarka Surowcami Mineralnymi/Mineral. Resour. Manag. **34**(1), 155–174 (2018)
10. Jensen, M.C., Murphy, K.J.: Performance pay and top-management incentives. J. Polit. Econ. **98**(2), 225–264 (1990)
11. Frydman, C., Jenter, D.: CEO Compensation. Working Paper No. 77. Rock Center for Corporate Governance, Stanford University (2010)
12. Taylor, L.A.: Why are CEOs rarely fired? Evidence from structural estimation. J. Financ. **65**(6), 2051–2087 (2010)
13. Kaplan, S.N., Bernadette, A.M.: How has CEO turnover changed? Int. Rev. Financ. **12**(1), 57–87 (2012)
14. Warner, J.B., Watts, R.L., Wruck, K.H.: Stock prices and top management changes. J. Financ. Econ. **20**, 461–492 (1988)
15. Boschen, J.F., Duru, A., Gordon, L.A., Smith, K.J.: Accounting and stock price performance in dynamic CEO compensation arrangements. Acc. Rev. **78**(1), 143–168 (2003)
16. Conyon, M.J., Schwalbach, J.: Executive compensation: evidence from the UK and Germany. Long Range Plan. **33**, 504–526 (2000)

17. Chen, J., Ezzamel, M., Cai, Z.: Managerial power theory, tournament theory, and executive pay in China. J. Corp. Financ. **17**(4), 1176–1199 (2011)
18. Nannicini, A., Ferraz, D.P., Lopes, I.T.: Relationship between top executive compensation and corporate governance: evidence from large Italian listed companies. Int. J. Discl. Gov. **15**(4), 197–209 (2018)
19. Sakawa, H., Moriyama, K., Watanabel, N.: Relation between top executive compensation structure and corporate governance: evidence from Japanese public disclosed data. Corp. Gov. Int. Rev. **20**(6), 593–608 (2012)
20. Banker, R.D., Darrough, M.N., Huang, R., Plehn-Dujowich, J.M.: The relation between CEO compensation and past performance. Acc. Rev. **88**(1), 1–30 (2013)
21. Lopes, I.T., Ferraz, D.P.: The value of intangibles and diversity on boards looking towards economic future returns: evidence from non-financial Iberian business organizations. Int. J. Bus. Excellence **10**(3), 392–417 (2016)
22. Fogg, C.D.: Implementing Your Strategic Plan: How to Turn "Intent" Into Effective Action for Sustainable Change. BookSurge LLC, Charleston (2006)
23. Fey, C.F., Furu, P.: Top management incentive compensation and knowledge sharing in multinational corporations. Strateg. Manag. J. **29**, 1301–1323 (2008)
24. Bebchuk, L.A., Fried, J.M.: Executive compensation as an agency problem. J. Econ. Perspect. **17**(3), 71–92 (2003)
25. Bebchuk, L.A., Fried, J.M.: Pay without performance. Harvard University Press, Cambridge, MA (2006)
26. Bebchuk, L.A., Friedwalker, D.I.: Managerial power and rent extraction in the design of executive compensation. National Bureau of Economic Research Working Paper No. 9068 (2002)
27. Baxamusa, M.: The relationship between underinvestment, overinvestment and CEO's compensation. Rev. Pac. Basin Fin. Mark. Policies **15**(3), art. no. 1250014 (2012)
28. Sun, J., Cahan, S.: The effect of compensation committee quality on the association between CEO cash compensation and accounting performance. Corp. Gov. Int. Rev. **17**(2), 193–207 (2009)
29. Gomez-Mejia, L., Wiseman, R.M.: Reframing executive compensation: an assessment and outlook. J. Manag. **23**(3), 291–374 (1997)
30. Iwasaki, T., Otomasa, S., Shiiba, A., Shuto, A.: The role of accounting conservatism in executive compensation contracts. J. Bus. Fin. Acc. **45**(9–10), 1139–1163 (2018)
31. Danilina, E.I., Mingaleva, Z.A., Malikova, Y.I.: Strategic personnel management within innovational development of companies. J. Adv. Res. Law Econ. **5**(19), 1004–1013 (2016)
32. Esaulova, I., Semenova, I.: The impact of the leadership on the employee proactivity. In: 26th Interdisciplinary Information Management Talks, Strategic Modeling in Management, Economy and Society, pp. 489–496. Kutná Hora, Czech Republic (2018)
33. Kirillova, M.M., Soboleva, E.V.: Features of motivation of top managers. Topical Issues Mod. Sci. **1**, 75–81 (2013)
34. Levanova, L.N.: Features of motivation of upper managers in Russia. News of the Saratov University. New series. Seri.: Econ. Manag. Right **17**(1), 50–55 (2017)
35. Ruslom agency. https://ruslom.com/wp-content/uploads/2019/02/spravka-po-rynku-loma-2018.pdf. Accessed 12 Mar 2019

Models and Methods of Identification of Threats Related to the Uncontrollability of Capital Flows

Gennady Ross[1]([envelope]) and Valery Konyavsky[2]([envelope]) [ORCID]

[1] Financial University, Moscow, Russia
ross-49@mail.ru
[2] Plekhanov Russian University of Economics, Moscow, Russia
konyavskiy@gospochta.ru

Abstract. We have formulated and formalized the task of managing capital flows in order to identify threats to economic agents (EA). The class of capital which a particular EA operates has been determined. It has been shown that some of the threats to economic agents are due to the fact that capital flows can be illegal and pursue illegal goals: legalization of capital acquired by illegal means (money laundering), withdrawal of money through offshore companies, and financing of terrorism. The characteristics of business transparency and financial attractiveness of EA, as well as methodical approaches to assessing the transparency of their activities have been identified.

Keywords: Financial and economic security · Capital flow control ·
The theory of equilibrium random processes ·
Evolutionary-simulation methodology · The decision support system

1 Introduction

Problems of the adequacy of existing and new indicators of threats to economic agents are associated with overcoming uncertainties and inaccuracies. One of the basic means to solve this problem, as shown in [1–4, 7], is the use of economic and mathematical methods, particularly evolutionarily simulative methodology and the decision support system. The article discusses the use of evolutionary-simulation models of capital flow implemented in the framework of the decision system to identify threats to economic agents. The uncontrollability of capital flows is the cause of numerous threats at the local and global levels for all the economic agents. The content of these problems is analyzed in detail in [3, 9]. The content of the threats posed by the uncontrollability of capital flows is revealed in Theorems 1–9 in [5, 9].

1.1 Setting the Task of Managing Capital Flows

The main ideas of capital flow modeling are formulated in [5, 9]. There are classes of capital, in particular, the capital which a particular economic agent operates, or the capital used for a certain type of operations, for example, shareholders' equity, production capital, financial capital, etc.

© Springer Nature Switzerland AG 2020
T. Antipova (Ed.): ICIS 2019, LNNS 78, pp. 242–250, 2020.
https://doi.org/10.1007/978-3-030-22493-6_22

Let $j = 1, ..., J$ be the number of the capital class. Among the factors that determine the characteristics of any class of capital, you can distinguish the costs associated with the capital transaction ($f_{1,j}$), operating profit ($f_{2,j}$), the number of operations per unit of time ($f_{3,j}$). To write down the simplest evolutionary-simulative capital class model, we introduce the following quantities:

Fa_j is the expected demand for class j capital;
PL_j is the equilibrium volume of class j capital;
C_j is profit per monetary unit of investments (price) of class j capital;
S_j is the fee for the use of financing sources when using class j capital (cost price)

An evolutionary-simulative capital class model can be represented by a combination of relations (1)–(5):

$$Fa_j = \rho_j \left(f_{1,j}, f_{2,j}, f_{3,j} \right) \tag{1}$$

$$F_{1,j} = S_j \left(PL_j - Fa_j \right), \, PL_j \geq Fa_j \tag{2}$$

$$F_{2,j} = \left(C_j - S_j \right) \left(Fa_j - PL_j \right), \, PL_j < Fa_j \tag{3}$$

$$\min_{PL_j} \left\{ \max_{\chi \in \{1, 2\}} \left\{ M\{F_{\chi,j}\} \right\} \right\} \tag{4}$$

$$P_j^0 = P\left(PL_j \geq Fa_j \right) \tag{5}$$

Wherein:

- ρ_j is the imitational model of demand for class j capital;
- $F_{1,j}$ stands for the costs of overstatement with excessive irrational investments;
- $F_{2,j}$ stands for the cost of understatement with a lack of demand;
- $M\{F_{\chi,j}\}$ is the expectation of the overstatement costs (the risk of overstatement) when $\chi = 1$ and understatement costs (risk of understatement) when $\chi = 2$;
- P_j^0 is the probability that the supply will exceed the demand;
- $3/3_j = \frac{S_j}{C_j - S_j}$ is the ratio of the risk of overpricing to the risk of understatement.

Let us call the set of values $\{PL_j, 3/3_j, P_j^0\}$ the main characteristics of class j. From theorem 3 in [5] it follows that in the presence of 2 capital classes, namely j and j' and $3/3_j > 3/3_{j'}$, there occurs a flow from class j to class j' in the volume $V_{j, j'} = PL_j - PL'_j$, where PL'_j is calculated according to model (1)–(5) for class j with the substitution $3/3_j = 3/3_{j'}$. This statement is true, if capital j is absolutely liquid, capital j' is able to absorb this volume instantly and the capital movement from j to j' is not associated with losses.

The necessary refinement has been made in [6] to more adequately reflect the capital flow. It is important that the $3/3_j$ indicator expresses in a complex not only economic, but also psychological peculiarities of the behavior of market entities.

To simulate the flow of capital, a graph G is constructed, where there is a vertex for each class of capital j. In graph theory, the *momentum at the vertex* at the moment is usually understood to mean the change of the parameter at that vertex at the moment. *A pulse process* is a change in the parameters at the vertices that occurs as a result of the transfer of a pulse along the arcs of the graph. Considering the capital flow through this peak as an impulse we assume that there is a connection between the peaks if the probability of the impulse passing between them is not negligible. In general, therefore, the capital flow model appears as a flow model on a graph.

1.2   Threats to Economic Agents Generated by Money Laundering and Terrorist Financing

Some of the threats of economic agents are related to the fact that capital flows can be partly illegal and pursue illegal goals: money laundering, money withdrawal through offshore companies, and terrorism financing. To identify these threats, it is necessary to take into account the specifics of illegal financial flows in the evolutionary simulation model (1)–(5).

Let us consider illegally laundered money as one of the capital classes, and operations with this capital as one of the sectors of the market in which we distinguish objects of exchange (goods and money) and subjects (economic agents). Herewith, for money, we take any financial assets, including money in any form (paper, electronic, debt warrant).

The Federal Financial Monitoring Service counteracts the legalization of EA's criminal income, which significantly hinders illegal operations. At the same time, the results of the Service activity indirectly affect the prices on the market of illegal incomes, which makes it less profitable compared to the legal movement of capital. The following factors affect the prices at the market of illegal money:

– the risk of possible detection of a crime scheme;
– the level of criminal responsibility for the crime;
– money laundering costs;
– the expenses that increase the cost of criminal operations.

The price of legalization of criminal capital is determined by the total of all risks, costs and profits of each participant in the criminal chain. The money laundering market is based on price, demand, supply and competition. In the money laundering market, an increase in the price while strengthening the state system of counteracting these crimes forces criminals to improve illegal schemes, or leads to higher prices. Demand in the money laundering market represents the need for legalizing criminal capital, in which, the higher the liquidity of "dirty" money, the lower the demand for money laundering, and vice versa.

The offer in this market is the amount of criminal capital that can be legalized at a given price for this service. The change in the ratio between supply and demand leads to fluctuations, which result in the establishment of the market equilibrium price and equilibrium quantity.

The purpose of illegal business operations is the same as legal ones, to maximize profits. This goal is achieved in a competitive environment among criminals working in the field of money laundering.

The mechanism by which the improvement of legislation and other forms of regulation leads to a reduction in money laundering is as follows. If the increase in risks reduces demand, the situation when supply exceeds demand arises. In a competitive environment, this forces a price reduction, which leads to the dropout of some of those who provide criminal services. As a result, the market contracts.

Through such a mechanism, both the measures increasing the risks and the cost of supply of criminal services lead to the market contraction.

However, the most difficult problem is to develop such a combination of measures to prevent money laundering, which achieves the goal without worsening legal activities. One of the central tasks of identifying threats to the economic security of economic agents is to identify the consequences of certain legislative, organizational, and other initiatives for legal and illegal business. This is the task that should be solved with the use of the capital flows model and the inclusion of the capital class conventionally called money laundering in this model. Let us turn to the construction of an evolutionary- simulative model of this capital class. In [3], an attempt has been made to explain the behavior of a criminal with purely economic motives, namely, on the basis of an assessment of costs and benefits. In this paper we propose the following formula for calculating the expected utility of criminal activity:

$$EU = (1 - p)U(Wi) + pU(Wi - F) = U(Wi - pF) \tag{6}$$

where

EU is the criminal's expected utility of the crime;

Wi is the profit from a crime (commission for cash withdrawal, or payment for fictitious contracts);

U is a function expressing the utility of a crime;

p is the probability that the crime will be solved;

F is the amount of lost profits from the punishment (the severity of the punishment).

Formula (6) can be used to calculate the overpayment costs (3), which is the basis for the development of an evolutionary simulation model of the class of capital under consideration.

In [3, 9] it is argued that the commission of crimes is more controlled by the probability of exposure than the severity of the punishment, and the one who has already chosen the criminal path, on the contrary, is more restrained by the expected severity of the punishment.

Possible governmental measures to prevent money laundering are rather diverse and are applied to a greater or lesser extent in practice. They include the following:

(1) strengthening personal responsibility of employees of credit institutions for compliance with the requirements of legislation in the field of money laundering and financial terrorism (ML/FT);

(2) strengthening supervision and sanctions for violation of the law in the field of ML/FT;
(3) strengthening administrative and criminal liability for ML/FT;
(4) automation of control of credit institutions;
(5) formation of a unified information and reference system on ML/FT and persons involved in the legalization of criminal proceeds.

1.3   Characteristics of Business Transparency and Financial Attractiveness of Economic Agents

Since the class of capital, on the one hand, is tied to economic agents and, on the other hand, it is associated with the top of the graph, one of the important characteristics can be the one we call "Transparency of business" and denoted by Pr. By Pr we mean that the probability that an economic agent and, therefore, the capital it controls, or capital circulating in a certain sector of the market, is legitimate.

Based on the Pr indicator, methodical approaches can be proposed for solving the following complex of interrelated and mutually additional tasks:

– identification of the most likely money laundering schemes;
– identification of the least transparent and powerful illegitimate channels of capital flow;
– identification and forecasting of threats to economic agents, generated by the lack of transparency in their activities or in the classes of capital;
– development of information systems for identifying threats from insufficient transparency;
– development of methodological approaches to the management of transparency and investment attractiveness of economic agents.

If we mean not the categories of economic agents, of which are only 4 altogether, but real agents, in particular, households and firms, their number is enormous. Considering that in order to calculate the Pr value, it is necessary to involve statistical and expert information, it becomes clear that to study any significant sets of economic agents it is necessary to have a fairly simple, low-cost, automated information system for calculating Pr.

In Automated Information Systems (AIS), for a characteristic of economic agents the following indicators are taken: fixed assets; working capital; the number of incoming and outgoing financial transfers; the frequency of remittances during the day; the number of transfers of sums below the threshold value; the number of large amounts transferred; the number of operations from offshore countries against which sanctions have been imposed; the number of transactions with foreign entities. The indicators of different categories of agents from the perspective of business transparency include [10]: signs of a high-risk state, signs of firm fictitiousness; the use of offshore money laundering operations; foreign economic operations of laundering or illegal compensation of VAT and reinsurance of funds for the purposes of money laundering.

Based on these data, various procedures can be developed with or without the participation of experts to calculate the Pr indicator. Indicators of different categories of agents in terms of business transparency are measured in points, which can be estimated in an expert manner if there is a scale. They can also be rated by assigning a "Yes /No" tag. It is assumed that if the expert indicates the value "Yes", the corresponding indicator is assigned the value 1, if "No", then 0.

1.4   Methodical Approaches to Assessing the Transparency of an Economic Agent

To calculate Pr, it is necessary to develop ways of estimating the values of indicators K_j^i, $\forall i$, $\forall j$ and methods for calculating Pr on the basis of these indicators. Let us turn to the consideration of the main methodological approaches for obtaining estimates of indicators K_j^i, $\forall i$, $\forall j$ and Pr.

A taxonomy is the decision making by choosing from a finite set of options based on precedents. Taxonomy methods allow classifying multidimensional observations described by a set of variables. The goal is to form clusters (sets). For classification, various measures of proximity or similarity are applied. One of such measures may be distance in hyperspace. Taxonomy allows you to select areas of condensation in multidimensional space. With the usage of taxonomy methods, an approach is possible based on identification of a number of economic agents for which there is sufficient direct data for calculating Pr (for example, based on judicial statistics). Consider the hyperspace with coordinates K_j^i, $\forall i$, $\forall j$ and assign appropriate Pr values to similar economic agents.

Another use of taxonomy can be based on ranking. There are various ranking algorithms. For example, the economic agent can be considered the least corrupt if it is better than any other according to the following majority rule: it has more inequalities of the type $K_{j,n}^i < K_{j,n'}^i$ (i.e., the greatest number of the best qualities) than other EA, where n and n' are the index numbers of economic agents. If the best economic agent is assigned $Pr = 1$, the transparency level of any economic agent can be normalized.

The elements of utility theory can be used to quantify already ordered transparency estimates. If $U(Wi)$ denotes the utility of the set of characteristics Wi, $i = 1, ..., m$ from the point of view of their use for evaluating the transparency Pr of an economic agent, and if Wi is preferable, than Wi' i.e. $W_i > W_{i'}$, for the utility function the following should be true by construction: $U(W_i) > U(W_{i'})$.

A special and simplified case of taxonomy is the method of standards, the feature of which is, on the one hand, the possibility of its use for the problems considered above and, on the other hand, that it provides a Pareto optimal solution. For multi-criteria tasks, this means that the improvement of a solution based on one of their criteria will necessarily entail a deterioration on some other.

The reference method is based on the fact that a reference object (or solution) is formed, that is, the best (or worst) one by all the criteria, regardless of whether such an object exists in the search area. Then, in the search area, an object is found, which in the space of the criteria is located at an extreme distance from the reference.

To automate the process of collecting information, to minimize the labor of experts, as well as to improve the adaptation of the technology for collecting and processing data for specific categories of economic agents, for example, to take into account the specialization of firms, their geographical location, their legal status, it is important to select K_j^i, $\forall i$, $\forall j$ corresponding to limited sets from all the characteristics.

For example, if it is necessary to examine some specific category of economic agents, you can choose the appropriate set of the most significant characteristics in 2 stages: at the 1st stage, some redundant, approximate set of samples W_1, ..., W_m is formed from the characteristics K_j^i, $\forall i$, $\forall j$ (i.e., each sample W_i contains only a part of the characteristics from the sample, and W_1, ..., W_m do not intersect), while m is not greater than 9; at the 2nd stage, among W_1, ..., W_m, several (not more than 3) of the most significant ones are selected. Only the characteristics of economic agents included in their association are used to obtain estimates of the transparency of economic agents.

Binary relations language can be used to obtain the most significant samples. Decision making technology based on the language of binary relations is based on pairwise comparison of alternatives. Herewith:

1. No quantitative characteristics of the preferences of alternatives W_i and $W_{i'}$ are required, in other words, no quantitative estimates of alternatives are required. It is enough to be able to say that W_i is better or worse than $W_{i'}$.
2. For each pair of alternatives W_i and $W_{i'}$ one of the 3 is true:
   (a) one alternative is preferable to the other;
   (b) alternatives are equivalent;
   (c) alternatives are incomparable.
3. The relations of preferences for any pair do not depend on the others.

There are 4 ways to set binary relationships: by a direct listing of pairs; using a matrix; in the form of a graph; task with sections.

There are fairly simple and efficient algorithms for solving ranking and classification problems [8].

One of the widely used methods of collective decision-making is voting. With the help of voting, you can organize the elements of a set without quantifying each element. It is assumed that there are electors and each of them is able to order the elements according to their own preferences. With the help of the voting procedure, it is possible to organize the elements, to some extent taking into account or coordinating individual preferences.

We explain the ways of summarizing the total vote using a conditional example. Suppose there is an array of economic agents, consisting of 3 EA: a, b, c. The record $a \rangle_i b \rangle_i c$ means that the i-expert believes that EA is better EA, EA than EA b, and EA b is better than EA c. Suppose there are Q experts. Suppose the number of experts is $|Q| = 13$ and experts can join in coalitions. The results of the experts' voting of each coalition for assessing the transparency of business for all the three EA are summarized in Table 1, in which:

- The 1st coalition of 2 experts established the preferences: a > b > c, which correspond to column 1;
- The 2nd coalition of 3 experts set the preferences: c > b > a, which corresponds to column 2;

Table 1. Voting table

	Coalition №1 (2 experts)	Coalition №2 (3 experts)	Coalition №3 (4 experts)	Coalition №4 (4 experts)
1	a	c	a	b
2	b	b	c	c
3	c	a	b	a

- The 3rd coalition of 4 experts set preferences: a > c > b, which correspond to column 3;
- The 4th coalition of 4 experts set preferences: b > c > a, which correspond to column 4;

In Table 1, the columns correspond to coalitions of 2, 3, 4, and 4 experts, respectively, and the row number is determined by the ratings (preferences) of experts to economic agents in these coalitions.

Summarizing by the rule of relative majority is implemented as follows. The number of votes that received the highest rating in each coalition (first row) is calculated and the option with the highest number of votes wins. The voting results given in Table 8 are interpreted as follows:

- EA a has 6 votes, as occurs twice in the first row (i.e., in coalition 1 it was given the 1st place by 2 experts, and in a coalition №3 4 experts, the total number of votes equals $2 + 4 = 6$);
- EA b has 4 votes, since it was assigned the 1st place only in coalition 4, which corresponds to the 2nd place;
- EA c has 3 votes, since coalition 2 gave it the 1st place, which corresponds to the 3rd place.

As a result, the highest rating of "business transparency" has EA a, the one following is EA b, and the business of EA c has the lowest rating, which may entail a more thorough verification of this business.

2 Conclusions

Based on the proposed integrated approach to capital outflows managing, it is possible to more accurately identify threats and identify the level of capital legitimacy based on an assessment of the "transparency of business" of economic agents. A significant part of the threats of economic agents is related to the fact that capital flows can be illegal and pursue illegal goals: money laundering, money transfer to offshore companies, and financial terrorism. To identify these threats the following measures are designed:

- evolutionary-simulation model, which allows to take into account the specifics of illegal financial flows;
- a model explaining the behavior of the criminal with purely economic motives, namely, based on the assessment of costs and benefits;

- a method for calculating the new characteristic of "business transparency" of an economic agent, which determines the level of legitimacy of capital of the economic agents.

Based on the indicator "business transparency" of economic agents we have proposed methodological approaches to address such tasks as identifying the most likely money laundering schemes and identifying the least transparent and powerful illegitimate channels of capital flow, as well as identifying and predicting threats to economic agents generated by the lack of transparency of economic agents or classes of capital.

A modern feature of capital flows is the digitization of the economy. It means that the above analyzed indicators, incl. of "business transparency" of economic agents, supplement methodical approaches to solve such tasks as identifying the most likely money laundering schemes and identifying the least transparent and powerful illegal capital flow channels, and identification of threats can be used for the digital economy.

References

1. Abalkin, L., et al.: Russia's economic security: threats and their reflection. Questions of the economy (1994). N 12
2. Burtsev, V.V.: Factors of financial security of Russia. Management in Russia and abroad (2001). N 1
3. Kaurova, N.N.: Financial and economic security in the conditions of openness of the national economy (theoretical and methodological aspect), Moscow (2013)
4. Senchagov, V.K.: Economic Security of Russia, 2nd edn. Moscow (2010)
5. Liechtenstein, V.E., Ross, G.V.: Equilibrium random processes: theory, practice, info business. Finance and Statistics, p. 423 (2015)
6. Avdiysky, V.I., Bezdenezhny, V.M., Liechtenstein V.E., Ross, G.V.: Economic justice and security of economic agents. Finance and Statistics, p. 272. Moscow (2016)
7. Liechtenstein, L., Ross, G.: Management of financial bubbles as control technology of digital economy. In: Antipova, T., Rocha, Á. (eds.) Information Technology Science, MOSITS 2017. Advances in Intelligent Systems and Computing, vol. 724. Springer, Cham (2018)
8. Ross, G.V.: Modeling of Production and Socio-Economic Systems Using the Apparatus of Combinatorial Mathematics, p. 303. Moscow, Mir (2001)
9. Zolotarev, E.V.: Improving the system of combating money laundering and control mechanisms in credit institution. PhD Thesis, Moscow (2014)
10. Information letter of Rosfinmonitoring dated 23.11.2018 No. 56 "On methodological recommendations for review by audit organizations and individual auditors when providing audit services to the risks of legalization (laundering) of criminally obtained income and financing of terrorism" (ML / FT Risks). http://www.consultant.ru/document/cons_doc_LAW_311943/45f1fce0324f79354ce78e13fb9ee2f676edb7e1

Matrix Planning for the Development
of Enterprises of the Fuel and Energy Complex

Kirill Litvinsky[1]([⊠]) [iD] and Elena Aretova[2] [iD]

[1] Kuban State University, Krasnodar 350040, Russia
litvinsky@econ.kubsu.ru
[2] Kuban State University, Stavropol'skaya St., 149,, Krasnodar 350040, Russia

Abstract. Due to the increasing role of hydrocarbon resources in the development of the Russian economic system, as well as taking into account the need to study the economic processes taking place in modern enterprises of the fuel and energy complex, the problem of modeling their activities, with the involvement of the mathematical apparatus of matrix planning, becomes now while an even more significant. Formulated in the article the concept of modeling the economy of modern enterprises of the complex, involving the mathematical apparatus of matrix planning. An algorithm for synthesis of economic-mathematical model of the complex's companies in the market conditions based on the matrix-differential equations. Defined by the differential equations of the matrix planning in the development of modern economy the complex. The results of studies of the relationship of factors of production, demand and consumer criteria on the example of the complex companies. In the study, the authors concluded that the use of differential equations planning matrix allows one side of the energy companies to more accurately produce the strategy and tactics of its development, and on the other - to identify options for dynamic interactions Energy with related industries.

Keywords: Fuel and energy complex · Matrix-differential equations ·
Total production per cycle of capital investment
Dependence of the level of consumption of time

1 Introduction

The aim of this study is the economic and mathematical modeling of the dynamics of the FEC using matrix-differential equations. The choice of the mathematical formalism in this research as a tool to describe the economy of production FEC allows to analyze the state of the industry in the preceding period of time, gives an estimate of the state of affairs at the moment, and also determines the prospects for the development and production of FEC in market conditions. Identified two economic situations, depending on the dynamics of change and the degree of satisfaction of demand; the level of total production in view of capital investments, as well as the level of consumption of time, which allows more efficient to simulate the development of the energy sector companies.

© Springer Nature Switzerland AG 2020
T. Antipova (Ed.): ICIS 2019, LNNS 78, pp. 251–262, 2020.
https://doi.org/10.1007/978-3-030-22493-6_23

1.1   Analytical Description of the Economy by Price Matrices: Foreign Market Case

In the context of fuel and energy complex (FEC) enterprises we will manipulate the economic numbers [3, 4, 9–11] of production in a single abstract enterprise. The description of development dynamics and main principles of economy of FEC production [12, 17, 19] will be carried out by differential Eqs. (5, 14) in accordance with the algorithm (see Fig. 1).

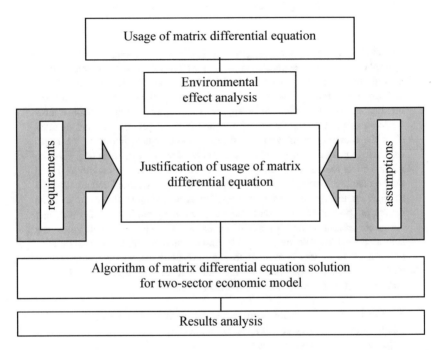

Fig. 1. Algorithm of synthesis of economic-mathematical model of FEC companies development on basis of matrix differential equations

It should be noted herein that the choice of mathematical apparatus as a tool to describe the economy of FEC production allows to analyze the condition of the industry in the preceding period of time, to estimate the present condition, and also to determine the prospects for FEC development and production in the market economy environment [1, 2, 6, 8, 13, 15, 16, 20].

Let us consider the matrix dynamic Eqs. [7, 18] of change in the level of gross i-th production in the j-th production of one weight category per cycle of capital investment:

$$y_i^{(t)} - A_{ij}^{(t)} - k_{ij}^{(t+1)}\left(y_j^{(t+1)} - y_i^{(t)}\right) = c_j^{(t)} \tag{1}$$

y: the matrix of the level of gross output per cycle of capital investment; *k:* capital investment matrix; *A:* technological coefficients matrix (input matrix); *C:* consumption level matrix; *t:* period (number of capital investment cycles).

If we introduce the identity matrix *I*, all elements of which (except diagonal) are equal to zero, and diagonal elements are equal to one, then the Eq. (1) can be represented in matric form as follows:

$$(1 - A)\bar{Y} - K \times \varDelta\bar{Y} - C = 0 \tag{2}$$

here $\varDelta Y$ is the difference in the level of gross production per cycle of capital investment *K*.

If we take the cycle of capital investment as a unit of time, the $\varDelta Y$ can be considered as the rate of behavior of the level of gross production in time: $\varDelta\bar{Y} = d\bar{Y}/dt$

In this case, the equation of the difference in the level of gross production per cycle of capital investment can be transformed to matrix differential equation:

$$K \times \frac{d\bar{Y}}{dt} - (1 - A)\bar{Y} + \bar{C} = 0 \tag{3}$$

Let us multiply this Eq. (3) by inverse matrix K^{-1}:

$$\frac{d\bar{Y}}{dt} - K^{-1}(1 - A)\bar{Y} + K^{-1}\bar{C} = 0 \tag{4}$$

The Eq. (4) includes *N* of differential equations for each of *N* of FEC linked industry.

Those equations intertwined by off-diagonal elements of technological (technical-and-economic) matrix *A* and capital investment matrix *K*, i.e. by mutual production costs.

The solution of matrix differential Eq. (4) is as follows. We introduce the designation of a new vector - the column *Z* of the matrix *Y* with an exponent function e^{Qt} of the matrix argument *Q*, then:

$$\bar{Y} = e^{Qt}\bar{Z}; \quad Q = K^{-1}(1 - A) \tag{5}$$

here *Z* is new variable.

Then the Eq. (4) can be transformed by inserting the formula (5) into Eq. (4) and differentiating the exponent function of the matrix argument *Q*.

There are several cases of dynamic interactions of FEC with linked industries [4, 9, 11]. The first case is realized when the parameters of the technological matrix *A* and the matrix of capital investments *K* are constant values and do not depend on time. The second case considers the components of Eq. (4) as variable values.

The implementation of these two cases is determined by the duration of the forecast for the development of the economy of FEC production. In the short-term intervals, the first case of dynamic interactions can be realized. In the long-term perspective, we should take into account changes in production technology over time: a decrease in production costs, the creation of new types of products, etc.

In the first case of dynamic interactions, only two values change: level of gross production and the level of consumption. The change in the level of gross production is carried out only by means of the capital investments cost K.

Let us differentiate the Eq. (4) by time. To do this, it is necessary to decide how to differentiate the exponential function of the matrix argument Q. The exponential function of the matrix argument can be expanded into a series.

An operation of division in the matrix form is replaced to multiplication operation by the inverse matrix. When multiplying matrices, it required to follow the order of multiplication as in matrices multiplication the permutation law of the factors does not work.

In the case of matrix multiplication, there is a combination law that is carried out without changing the order of multiplication of matrices. The powers of matrices are defined as the product of the matrix itself for itself, for which the law of permutation of the factors works. Therefore, the power series expansion of the exponential function of the ordinary argument and exponential function of the matrix argument are the same. So the exponential function of the matrix argument can be expanded into a power series:

$$e^{Qt} = E + \frac{t}{1!}Q + \frac{t^2}{2!}Q^2 + \frac{t^3}{3!}Q^3 + \dots$$

and inverse technological matrix can be expanded into a power series:

$$(1 - A)^{-1} = \sum_{p=0}^{\infty} A^p$$

here E is zero-unit matrix,

The conditions of this expansion are reduced to the fact that the absolute value of the eigenvalues of the matrix A is less than unity. This condition is satisfied for most technological matrices A, since the coefficients of matrix A are positive and less than unity. The coefficients of matrix A determine the costs of production and must be positive values and their values must be less than unity, because these costs should be less than the level of the total production of FEC.

Since the sum of the eigenvalues of the matrix is equal to the sum of the diagonal elements, the eigenvalues of the technological matrix A are also less than unity. However, the exact fulfillment of this condition is determined by the value of the off-diagonal elements of the matrix A and must be checked additionally. The differentiation of matrices is carried out in the usual way, as the differentiation of all components of the matrix.

After differentiating the exponential function of the matrix argument, the derivative of the column vector of the level of gross production will be:

$$\frac{d\bar{Y}}{dt} = Q \times e^{Qt} \times \bar{Z} + e^{Qt} \times \frac{d\bar{Z}}{dt}$$

The Eq. (4) can be transformed to the following equation:

$$\frac{d\bar{Z}}{dt} = -\left[K^{-1}\bar{C}\right]e^{-Qt} = F(t)e^{-Qt}; \quad F(t) = -\left[K^{-1}\bar{C}(t)\right] \tag{6}$$

here F(t) is column vector of consumption matrix C.

The solution of Eq. (6) shall be found in the usual way:

$$\bar{Z}(t) = \bar{q} + \int_{t_o}^{t} e^{-Q\tau}\bar{F}(\tau)d\tau, \tag{7}$$

here q is the integration constant determined from the initial condition at $t = t_0$, τ is the time since the start of capital investments K.

Taking into consideration the equality (7) the solution for level of gross production is:

$$\bar{Y}(t) = e^{Qt}\left[\bar{q} + \int_{t_o}^{t} e^{-Q\tau}\bar{F}(\tau)d\tau\right] = e^{Qt}\bar{q} + \int_{t_o}^{t} e^{-Q(t-\tau)}\bar{F}(\tau)d\tau \tag{8}$$

The initial conditions can be taken as the start of the zero cycle of capital investments K:

$$\bar{q} = e^{-Qt_o}\,\bar{Y}_0 \tag{9}$$

Then the solution to Eq. (8) will be:

$$\bar{Y}(t) = e^{Q(t-t_o)}\bar{Y}_0 + \int_{t_o}^{t} e^{Q(t-\tau)}\bar{F}(\tau)d\tau. \tag{10}$$

The integral in formula (10) determines the total costs for consumption over the entire period of capital investments K, from moment t_0 to moment t. These costs depend on the consumption column vector (C): $F(t) = -\left[K^{-1}\bar{C}(t)\right]$ and its time dependence.

1.2 Dependence of Consumption Level on Time with Constant Demand for FEC Products

If the demand level in the process of capital investments K does not change, then we assume that $C^{(t)} = C$ and the equality (10) will be:

$$\bar{Y}(t) = e^{Q(t-t_o)}\bar{Y}_0 + \left[\int_{t_o}^{t} e^{Q(t-\tau)}d\tau\right] \times \bar{F}. \tag{11}$$

Now we need to calculate the integral of the exponential function of the matrix argument Q. We use the relation between the integral and the differential:

$$\frac{d}{dt}\left[\int_{t_o}^{t} e^{Q(t-\tau)}d\tau\right] = e^{Q(t-\tau)}$$

Since the derivative of an exponential function with a matrix argument is equal to the exponential function itself, multiplied from the left by the matrix of the argument, it can be stated that the integral is equal to the exponential function multiplied by the inverse matrix from the left. Hence the following equality is true:

$$\frac{d}{dt}\left(Q^{-1} \times e^{Qt}\right) = e^{Qt}$$

Then the solution of the differential equation will be as follows:

$$\bar{Y}(t) = e^{Q(t-to)}\bar{Y}_0 - Q^{-1} \times e^{Q(t-\tau)}\Big|_{t}^{t_o} \times \bar{F} \tag{12}$$

Substituting the integration limits into this Eq. (12), we obtain:

$$\bar{Y}(t) = e^{Q(t-t_o)}\bar{Y}_0 - Q^{-1} \times e^{Q(t-\tau)}\Big|_{t}^{t_o} \times \bar{F}. \tag{13}$$

The exponential function of the matrix argument Q commutes with the matrix Q^{-1}. Indeed, an exponential function can be expanded in a series in powers of Q^n, but any matrix commutes with its inverse matrix. Therefore, e^{Qt} can be interchanged with Q^{-1}, merge with the first component (13) and get the ratio:

$$\bar{Y}(t) = e^{Q(t-to)}\left(\bar{Y}_0 + Q^{-1} \times \bar{F}\right) - Q^{-1} \times \bar{F} \tag{14}$$

The inverse matrix of Q is equal to the product of inverse matrices taken in the reverse order:

$$Q = K^{-1}(1 - A); \quad Q^{-1} = (1 - A)^{-1}K.$$

Consider now the product of two matrices:

$$\begin{aligned} Q^{-1} &= (1 - A)K^{-1}; \\ F(t) &= -\left[K^{-1}\bar{C}(t)\right]; \\ Q^{-1} \times F(t) &= -(1 - A)^{-1}K \times \left[K^{-1}\bar{C}(t)\right] \end{aligned} \tag{15}$$

Opening the brackets in this equality (15), we get:

$$Q^{-1} \times F(t) = -(1 - A)^{-1}\bar{C}(t) \tag{16}$$

Substituting (14) into (16), we get:

$$\bar{Y}(t) = e^{Q(t-t_o)}\left[\bar{Y}_0 - (1-A)^{-1}\bar{C}\right] + (1-A)^{-1}\bar{C} \qquad (17)$$

Solution (17) satisfies the initial condition: $t = t_0 \quad \rightarrow \quad Y = Y_0$

The value of $(1-A)^{-1}\bar{C}$ is equal to the level of gross production in the absence of capital investments. Therefore, the value of $\bar{Y}_0 - (1-A)\bar{C}$ determines the level of capital investments K. The contribution of capital investments K to the level of gross production depends exponentially on time, and the decrement of the time dependence is the matrix Q. Therefore, the next task is calculating an exponential function with the matrix argument Q.

2 The Decrement of the Time Dependence of the Level of Gross Production of FEC Enterprise

2.1 Initial Section of the Time Dependence

The exponent $e^{Q(t-t_o)}$ for the initial sections of the time dependence can be expanded in a series in powers $t - t_0$:

$$e^{Q(t-t_o)} = E + \frac{(t-t_0)}{1!}Q + \frac{(t-t_0)^2}{2!}Q^2 + \dots,$$

here E is the identity matrix and the matrix $Q = K^{-1}(1-A)$.

The equation for level of gross production will be:

$$\bar{Y}(t) = \left[E + \frac{(t-t_0)}{1!}Q + \frac{(t-t_0)^2}{2!}Q^2 + \dots\right] \times \left(\bar{Y}_0 - (1-A)^{-1}\bar{C}\right)$$

$$+ (1-A)^{-1}\bar{C} = \bar{Y}(t) + \left[\frac{(t-t_0)}{1!}Q + \frac{(t-t_0)^2}{2!}Q^2 + \dots\right] \qquad (18)$$

$$\left(\bar{Y}_0 - (1-A)^{-1}\bar{C}\right)$$

If the initial level of gross production is equal to its static value $\bar{Y}_0 - (1-A)^{-1}\bar{C}$, then no change in the level of gross production occurs, i.e. $Y(t) = Y_0$ and $\Delta Y = 0$. It is necessary for the initial level of gross production to exceed the static level, only then an increment or change in the level of gross production will occur. This means that it is necessary to create capital assets that are not supplied on demand to consumers, but are invested in the next FEC production cycle.

2.2   Clarification of Initial Conditions of the Level of Gross Production of FEC

The definition of the initial condition for solving a differential equation was made formally and requires clarification.

If the entire level of gross production (Y_0) after deduction of costs (AY_0) goes to consumers (C), then there will be no increase in production. It is required that a part of the FEC products shall not be delivered to consumers creating an excess, which goes to increase production. This follows from the definition of the derivative of the level of gross production in time at the initial moment when $t = t_0$:

$$K \frac{d\bar{Y}}{dt} |t = t_o = (1 - A)\bar{Y}_0 - \bar{C}. \tag{19}$$

If the right-hand side of (19) is zero, then the derivative of the level of gross production is zero. The capital investment tensor K is the coefficient of proportionality between the growth rate of the level of gross production and the difference in FEC production and supply levels.

To determine the initial level of gross production Y_0, let's compare (19) with the change in the level of gross production after one cycle of capital investments K, obtained from the analysis of the dynamic matrix:

$$\Delta \bar{Y} = \left[D^{(t)} \right]^{-1} \left(\bar{C}^{(t+1)} - \bar{C}^{(t)} \right) - X^{(t)} \left[D^{(t+1)} \right]^{-1} \bar{C}^{(t+1)}$$

here D is the matrix of combinations of the technological matrix A and the capital expenditure matrix K, X is the product of the matrices A and K.

For a constant level of consumption, we should set $C^{(t)} = C^{(t+1)} = C$ and $C^{(t)} - C^{(t+1)} = 0$, then for incrementing the level of gross production we can write:

$$\Delta \bar{Y} = -X^{(t)} \left[D^{(t+1)} \right]^{-1} \bar{C}. \tag{20}$$

In Eq. (19), we may transfer the tensor K to the right side and make the replacement:

$$\frac{d\bar{Y}}{dt} |t = t_o = K^{-1}(1 - A)\bar{Y}_0 - K^{-1}\bar{C} = Q\bar{Y}_0 - K^{-1}\bar{C}. \tag{21}$$

Comparing (20) and (21), we obtain:

$$Q\bar{Y}_0 - K^{-1}\bar{C} = -X^{(t)} \left[D^{(t+1)} \right]^{-1} \bar{C}. \tag{22}$$

Hence the initial level of gross production will be equal to:

$$\bar{Y}_0 = Q^{-1} - \left\{ K^{-1} - X^{(t)} \left[D^{(t+1)} \right]^{-1} \right\} \bar{C}. \tag{23}$$

Matrix X is equal to:

$$X^{(t)} = \left(1 - A^{(t)} + K^{(t+1)} \right)^{-1} \left[K^{(t+1)} \right] = \left(\left[K^{(t+1)} \right]^{-1} - \left[K^{(t+1)} \right]^{-1} A^{(t)} + 1 \right)$$

and the combination of the technological matrix A and the capital investment matrix K will be expressed as:

$$D^{(t)} = \left(1 - A^{(t)} + K^{(t+1)} \right).$$

The product of these matrices will be:

$$X^{(t)} = \left[D^{(t+1)} \right]^{-1} = (1 - A + K)^{-1} K (1 - A + K)^{-1}.$$

Substituting these matrices into (23), we obtain:

$$\bar{Y}_0 = (1 - A)^{-1} \bar{C} - (1 - A)^{-1} \left[(1 - A)K^{-1} + 1 \right]^{-2} \bar{C}. \tag{24}$$

The difference with the static level of gross production will be:

$$\bar{Y}_0 - (1 - A)^{-1} \bar{C} = -(1 - A)^{-1} \left[(1 - A)K^{-1} + 1 \right]^{-2} \bar{C}. \tag{25}$$

Then Eq. (21) can be changed as follows:

$$\frac{d\bar{Y}}{dt}\big|t = t_o = -(1 - A + K)^{-1} K (1 - A + K)^{-1} \bar{C}. \tag{26}$$

The solution of the differential equation of the level of gross production (26) with constant demand with regard to the initial conditions will be:

$$\bar{Y}(t) = (1 - A)^{-1} \bar{C} - \left[1 + \frac{(t - t_0)}{1!} Q + \frac{(t - t_0)^2}{2!} Q^2 + \cdots \right]$$
$$\times \left\{ (1 - A)^{-1} \left[(1 - A)K^{-1} + 1 \right]^{-2} \bar{C} \right\} \tag{27}$$

and to increment the level of gross production we will have:

$$\Delta \bar{Y} - \bar{Y}(t) - \bar{Y}_0 = - \left[\frac{(t - t_0)}{1!} Q + \frac{(t - t_0)^2}{2!} Q^2 + \dots \right]$$
$$\times (1 - A)^{-1} \left[(1 - A)K^{-1} + 1 \right]^{-2} \bar{C} \tag{28}$$

From the square brackets of equality (28), we can take out the matrix Q and multiply it with other matrices:

$$\Delta \bar{Y} = - \left[\frac{(t - t_0)}{1!} Q + \frac{(t - t_0)^2}{2!} Q^2 + \dots \right] \times (1 - A + K)^{-1} K (1 - A + K)^{-1} \bar{C} \tag{29}$$

It is possible to clarify the time dependence of the level of gross production. Let us add equality (24) to (29) and obtain a new equality for the time dependence of the level of gross production at constant demand:

$$\bar{Y}(t) = (1 - A)^{-1} \left\{ 1 - \left[(1 - A)K^{-1} + 1 \right]^{-2} \right\} \bar{C}$$
$$- \left[(t - t_0) + \frac{(t - t_0)^2}{2!} K^{-1}(1 - A) + \frac{(t - t_0)^3}{3!} \left[K^{-1}(1 - A) \right]^2 + \dots \right] \tag{30}$$
$$\times (1 - A + K)^{-1} K (1 - A + K)^{-1} \bar{C}$$

Hence, the level of gross production for the initial capital investment cycles is equal to:

$$\bar{Y}(t) = (1 - A)^{-1} \left\{ 1 - \left[(1 - A)K^{-1} + 1 \right]^{-2} \right\} \bar{C}$$
$$- (1 - A + K)^{-1} K (1 - A + K)^{-1} \bar{C}(t - t_0). \tag{31}$$

The increase or decrease in the level of gross production depends on the sign of the coefficients of the capital investment matrix K. If these coefficients are negative ($K < 0$), then the level of gross production increases with time, if these coefficients are positive, then the level of gross production decreases with time.

This corresponds to two economic situations. While demand is constant, production may decrease, since this demand can be met with a lower level of gross production ($K > 0$) taking into account capital investments K. The second situation is associated with an increase in the level of gross production, which means that with constant demand, the entire increase in the level of gross production is spent on capital investments K of the next cycles.

2.3   The Initial Stage of Capital Investment

Formula (31) allows to calculate the level of gross production, provided that the exponential function of the matrix argument is expanded into series:

$$e^{Q(t-t_0)} = E + \frac{(t-t_0)}{1!}Q + \frac{(t-t_0)^2}{2!}Q^2 + \cdots \qquad (32)$$

This expansion is valid for any matrices Q. Under the condition of a small argument value of the exponential function $Q(t-t_0) << 1$, one or two elements can be kept in the expansion (32). This condition can be converted as follows:

$$K^{-1}(1-A)(t-t_0) << 1$$

Multiplying from the left the two parts of this inequality successively by the matrix of capital investments K and the inverse "input-output" matrix $(1-A)^{-1}$, we obtain:

$$(t-t_0) << K(1-A)^{-1} \qquad (33)$$

The right side of this inequality (33) is the value that can be interpreted as the relative share of capital investments K in relation to the level of FEC production $(1-A)$. This value, by definition, should not exceed the unit. Time in our model is scaled by cycles of capital investments K. Therefore, considered development model refers to one cycle of capital investments K. For a larger number of capital investment cycles K, inequality (33) is not satisfied and in expansion (32) it is necessary to take into account elements with large powers. In this case, it is reasonable to solve the equation completely without expanding into series the exponential function of the matrix argument Q.

3   Conclusion

The study showed that the use of the differential equations of matrix planning allows, on the one hand, enterprises of the fuel and energy complex to more accurately develop the strategy and tactics of their development, and on the other hand, determine the options for dynamic interactions of the fuel and energy complex with related industries. The features of modeling the activities of enterprises of the fuel and energy complex, the purpose of which is to improve the efficiency of the fuel and energy complex management, which will improve energy and economic efficiency and improve their investment and production attractiveness, have been identified. In addition, the article identifies economic situations depending on the dynamics of change and the degree of satisfaction of demand; the level of gross production, taking into account capital investments, as well as the level of consumption over time, which make it possible to more effectively model the development of enterprises in the fuel and energy complex.

References

1. Ackoff R. The Art of Solving Problems. Moscow (2012). Book on request
2. Ackoff R. Creating the Corporate Future. Moscow (1985). Economika
3. Yegorkina, T.A.: Methods of planning expenditures of the main type of activity of an industrial enterprise. Bus. Inform. **7**, 247–252 (2013)
4. Yekutech, A.D., Yekutech, T.D., Loginov, A.Y., Sinkovetz I.A., Fedoseeva, Y.N.: The system's problems in management of enterprises of fuel and energy complex Russia. Econo. Math. Meth. **44**, 114–118 (2008)
5. Zatonskiy, A.V., Sirotina, N.A.: Prediction of economic system based on regression model with differential equation. Econo. Math. Meth. **50**, 91–99 (2014)
6. Kalaydin, E.N., Litvinsky, K.O.: Methodology of construction management models of actors of nature. Reg. Econ. Theo. Pract. **25**(352), 40–47 (2014)
7. Kirillov, I.V., Nazimko, E.N.: Using of differential equations for the financial mathematics problems solution. Financ. Analytics Sci. Experience **20**(206), 58–64 (2014)
8. Kolpakov, V.F.: Modeling of dynamic processes in economy. Financ. Analytics Sci. Experience **3**(189), 31–36 (2014)
9. Kumaritov, A.M., Sokolova, E.A.: Development of the system of data analysis and processing on strategic management of fuel and energy complex. Sci. Bus. Ways Dev. **5**(35), 113–116 (2014)
10. Litvinsky, K.O.: Methodological approaches to the management of ecological-economic systems. Terra Economicus **11**(2). Part 3, 40–44 (2013)
11. Litvinsky, K.O.: The model of preference structure of business-nature in fuel and energy complex. J. Econ. Theor. Pract. **3**(35), 43–47 (2014)
12. Litvinsky, K.O., Shevchenko, I.V.: Basis of the Production and Consumption of Goods: Theory and Practice. Krasnodar, Ecoinvest (2010)
13. Serbulov, Y.S., Glukhov, D.A.: Market situation mathematical simulation of resource interaction of industrial and economic systems. Inf. Control Syst. **4**(65), 27–29 (2013)
14. Serbulov, YuS, Gluhov, D.A., Malyshev, V.A.: Model partition neutralize external effects on industrial and economic systems in a competitive environment. Inf. Technol. **4**(65), 27–29 (2013)
15. Serbulov, Y.S., Kurchenkova, T.V., Kurchenkov, O.A., Lemeshkin, A.V.: The models of resource planning of interaction in technological systems. Control Syst. Inf. Technol. **1**(27), 275–279 (2007)
16. Sinkovets, I.A., Shevchenko, I.V., Fedoseeva, E.N.: The problems of management of development of Russian's fuel and energy complex. Finan. Credit **5**(245), 44–49 (2007)
17. Trusov, V.A.: Model of the shaping the system information production industrial enterprise and enterprise fuel-energy complex. Energetic. Innovative Dir. Energ. CALS-Technol. Energ. **1**, 74–80 (2013)
18. Wilks, S.: Mathematical Statistics. Nauka, Moscow (1967)
19. Shevchenko, I.V., Gakame, A.K.: The perfection of investment policy in Russian fuel-energy complex corporative structures. Reg. Econ. Theor. Pract. **29**, 20–23 (2010)
20. Shevchenko, I.V., Nikolaev, N.V.: Analysis of development of the Russian ferrous metallurgy. Reg. Econ. Theor. Pract. **8**, 2–5 (2008)

Educational Sciences in Digital Age

Digital Transformation in Education

A. A. Bilyalova[1]([✉]) [iD], D. A. Salimova[1] [iD], and T. I. Zelenina[2]

[1] Kazan Federal University, Kazan 420008, Russia
abill71@mail.ru
[2] Udmurt State University, Izhevsk 426034, Russia

Abstract. Digital technology in the modern world is not only a tool, but also a living environment that opens up new opportunities: learning at any convenient time, continuing education, etc. This article aims to describe the specificity of digital education, the current state of its implementation, the expected results and concerns in this respect. Having shown the core of the digital education and the state of its implementation in modern society, this type of education must be critically analyzed in terms of advantages and risks with reference to contemporary students and the effectiveness of the teaching – learning process, in which they participate. In the study pros and cons of digital learning are revealed.

The paper concludes information about advantages of using electronic educational resources in teaching a foreign language based on the experimental work which was done in Naberezhnye Chelny Institute of Kazan Federal University. The success of the experiment presented in this paper is demonstrated by comparing the results of the test group who were taught using electronic educational resources with the reference group who were taught in a common traditional way. The statistical analysis shows that the test group students had better achievements compared to the reference group.

Keywords: Digital education · Electronic educational resources ·
Digital technologies · Informatization

1 Introduction

The modern world is constantly changing. Innovations are being introduced into various spheres of human activity, which orients people towards continuous development, improvement of their knowledge, skills, competencies, mastering new types of activities in related industries. Moreover a person needs creativity, willingness to cooperate with colleagues in finding new solutions, and - most importantly - the ability to critically evaluate the information offered, both in terms of reliability and in terms of its logical integration into the current task [1].

The modern society moved to the next level of development of new technologies. The first one was the creation of a steam engine; the second is electrification; the third is informatization; the fourth is digitalization, that is, the era of big data and the technologies based on them. It should be noted that digital technologies, on the one hand, help to further increase the volume and production efficiency, on the other hand, they allow implementing an individual approach in various fields. Today we can talk about the need for a modern person to have an information culture as an element of

T. Antipova (Ed.): ICIS 2019, LNNS 78, pp. 265–276, 2020.
https://doi.org/10.1007/978-3-030-22493-6_24

human culture and as an indispensable condition for comfortable existence in society. Purcell K. states: "Be ready to face the fact that after a while the majority of work places will require the knowledge of the latest technologies, which are extending and progressing on a going basis. Irrespective of the profession or industry, which is currently covering major scope of human resources, the companies and organizations wish to employ and keep the staff knowing these technologies and ready to advance their skills" [12].

The developing of information culture is one of the most important tasks of the education system nowadays. Requirements for students' skills have changed, since it is necessary not only to read, write and count, but also to be able to organize data resources, cooperate productively, collect, evaluate and use information.

The digital resources used today in daily human activities allow us to overcome the barriers of traditional learning: the pace of program development, the choice of a teacher, the forms and methods of teaching.

In this regard, adaptation to changing conditions and requirements is necessary, which entails the digitizing education. This type of education would not be therefore possible without rapid development of computers and the Internet. The prevalence of computers and broadband Internet has given a very strong impulse to use them in the educational activity. This article aims to describe the specificity of digital education, the current state of its implementation, the expected results and concerns in this respect.

2   Methodology

To solve the problems set in accordance with the purpose of the study, the following methods of research were used: theoretical methods – the analysis of scientific and methodical literature on the digitizing of education and development and application of electronic educational resources in teaching; empirical methods – questionnaire of students, observation, analysis of the students tests; statistical methods – adapted to the objectives of this study.

Firstly, the study identifies the specificity of digital education, its pros and cons, the main fields of application of IT, digitizing of Higher education in Russia. Also the article includes the analysis and synthesis of experimental work on the use of EER at the lessons of a foreign language, assessment of EER as a means of improving the efficiency of the process of teaching foreign languages. The experimental work was done on the basis of Naberezhnye Chelny Institute of Kazan Federal University.

3   Digital Education as a Global Trend

The term "digitalization" deals with the intensive development of information and communication technologies. Klaus Schwab believes that the first digital revolution was in the 1960s and 1980s. He called it "industrial", and considered that its catalyst was the development of semiconductor computers, then personal computers, and in 90-x- it was the Internet [13]. The author has predetermined the approach of the fourth industrial revolution, which will also be digital in connection with the mobile Internet, miniature devices, and the development of artificial intelligence.

With the advent of the Internet in 1982, a virtual world is being formed, filled with new connections, such as online games, social networks, linking it with the real world. The real and virtual worlds are interdependent, and according to one of them, you can identify the person [8]. Their merger forms a hybrid world, through which the vital actions of the real world are accomplished with the help of the virtual one. Prerequisite for this process is the effectiveness of information and communication technology and the availability of digital infrastructure.

The digital revolution that has gripped the global world is impressive in its pace and scope. The transition from electronic computers to personal computers lasted for decades, but now similar global technology changes occur in months. Initially, digitalization was limited to the automation of technologies, the spread of the Internet, mobile communications, social networks, the emergence of smartphones, and the growth of consumers using new technologies. However, very quickly digital technologies become part of the economic, political and cultural life of a person.

The terms "digital ecosystem", "digital environment", "digital community", and "digital economy" are introduced in various areas of the economy. Currently, digitalization has penetrated into education.

There is no well-established definition of the term "digitalization".

Wiktionary reveals the content of the concept of "digitalization" as "a digital method of communication, recording, data transmission using digital devices". Marey considers digitalization as a change in the paradigm of communication and interaction with each other and society [9]. Vartanova clarifies the content of this concept: this is not only a translation of information into digital form, but a complex solution of an infrastructural, managerial, behavioral, cultural character [15]. That is, it can be concluded that the development of the Internet and mobile communications are basic digitalization technologies.

The Russian pedagogical encyclopedia considers the digitalization of education in a broad sense as a complex of socio-pedagogical transformations associated with the saturation of educational systems with information products, tools and technologies; and in the narrow sense it is the introduction of information tools based on microprocessor technology into educational institutions, as well as information products and pedagogical technologies based on these tools.

So, despite the fact that there is no single definition of the term "digital education" almost every above mentioned includes the following: the use of large data in teaching in the process of mastering individual students of certain disciplines and automatic adaptation of the educational process based on them; use of virtual and augmented reality and cloud computing, and many other technologies.

In the article, the term "digital education" means a set of measures to transform pedagogical processes through the introduction of information products, tools and technologies (IT) in education and training.

The main fields of application of IT in education are:

- development of pedagogical software for various purposes;
- development of educational websites;

- development of methodological and didactic materials;
- management of real objects;
- organization and conduct of computer experiments with virtual models;
- targeted information retrieval.

Digitalization of education leads to changes in the labor market, in educational standards, identifying needs in the development of new competencies of the population and it is also focused on the reorganization of the educational process, rethinking the role of the teacher. On the one hand, digitalization undermines the methodological basis of school inherited from the past, on the other hand, it generates accessibility of information in its various forms, not only in text, but also in sound, visual. The availability of information will require constant search and selection of relevant and interesting content, high processing speeds. Consequently, the digitalization of education leads to its radical, qualitative restructuring. The teacher must learn to apply new technological tools and virtually unlimited information resources. It is necessary to teachers to form the ability to navigate the flow of digital information, to be able to work with it, to process and embed it in a new technology.

It should be noted that digital education system includes the following: information resources, telecommunications, management system. Information resources: hyper-collections (media, video, audio, biblio, photo, graphics, animation), informational data files, educational portals, Internet sites. Telecommunications: network and mobile environments, media, television, telephony, teleconference, hosting, postal services. Management system: user authorization, testing, content, ratings, personal and collective information space (website, blog, chat, forum, mail, database).

Thus, digitalization transforms the social paradigm of people's livelihoods, opens up the possibility of obtaining and improving knowledge, and surely expands horizons. Digital technology in the modern world is not only a tool, but an environment of existence that opens up new opportunities: learning at any convenient time, lifelong learning, the ability to design individual educational routes, from consumers of electronic resources to become creators of them.

4   Advantages and Disadvantages of Digital Education

Having shown the core of the digital education and the state of its implementation in modern society, this type of education must be critically analyzed in terms of advantages and risks with reference to contemporary students and the effectiveness of the teaching – learning process, in which they participate.

There is no doubt that the digitizing of education is one of the effective ways to improve the quality of teaching. Digital learning appears to be something extremely beneficial to all schools and students. So, there are a number of advantages of digital learning, such as:

- *Technology gives an opportunity to experiment more with pedagogy and get instant feedback.*
   Modern technologies allow students to become more active participants in the educational process, and teachers to create new approaches, methods, models of

training. For example, a teacher can conduct an online survey at any stage of a lecture to determine the level of mastering the material being studied. Or teachers have the opportunity to implement new models of the organization of the educational process, for example, the "flipped classroom" when a teacher first gives students the opportunity to independently study the new material at home, and then in the classroom organizes the practical application of this material.

Thus the learning process becomes more dynamic with the use of digital textbooks, when the student can use the links to relevant materials or resources. Students can search for answers to the questions asked, form their position, and then defend it.

- *Technology helps to ensure the active involvement of students in the learning process.*

Online surveys and other digital tools help to engage all students in the learning process, including those who are shy, not confident in their abilities, and usually not taking the initiative. Online systems can receive regular feedback, including students' feedback on the availability of training materials and assignments. Data analysis allows the teacher to easily and quickly identify the difficulties of each student and assist in time, identify areas where students can compete, and therefore easily adjust each student's work or group work. For example, technology can significantly improve the efficiency of using such an active teaching method as a quiz. At the beginning of the lesson, the teacher can conduct a quiz using technical devices and quickly assess the starting level of students, spending only a few minutes to obtain reliable information and its analysis. Further, the teacher can make adjustments to the organization of the educational process objectively understanding where it is worth to direct its efforts and how to organize the work of students. Conducting the same quiz at the end of the lesson will again allow you to get feedback with minimal time, and allow students to evaluate the results and success of their training.

- *There are many resources for organizing students' productive learning activities.*

In applications of mobile platforms and electronic textbooks there is no shortage of tools that significantly change the organization of training activities. Some technical devices use various types of incentives and help to assimilate information in the study process, use competitive scenarios for the distribution of points and awards to make the learning process more exciting and attractive. Some mobile platforms and e-books include role-playing games in which students are given the opportunity to present facts and their arguments in favor of, for example, historical figures or scientific concepts. In addition, gaming technology contributes to the introduction of healthy competition in the educational process. Modern automated learning systems can significantly help in the organization of productive learning activities and realistically assess the achievements of each student. It's important to underline here that significant condition for the use of such technical devices is the achievement of learning objectives.

- *Technology will help the teacher to automate or simplify the implementation of a number of tedious duties.*

Automation can simplify and reduce time for routine but time-consuming tasks such as checking student attendance and learning performance. Modern technological tools simplify the systematization and selection of individual tasks for students, help to track the activity of their participation in the discussion, etc. The ability of modern technological means to visualize difficult for perception and understanding of educational material, reduces the time and effort of the teacher to explain. For example, augmented reality technology allows students from atoms to create molecules of complex chemical compounds in a virtual environment with their own hands. Due to the effect of the presence which is created by the impact on the human senses, the technology allows students more effectively study the process of creating a molecule or substance than a presentation on the screen or a picture on paper.

- *Technology provides instant access to the necessary information and brings up important skills to work with different sources.*
  Modern technology expands communication and creates a more productive learning environment. By joining groups on the Internet, students can share information, work together on group projects and interact with the teacher. For example, a free Community resource Scratch allows students to effectively use not only the media capabilities (working with music, graphics, etc.), but also the ability to work together on tasks, networking, self-organization and other system effects. The work of students in the group allows to fill the gap between classical and digital culture by the fact that cultural institutions (libraries, museums, centers of contemporary art) provide source materials (texts, illustrations, musical records) for further use by the students in the creating of games, cartoons and digital stories.

- *The ability to use technology is a life skill and an important type of literacy in modern world.*
  To be digitally literate is more than having "separate technological skills". Today we are talking about a deep understanding of the digital environment, which provides an intuitive adaptation to new contexts and co-creation of content with other students. Making presentations, learning how to find reliable sources on the Internet, supporting proper online etiquette, etc., these are life skills that students can learn in the study process, and they will be useful to them throughout their lives. Digital literacy can help educational organizations not only improve the quality of learning, but also allow learning outcomes to always be relevant. Thus, users of electronic courses: both teachers and students develop their skills and knowledge in accordance with the latest modern technologies and standards. Electronic courses also allow for timely and prompt updating of training materials.

Summing up, it should be especially noted that important advantages were appreciated by the students themselves. SkillSoft (the modern cloud-based content delivery platform that provides e-learning) surveyed students at 16 major foreign universities that use e-learning technologies as a supplement to traditional education. The following advantages of e-learning courses were especially noted by students: flexibility, time saving, ease of returning to the completed study material. Making this paper objective, apart from the advantages of digital education presented above, one needs to indicate the threats which it carries (with its implementation). So in the next part of the article we intend to point to disadvantages of digital education.

- *Loss of writing skills and as a result loss of creativity.*
If a person loses writing skills, his motor skills and coordination will suffer. When handwriting is involved areas of the brain are responsible for the sensory inter- pretation and the development of speech. And for those who do not write with a hand, these sections are included much less frequently. We have a so-called Brock center in our head, it is a section responsible for folding letters into words and recognizing them. In other words, Brock center is responsible for the ability to read and write. By hand writing, this center activates its work. According this scientists from the Norwegian University of Stavanger concluded that people who write quickly read better. And vice versa: people who are slow to read and hard to understand the text do not write well [2].

In addition, students will learn less spelling, punctuation and grammar, because almost every gadget and browser has auto-correction function. Therefore, a person who will not be able to write by hand, most likely, will not be able to write correctly. Also, students will have a poor imagination. People who write a text by hand have a better idea in their minds what it is about. It was found out, forcing people to record lectures in the tomograph. Experts point out that the use of electronic textbooks significantly reduces the imagination of young people, making them "reproductive recipient of an extremely attractive minimum of information" [6].

- *Gadget addiction.*
Gadget addiction can be described as is an obsession with various gadgets, also with the Internet, social networks and online games, and their abusive usage. Gadget addiction leads to attention disorders. Students with this problem experience the lack of concentration or an ability to focus on something for a long time; they also have problems with long-term memory, their decision-making capabilities are in general poorer than among people who have no gadget addiction. Moreover physiological problems directly connected to a prolonged exposure to gadgets include the development of shortsightedness, regular headaches, and aches in the neck and back (because of constantly leaning above the screen of a gadget). Some researchers also connect gadget addiction to problems with fertility, supporting their point with the over-exposure to electro-magnetic fields emitted by gadgets, but this thesis needs to be checked. Alarming is that recently gadget addiction started to develop among children of a rather young age; if a couple of years ago psychol- ogists were talking about teenage addiction to gadgets, nowadays specialists tend to believe a child may develop this addiction earlier – starting from the age of 11. According to recent research, in which around 2,200 young people participated, approximately 65% of children aged between 11 and 17 take their gadgets to bed to be able to browse the Internet before sleep or to play games [14].

According to experts, soon in developed countries gadget-dependence will come out on top in the number of people who suffer from it, ahead of smoking and alcoholism.

So, technology is increasingly penetrating our lives: with their help we communi- cate, learn, work and have fun. However, the paradox is that numerous gadgets, being invented in order to save time and facilitate communication between people, often have the opposite effect. In developed countries, there has long been and rapidly developing people's dependence on their gadgets. For example, in the United

States, about 10% of the population is dependent on smartphones, and thousands of people have applied for medical care in connection with this problem [7].

Now, most children in different degree suffer from screen addiction. As stated above, there is only one way out - no gadgets. And it is also clear that in the digital school it will be completely impossible, which means screen dependency will only progress.

- *A decrease in social skills.*

Most students suffer from loneliness and can not live without social networks. These are the results of a survey conducted by the Russian popular front. The study affected almost 80 regions of Russia. It is wellknown that the personality of the child is developed in the process of interaction with the surrounding world. Here, the emotional participation of the environment where a person character is formed, is of great importance. Internet communication is something like a substitute, surrogate form of communication. It will be difficult for students who communicate on the Internet to build real human rather than virtual relationships. Spiritual wealth, adherence to moral principles, responsibility for their own actions, emotional stability, the ability to respond to conflicts, the ability to sacrifice something – these are just some of the psychological characteristics that are necessary for a harmonious personality for the development of social skills. However, students who spend most of their time online become more authentic and unable to interact with society.

- *Negative influence on health.*

The growing risk of health loss caused by the use of electronic devices, especially by children. This mainly concerns the eyesight and the risk of myopia, because the use of e-course books leads to the fact that children intensely stare at a computer screen for 'at least 9 h of lessons' a week. The time spent in front of the computer at home, which is difficult to estimate, adds to the above problem. According to American studies, even "over two hours a day is harmful for children. It increases the risk of psychological problems" [10]. These can include anxiety, depression, insomnia, dizziness, memory loss, etc. The negative influence of the light emitted by computer screens on the ability to fall asleep has also been proven. This is due to the decrease in the level of melatonin, which regulates the biological rhythm of human life. Also spine defects arising from the faulty posture in front of the computer and from spending many hours in a sitting position are a serious condition stimulated by digital education.

Thus, it is worth underlining that digitization programmes is not mere transfer from the traditional paper version into the electronic one. Digitization of education requires a separate original programme, and most of all long-term studies. The University of Durham completed the studies of digital technologies impact on education and made several interesting conclusions. According to its experts, digital technologies should only complete, but not supersede traditional teaching methods, with their full potential to be used by slow-learning students or students with special needs. The best results can be achieved if IT are used at certain intervals, approximately three times a week, since frequent use of innovative and primarily digital methods can gradually decrease the students' information processing efficiency. And finally the scientists stress the importance of holding professional development seminars for teachers, who will use digital technologies in their work [5].

5   Digital Education in Russian Federation

Presently education digitalization in the Russian Federation is closely related to implementation of Federal state education standards (FSES), developed for all education levels. One of the key competences of FSES is "to generate and develop major competences in the area of information and communication technologies use; to motivate students to actively use dictionaries and other search systems". One of the tools to reach the stated objectives is "access to school library and Internet information resources, course books and popular literature, media-sources on electronic media, copy machines for replication of tutorials, guidance materials, texts and graphics, audio and video information, creative, research and development content as well as the projects of students".

Another important state document promoting education digitalization concept in Russia is priority project "Modern Digital Educational Environment in the Russian Federation". This project was approved by the Government of the Russian Federation on October 25, 2016 as part of the implementation of the state program "Development of Education" for 2013–2020. Within the framework of this project, it is planned to "modernize the education and training system, bring educational programs in accordance with the needs of the digital economy, widely introduce digital tools for learning activities and integrate them integrally into the information environment, and ensure that citizens can learn about the individual curriculum throughout their lives - at any time and in any place" [9]. So, its core objective is to create conditions for improving the overall quality and broadening options for continuing education of all citizens by developing Russian digital educational environment and raising the number of online students up to 11 mln. people by the end of 2025.

The Government of the Russian Federation approved the passport of the project "Modern Digital Educational Environment in the Russian Federation", which is aimed at creating the conditions for systematic quality improvement and expansion of lifelong learning. The project will be implemented due to the digital educational space, the availability of online learning and is aimed at the possibility of organizing blended learning, building individual educational learning routes, self-education, family and non-formal education.

Currently Russian schools are deeply involved in education digitalization projects. Most of them follow the concept that digital education must go hand-in-hand with traditional education models and techniques. The school offers video lessons of the best teachers (including foreign), online seminars, distant learning, discussions and conferences. One of the objects of stage-by-stage reform and subsequent modernization of Russian education is electronic journal and a pupil's grade book that, for example, allows students to assess the activity by selecting a certain scale that best corresponds to the individual characteristics of each student.

Top universities of Russia such as Moscow State University, St. Petersburg State University, National Research University "Higher School of Economics", Moscow Institute of Physics and Technology and some others established the National Platform "Open Education" [11]. It offers online courses on basic subjects trained in Russian universities. All courses hosted on the platform are available free of charge. And those who want to set off the completed online course, while mastering the educational program of undergraduate or specialist programs, can receive certificates. According to the latest data currently more than 197 thousand students have been trained in the online courses of the National Open Education Platform.

At Kazan Federal University the programmes of training and introduction of electronic educational resources (EER) in the teaching process are successfully realized in recent years. Today the electronic versions of working programmes, methodical instructions on all studied disciplines, provided practically for each training direction by the FSES, are available to any student. The use of EER in KFU is realized in different ways, including the use of LMS MOODLE (Learning Management System Module Object-Oriented Dynamic Learning Environment). It is a learning management system that allows to create distance learning courses (e-learning courses), which includes all the necessary training, support and control materials (or links to them) as well as methodological instructions (for both the teacher and the trainee) in accordance with curriculum [3, 4].

We suggest the results of experiment that was held on the basis of Naberezhnye Chelny Institute of Kazan Federal University.

The group of the first year students of the Department of Philology were chosen randomly to be the sample of the experiment. The total number of the students was 30. The students of this group were appointed randomly into two groups; the control group which had (15) students, and the other experimental group which had (15) students. In order to achieve the purpose of the study, a set of research instruments was developed: Vocabulary Test and Listening Test. Both Tests consisted of 17 multiple choice questions. Students were given twenty minutes to answer the questions. Test items had variants, only one of which was correct.

To ensure the equivalence of the two groups before starting the study, a pre-test was applied. Table 1 shows the results:

Table 1. Results of pre-test of the equivalence of the two groups

	Group	Students	Mean score (vocabulary test)	Mean score (listening test)
Pretest	Control	15	4.1	4.2
	Experimental	15	4.0	4.3

Table 1 shows that there were no statistically significant differences between the experimental and control group at the pre-test of reading comprehension and pre-test of listening skills. This means that the control group and the experimental group were equivalent before starting the experiment.

During the second term students of the experimental group were taught a foreign language using EER, students of control group were taught a foreign language without using EER. Are there significant differences in the students' reading comprehension and listening skills due to the strategy of teaching with the help of EER? To answer this question mean score in the post tests was calculated. Table 2 presents the results in comparison.

Table 2. Results of using EER in teaching a foreign language for Each Group (Control and Experiment)

	Group	Students	Test	Mean score
Vocabulary test	Control	15	Pretest	4.1
	Control	15	Posttest	4.0
	Experimental	15	Pretest	4.0
	Experimental	15	Posttest	4.5
Listening test	Control	15	Pretest	4.2
	Control	15	Posttest	4.3
	Experimental	15	Pretest	4.3
	Experimental	15	Posttest	4.6

Table 2 shows the mean scores of the two groups (Control and Experimental) in the pre Vocabulary Test were (4.1) and (4.0) respectively while in the post test were (4.0) and (4.5) respectively; in the pre Listening Test were (4.2) and (4.3) respectively while in the post test were (4.3) and (4.6) respectively. This means that the students' vocabulary and listening skills scores in experimental group were improved in the post test because of the use of EER.

Thus, the study investigated the effectiveness of using EER in Foreign Languages Teaching at University. The results showed that there were statistically significant differences in students vocabulary and listening skills due to the strategy of teaching (teaching using EER and teaching without using EER) in favor of the EER - teaching strategy.

The results showed that using EER at the lessons of foreign languages might have contributed to the improvement of students' achievement in vocabulary and listening. Moreover students might have the opportunity to receive information from different resources, share their knowledge and skills in an appropriate manner and interact more effectively.

6 Conclusion

The digitization of education in Russia and all over the world is spreading into more and more areas.

Therefore, it becomes necessary to analyze all its aspects, both advantages and disadvantages. In the performed study an attempt to present the current state of the access to computers, the Internet and electronic educational courses in Russian schools has been made. Pros and Cons of digitizing of education were the axis for the reflection undertaken. Many arguments in favour of teaching based on modern methods and techniques have been presented; however, at the same time, a number of risks which this technology carries have been identified.

It is obvious that digitalization of education involves the use of mobile and Internet technologies by students, expanding the horizons of their knowledge, making them limitless. The productive use of digital technologies, the inclusion of students into an independent search, the selection of information, and participation in project activities form the competences of the 21st century.

It is clear that the benefits outweigh the cons. Technology can be a very effective tool, but it is just a tool. Technologies are not intended to replace the teacher, rather, the idea is to create a learning environment that will allow you to switch the organization of the educational process from one-actor theater to cooperation and interaction.

Digitization has no doubt changed our education system, but we cannot say that it has diminished the value of traditional classroom learning. The best part about the digitization of education in the 21st century is that it is combined with the aspects of both; classroom learning and online learning methods. Walking hand in hand both act as a support system to each other, which gives a stronghold to our modern students

The implementation process can upset someone, annoy, take a lot of time and effort, but ultimately technology can "open doors" to new experiences, discoveries, ways of learning and cooperation of students and teachers.

References

1. Aksyukhin, A.A., Vyzen, A.A., Maksheneva, Z.V.: Information technologies in education and science. Mod. High Technol. **11,** 50–52 (2009)
2. Better learning through handwriting [Electronic resource]. https://www.sciencedaily.com/releases/2011/01/110119095458.htm
3. Bilyalova, A.: ICT in teaching a foreign language in high school. Educ. Health ICT Transcult. World **237,** 175–181 (2017)
4. Bilyalova, A., Sharypova, N., Akhmetshina, A.: Electronic educational resources in foreign languages teaching. In: Proceedings of the international conference on the theory and practice of personality formation in modern society (ICTPPFMS 2018), vol. 198, pp. 173–177 (2018)
5. Enhancing Learning and Teaching with Technology [Electronic resource]. https://www.ucl-ioe-press.com/books/schools-and-schooling/enhancing-learning-and-teaching-with-technology/
6. Internet Addiction Disorder - What Can Parents Do for Their Child? [Electronic resource]. https://www.webroot.com/ca/en/resources/tips-articles/internet-addiction-what-can-parents-do
7. Is Society Today too Dependent on Technology? [Electronic resource]. http://theessayblog.com/2016/05/17/is-society-today-too-dependent-on-technology
8. Keshelava, A.V.: Introduction to the "Digital" Economy - VNII Geosystems, p. 28 (2017)
9. Marey, A.: Digitalization as a paradigm shift [Electronic resource]. https://www.bcg.com/ru-ru/about/bcg-review/digitalization.aspx
10. Moore, A.: Warning to Teens: Gadgets Can Disturb Your Sleep [Electronic resource]. http://www.medicaldaily.com/warning-teens-gadgets-can-disturb-your-sleep-242294
11. Open education [Electronic resource]. https://openedu.ru
12. Purcell, K.: How Teens Do Research in the Digital World. "The Pew Research Center's Internet and American Life Project Online Survey of Teachers", 1 November 2012
13. Schwab, K.: Die Vierte Industrielle Revolution. — Эксмо, 208 с (2016)
14. The online generation: four in 10 children are addicted to the Internet [Electronic resource]. https://www.independent.co.uk/life-style/gadgets-and-tech/news/the-online-generation-four-in-10-children-are-addicted-to-the-internet-9341159.html
15. Vartanova, E.L.: The Russian Media Industry: A Digital Future: An Academic Monograph, p. 160. MediaMir, Moscow (2017)

Scientific Researches Analysis in Digital Multitasking Field of Educational Process in Modern Universities and Determination of New Conceptual Boundaries of It

Tatyana Vasilyevna Lyubova$^{(\boxtimes)}$ ⓘ
and Gulnara Tavkilievna Gilfanova ⓘ

Kazan Federal University, Kazan 420008, Russia
lubovatv@bk.ru

Abstract. This paper is a review of recent researches on digital multitasking. Claiming that digital technologies are changing the traditional roles of a teacher and a student, we must clearly understand a kind of effects arised at the moment digital technologies are used in the classroom. Moreover, the methods of applying digital technologies for the success of students' academic progress are not always obvious to the teacher himself. The goal of the research analysis on digital multitasking in education problem is to identify shifts in settings and accents from the operationalism methodology to attempts of application of analytical philosophy of consciousness methodology and/or postmodern philosophy. This allows us to analyze the intentions, motives of the educational process participants and thereby define new conceptual boundaries of digital multitasking.

Keywords: Digitalization · Multitasking · Digital technologies ·
Cognitive overload · Prefrontal cortex · Flexibility ·
Syndrome of the limbic system · iGeneration

1 Introduction

Around the middle of the last century, the digital revolution began as a multifactor transition from analog to digital processing, storage and transmission of data, and, accordingly, the rapid development of the hardware and software serving these processes. Changes begin with the research of innovations, which having a successful set of socio-economic circumstances, reach industrial developments and mass usage. Some digital innovations overcome this path significantly faster, such as, for example, a tablet personal computer. Other examples of important theoretical developments and their successful implementations are mathematical substantiation of the method of converting an analog signal to a digital one and vice versa by V. A. Kotelnikov (the reading theorem, 1933), the formalization of the principle of algorithmic processing for construction a digital computer by A. Turing (*a Turing machine*, 1936), principles and invention of an electromechanical programmable digital machine K. Zuse (*Z3*, 1941), an electronic digital computer J. Atanasova and K. Berry (*ABC*, 1942), Harvard digital

© Springer Nature Switzerland AG 2020
T. Antipova (Ed.): ICIS 2019, LNNS 78, pp. 277–288, 2020.
https://doi.org/10.1007/978-3-030-22493-6_25

computing Mark-1 machines (G. Aiken, *IBM*, 1944), British Colossus (Mark 1, Mark 2, 1944), American (D. Mauchly, P. Eckert, *ENIAC*, 1946) and the Russian digital computer (*SECM*, 1950, *BECM*, 1952), etc. which became the forerunners of modern digital technology.

Most often, digital technologies and systems are mentioned in connection with multimedia formats of information representation and processing (for example, digital television, photo, video and audio equipment, as well as in connection with digital communication systems (for example, digital networks based on ISDN and xDSL. This is not surprising, since computers were called digital computers since the 1940s, the term (digital) electronic computer (computer) became common in the late 1960s, and the first personal computers appeared in the 1970s we continued to call as ECM by the existing Russian standards.

Around the same years methods of digitizing or digital image and sound processing began to be actively investigated and developed. Currently, the term computer has lost its original meaning and more popular term "computer" has appeared. Regardless of the name, all types of computers to varying degrees represent and process data (usually analog - texts, documents, numbers, tables, images, maps, drawings, image sequences/animation, three-dimensional models, signals or sounds) in digital (discrete) format. It is digital technology that allows you to manipulate data at high speed, including transmission via analog (continuous) or digital communication channels (analog-digital/digital-analog conversion, coding, modulation/demodulation of the signal). Computers, telecommunications, Internet network services have the ability to process these digital data, which are there thanks to the conversion (digitization, digitalization) of various types of analog signals. Then, in digital form, these data are combined by devices and programs into new formats, undergoing convergence or *mediaconvergence*.

According to our research, the density of the emergence of new digital systems and devices has increased significantly, approximately since the early 1970s. Their growing popularity is gradually leading to the displacement of traditional analog devices and systems. A vivid example is the intensive development of mobile (mobile) cellular communication, which surpasses fixed-line analog/digital telephony by the number of subscribers. According to a study published in Science Express, digital technologies began to dominate in the 1990s and in 2007 99.9% of the information was transmitted in digital format. Scientists believe the beginning of the digital revolution is 2002, when the volume of digital data stored in the world for the first time exceeded the total volume of analog data. Most of these innovations Clayton Christensen and his followers refer to "disruptive" technologies, which unlike "supported" ones are capable to generate socio-economic changes in all spheres of human activity.

As for education, many of the revolutions predicted by optimists under the influence of innovative technologies of their time did not take place. Nevertheless, in the coming years, one can estimate the intensity with which radical digital transformations in technology influence the "digitalization"/digitization of education, as we see it today in culture, telecommunications, media, on TV. For example, we can note a significant increase in the volume of Russian-language educational digital resources, many of which are becoming available on the Internet.

Positive changes in education towards digitalization should probably be linked to the state ICT policy, which aims to provide universities with broadband Internet access and to form an IT-rich environment.

One of the important tasks of educational programs and projects is to achieve digital literacy, as new skills in evaluating and effective using programs and files in various digital formats (for example, the most common: text - *MS Word, PDF, .txt, XML; graphic - BMP, GIF, JPEG, TIFF; sound - MP3, WAV; video - AVI, FLV, MPEG*, etc.), which are supported by computers and Internet services. Thus, digital (electronic) books or *ebooks* are massively created and distributedwhich based on the availability of a printed source, can be divided into two types: electronic versions (copies) of a printed publication and books, originally produced in digital format, therefore they have different quality as well as educational value. Revolutionary transformations in publishing practice have been associated with the proliferation of online bookstores, such as Amazon.

Sociological studies show that in recent years the volume of digitized visual and audio information has been significantly highlighted in the increasing flow. The predominance of non-textual visual elements in the information received by users leads to an increase of the need to form visual literacy. Such training implies the development of visual perception of signs, symbols and their systems, the basic techniques of typography, text message design (cognitive-oriented, semantic, text and graphic accents), understanding of hidden meanings, codes and metaphors that are saturated with computer graphics or infographics, as well as photography, video, television image and advertising on the computer and Internet. According to some researchers, visual, multimedia components of the educational material or educational website allow to increase the rate of their development significantly.

Undoubtedly, these changes in technology, which began in the second half of the 20th century, entail the need for technological re-equipment of the education system, which can no longer be described as the only processes of *computerization or informatization*. Current trends of globalization, transformation and modernization affect all levels of the education system - from pre-school to additional. The digital revolution, understood in the narrow sense, as the transition from analog to digital devices and signal transmission technologies in the period of the middle of the last century, at the beginning of the XXI century, entered a new, higher level of implementation, encompassing more and more fields of activity, including education.

However, in this article we would like to consider one of the serious problems of digitalization of education, namely, the problem of multitasking. The keen interest of psychologists and teachers to this problem is inspired by the influence that digital technologies have on the achievements of students in the educational process. Nowadays, among the parents and some scientists, there is a widespread view that digital technologies are changing our lives and our brains. These conclusions followed the assumption of a fundamental change in the brain structure of children and adolescents under the influence of online interactions: "If the young brain is exposed to multitasking, triggered by the child's constant interaction with digital media, flickering images on the monitor or TV, instant switching attention by pressing a button only then such a fast alternation of images can teach the brain to work in the mode of fast actions and superreactions" (Greenfield 1984).

Some authors call the actions of students in multitasking mode a global condition for increasing productivity in the new millennium (Prensky 2012; Gasser and Palfrey). Others mark the generation *iGeneration*, for which digital technologies are not "tools", but part of the environment. For the generation of *iGeneration* WWW docs not mean the World Wide Web, but Whatever, Whenever, Wherever, i.e. anything, anytime, anywhere. Many authors indicate that students of the *iGeneration* generation have a special need for multitasking, since their short-term memory has a large volume, speed, and efficiency (Rosen 2007).

2   Methodology

Let us try to present an overview of some psychological and pedagogical research on multitasking in education, generated by the use of digital technologies. A review of research will help teachers make a well-considered decision about the possibilities and consequences of the use of digital technologies in the classroom, both directly by themselves and by students. Claiming that digital technologies are changing the traditional roles of a teacher and a student, we must clearly understand a kind of effects arised at the moment digital technologies are used in the classroom.

Smartphones and other mobile devices that students use during a class create a digital multitasking mode. The situation in Universities that has emerged after the widespread dissemination of digital technology is such that it is methodologically unproductive to place the observed effects of multitasking in the previous theoretical framework. New analytical models and new theories are needed. There is no doubt about the need for the emergence of new theoretical frameworks of interpretations when analyzing any phenomena that possess fundamental novelty. But in the case of digital multitasking it is not simple. On the one hand, multitasking itself is not a new phenomenon in the educational process. In the epistemological sense, the learning problem can be interpreted as multitasking planned by the teacher. Therefore, in its analysis and description, an ontologically unfamiliar field cannot arise (Bilyalova et al. 2017).

That is why the first pedagogical research of digital multitasking extrapolated the meanings and conclusions obtained in the decision making theory to the educational process. Most often, economists proceed from the following view: each person solves the problem of switching attention, i.e. determines the method and speed of information transfer at the moment when he is forced to make a decision or prioritize actions quickly (Speier 2000). Simultaneously with the concept of "multitasking", such studies use the concepts of "information overload", "background multitasking", "switching attention" (McPeak et al. 2001), "attention interruption" (Speier et al. 2007) "cognitive overload" (Kirsch 2000).

It is generally accepted in decision making theory that interruption of attention can increase productivity if interrupted tasks are simple and interruption forces a person to turn to a task that is substantially simpler than the primary one by the level of complexity. Setting and solving switching tasks (commutation) requires not only a large amount of time from a person for researching resources and planning resources, but also special skills, especially when the task has a critical moment of novelty.

Cognitive theory in the middle of the 20th century concluded that parallel processing of information can improve the efficiency of routine actions. Each shift of attention involves psychological costs. In most contexts, overall performance decreases in proportion to the number of switches. Conversely, a quick switch from one mental task to another can reduce productivity, especially if both tasks are complex.

Psychologists Meyer, D., Small, G., Jones, Q. and others reduce research to the search for a multitasking substrate in the human brain. It is argued that each shift of attention from one task to another requires the activation of various neural circuits and is probably coordinated by the frontal lobes of the brain. The prefrontal cortex of the brain regulates a person's cognitive abilities, problem solving, state of attention, and suppression of emotional impulses. For those students who are not able to learn from past mistakes, the function of the prefrontal cortex is usually impaired. Their actions are not based on experience, but on what they want at the moment. We conclude that a student with a good work of the prefrontal cortex is able to learn the following - starting to work in a complex project as soon as possible, he will provide himself more time and he will have less reason to worry that he will not have time to pass it on time. A student with a reduced function of the prefrontal cortex does not take into account past troubles and failures and will constantly postpone everything to the last moment. Within the framework of such a theoretical approach, the possibility of a teacher's other actions is denied, besides his influence on the students' abilities or innate qualities (Sanbonmatsu et al. 2013). Here one can identify the range of judgments from the complete negation of the very possibility of multitasking (Medina 2008) to the statement of the inevitability of a change in the brain - the "digital psychomorphosis" under the influence of human interaction with digital technologies (Prensky 2012; Sandomirsky).

Some researchers (Medina 2008; Strayer 2012) insist that evolutionary brain is not adapted to multitasking. Man is arranged in such a way that in conditions of a constantly unstable external environment, the problem of survival always comes to the fore. The flexibility of the brain, its ability to solve arising problemsinstantly can be considered an evolutionary effect. At the same time, the ability of the brain to unstable and changing in its main parameterswork evolved evolutionarily the loss of those qualities that would allow it to process incoming information in parallel. Solving one problem is possible by minimizing the value of another problem. Every time the brain switches to solve a new problem, a large amount of energy and time is consumed. Conclusion of Medina, J. is ambiguous: training in a peer group is one of the types of multitasking and entails a nervous strain generated by cognitive dissonance. Multitasking in the educational process is therefore harmful. Since educational institutions actually encourage multitasking, students must be transferred to home schooling.

Nervous overstrain is exacerbated if the student uses different digital technologies and mobile devices in the classroom. Jones, Q., Chang, S., Viadero, D., Ophir, E. indicate that digital multitasking is potentially disruptive to students. Students become lethargic and boring people. They do not enjoy reading, they have problems using words, and social networks use to certify their own identity. Jones Q. and others believe that the multitasking mode also affects the attitude of students towards themselves. Programming the brain on the background of multitasking, video and audio stimulation upsets the balance of operational and long-term memory, the syndrome of the limbic system develops (Small 2011).

Constant immersion in the redundant information environment leads to the fact that the student uses an individual tactic of avoiding information ("the phenomenon of information output"), the essence of which is that a person ignores relevant and useful information, because there is too much of it to understand it and adopt it (Savolainen and Mannering 2008). Sometimes, there is a rejection of the comprehension of information and fixation on a quick decision, the so-called "paradox of choice".

Student experiences paralysis of the will facing to grandeur and the volume of the tasks assigned to him, therefore, is fixed on the first solution that came to mind. There is a well-known fact from teaching practice: despite information literacy, a student cannot select the necessary and sufficient number of information sources. This often leads to a random selection of sources and materials.

The aggravation of the information overload problem occurred after the advent of *Web 2.0* tools. There was a situation in which a large amount of the most significant information comes in simultaneously with insignificant information, both of which are available in digital rather than printed form. The nature of the tools *Web 2.0* contributes to the expansion of the information aspect of the educational process. We are talking about the perceived impermanence and variability of educational content, redundant information environment with information resources of various formats and types available to users (blogs, wikis, *RSS* feeds, podcasts, social bookmarks). Due to the fact that *Web 2.0* tools provide and encourage rapid updating of the material, the student has an expectation of constant novelty, which can be satisfied by the generation of surface and ephemeral changes, recomposing the existing material when it is reused. There are questions: if an encyclopedia or textbook is presented in a wiki format, can we talk about a "classical textbook", "educational standard", and "final" form of a document? The problem of novelty of knowledge and information loses its former meaning (Keen 2007).

In contrast tooverload and switching attention information, the concept of "digital multitasking" has a fundamental novelty and therefore must be analyzed in another conceptual field. Since the beginning of the XXI century under the influence of the philosophy of consciousness D. Chalmers, some attempts are being made to form a new theoretical base for the research of digital multitasking. The new research approach proceeds from the position that consciousness originates in any information system; therefore, a person is a system for processing information. An additional statement of the principles of coherence, the correlation of consciousness with a certain type of information system, makes it possible to imagine a student acting in digital multitasking mode as an organization having common and structural coherence. It is not the mental qualities of a person that are investigated, but his conscious (phenomenal) experience on indirect signs—behavior change (Hembrooke and Gay 2003). There are many opportunities for pedagogical activities. However, several articles state that the ability of the generation iGeneration, born in the mid-1990s, to integrate information in a special way is nothing more than a delusion (Willingham 2009). The experience of multitasking does not give young people advantages and does not bring benefits in comparison with adults, consider Verhaeghen and Salthouse (1997).

In general, attempts to develop new theoretical frontiers of digital multitasking analysis can hardly be considered as successful so far. Currently, among teachers and parents of different countries there is a widespread desire to save schoolchildren and students from multitasking. In the opinion of the author, this position is naive. It seems that the key may be a statement of facts and a discussion of strategies. In our opinion, training can include a structured conversation with students about the possibilities and consequences of multitasking. Students can share alternative coping strategies that they use to cope with information overload. The teacher can help students cope with the effects of multitasking, namely: (1) to learn to focus on only one activity; from time to time to offer students to close laptops, smartphones and participate in classroom discussions. Some teachers go further: close access to Wi-Fi in classrooms, at least during the test and verification work; (2) limit the amount of information. This means the advisory nature of the choice of sources of information; (3) to adopt and master digital technologies, while at the same time ensuring the limits of their use from time to time to teach students to reflect, plan and organize. It is important to find the direction in which invisible shifts and changes take place in students.

3   Results

Digital multitasking in the educational process occurs when digital technologies are used in the classroom. So, let's take a closer look at the data that was obtained by researchers in various circumstances:

1. Digital technologies are used by teachers. Specialists in the cognitive sphere pay a lot of attention to analyzing the consequences of the use of digital technologies by the teacher, for example, they learn the ways of interpreting and the effectiveness of memorizing training text by the students in multimedia classes. They point out that sometimes a multimedia lesson consists of two sequences of unrelated messages - a verbal and a video ones. Teachers, due to inexperience or lack of understanding of the consequences of their actions, can separate the studied text and the accompanying it graphics in time and/or space. It is assumed that the student should simultaneously understand both the text and the video sequence and coordinate the process of understanding both. The student is experiencing a heavy load, he must keep both messages in his memory at the same time. There is a possibility that he will not be able to do it successfully. Analysis of posts in social networks shows that users open and read short and simple messages, preferring not to open long and overloaded ones. This implies, as Rubinstein and Meyer (2001) show that in a case of a conflict of text and graphics, multimedia technologies are ineffective. Willingham (2009) states that the teacher's use of digital technology does not mean that students will learn better. Moreover, the methods of applying digital technologies for the benefit of students' academic performance to the teacher himself are not always obvious. Sometimes different learning outcomes are rooted not in the strengths or weaknesses of multimedia, but in differences between students. The amount of RAM varies from person to person. So a multimedia lesson that will benefit a student with a large amount of RAM can be detrimental to a student with a smaller volume. In a multimedia environment, the understanding and interpretation of content occurs differently than in traditional conditions. The transition from a two-dimensional picture to a three-dimensional moving video sequences can also create problems for the perception of content in some students.

2. Digital technologies are used by students as directed by the teacher, which implies a local network and/or free Internet access (Jungo 2012; Ophir et al. 2009). Under hypertext conditions, if a student uses related links to see the definition of a word, the reading comprehension is violated. The decision to follow the link, and then returning to the main text destroys the continuity of reading and endangers the possibility of building logic. The degree of influence of hyperlinks on reading comprehension depends on the volume of the student's memory and on the availability of prior knowledge. At the same time, students who have a greater amount of RAM or prior knowledge about the subject of the text, better assimilate the material, despite the hyperlinks. Studies of online learning mode have shown that students finally print online materials to avoid overloading (Ophir 2006) or limit the number of websites they visit (Hartmann 2003).

3. Digital technologies are used by students furtively on the contrary to the teacher's instructions along with the performance of classroom work. Researchers have repeatedly analyzed the functions of attention and memory of students at the moment when students read SMS messages or make entries in social networks (Gantz 2010; Jones and Shao 2011; Foehr 2006). Students were asked about the mode (frequency and duration) of using smartphones, instant messaging services in the classroom, volume and number of messages sent to each other.

4 Discussion

Most pedagogical studies are conducted from the standpoint of the methodological dogma of operationalism: researchers observe the actions of students, sometimes they transcribe their written texts, and keep diary. Operationalism is a common methodological setting in anthropology, psychology, and pedagogy, widespread in the 20th centuryand it consists in the fact that the researcher captures the actions of people in different ways (photo, audio, video), and then analyzes them, using certain approaches. Operationalism had its own strong point, and it is still relevant: it will be impossible to imagine the consequences of the use of digital technologies in the audience, if not to analyze the actions of students. In early pedagogical studies, the automatic use of a certain set of old and rather annoying concepts, sounding banal thoughts, which were formed almost a century ago, should be noted. The real picture of digital multitasking is simplified or distorted. There is a feeling of pettiness or unimportance of observations and conclusions. It is not by chance that digital multitasking appears in articles in the paradigm of collecting exotic technologies and devices:

In recent articles, there is a concern about the idea of representativeness of a study to legitimize a chosen research approach. This is manifested in an increase in a sample size. Researches in the 90-th were conducted with 7–9 students butin 2014 - 1,479 people took part in the studies (Ravizza et al. 2014). The object of research is changing: instead of objective indicators - speeds of cognitive processes, for example, it describes how multitasking is experienced by a student (Lavie 2005). In general, attempts to develop new theoretical boundaries of digital multitasking analysis can hardly be considered as successful so far.

Currently, among teachers and parents of different countries there is a widespread desire to save schoolchildren and students from multitasking. In the opinion of the author, this position is naive. It seems that the key solution may consist of a statement of facts and a discussion of strategies. In our opinion, training can include a structured conversation with students about the possibilities and consequences of multitasking (Crenshow 2010). Students can share alternative coping strategies that they use to cope with information overload.

The teacher can help students cope with the effects of multitasking, namely:

(1) teach to focus on only one type of activity; from time to time to offer students to close laptops, smartphones and participate in classroom discussions. Some teachers go further: close an access to *Wi-Fi* in classrooms, at least during the control and verification work;
(2) limit the amount of information. This means the advisory nature of the choice of sources of information;
(3) adopt and master digital technologies, while at the same time ensuring the limits of their use from time to time to teach students to reflect, plan and organize. It is important to find the direction in which invisible shifts and changes take place in students.

Thus, in the analyzed monographs and articles it is stated:

- digital multitasking in the educational process occursat the time of use of digital technology in the audience;
- Digital multitasking does not create major obstacles to learning. Despite this, researchers confidently conclude that switching attention when using digital technology significantly increases the amount of time needed to complete a learning task. Distraction of attention and time spent on switching from task to task will have a negative impact on the academic performance of students;
- The use of digital technology by the teacher creates a multitasking mode for the student. The specific use of digital technology at one time or another class is determined by the goals that the teacher sets;
- student use of mobile devices (devices) in the classroom may entail a negative effect on academic performance;
- Digital multitasking can qualitatively change learning, because the student relies on different information processing systems that differ in their degree of flexibility, that is, he chooses whatinformation structure to use in a particular case.

5 Conclusions

1. In the analyzed studies of digital multitasking in education, there is a shift in attitudes and emphasis from the methodology of operationalism to attempts to apply the methodology of an analytical philosophy of consciousness and/or postmodern philosophy. This allows you to analyze the intentions, motives of the participants in the educational process and thereby define new conceptual boundaries of digital multitasking.

2. The object of research is changing: instead of the speed of various cognitive processes, the authors explore the motives for using digital technologies or experiencing (qualia) participants in the educational process.
3. Preoccupation with the representativeness of the study forces the authors to expand the sample size.
4. In general, attempts to develop new theoretical boundaries of digital multitasking analysis can hardly be considered as successful so far.

References

Allen, D., Shoard, M.: Spreading the load: mobile information and communications technologies and their effect on information overload. Inf. Res. **10**(2), 227 (2005)

Bawden, D., Robinson, L.: The dark side of information: overload, anxiety and other paradoxes and pathologies. J. Inf. Sci. **XX**(X), 1–12 (2008)

Bilyalova, A.A., Lyubova, T.V., Valeeva, A.R.: Multicultural education and its targets in the process of intercultural competence formation: (based on foreign languages teaching experience). Int. J. Sci. Study **5**(6), 93–96 (2017)

Chang, S., Ley, K.: A learning strategy to compensate for cognitive overload in online learning. J. Interac. Online Learn. **5**(1), 104–117 (2006). [Electronic resource] Access mode: http://www.ncolr.org/jiol/issues/PDF/5.1.8.pdf

Castaño-Muñoz, J., Duart, J.M., Sancho-Vinuesa, T.: The Internet in face-to-face higher education: can interactive learning improve academic achievement? Br. J. Edu. Technol. **45**(1), 149–159 (2014)

Foehr, U.: Media Multitasking Among American Youth: Prevalence, Predictors and Pairings. Henry, J.: Kaiser Family Foundation, Menlo Park (2006). [Electronic resource] Access mode: http://www.kff.org/entmedia/upload/7592.pdf

Gasser, U., Palfrey, J.: Mastering multitasking. Educ. Leadersh. **66**(6), 14–19. [Electronic resource] Access mode: http://cf.linnbenton.edu/wed/dev/hakek/upload/mutli1.pdf

Gantz, J.F.: The Expanding Digital Universe: A Forecast of Worldwide Information Growth Through 2010. IDC, Framingham (2010). [Electronic resource] Access mode: http://www.emc.com/collateral/analyst-reports/expanding-digital-idc-white-paper.pdf

Greenfield, P.: Mind and Media: The Effects of Television, Video Games and Computers. Harvard University Press, Cambridge (1984). [Electronic resource] Access mode: http://www.cdmc.ucla.edu/Mind_and_Media_files/MMCHP1.pdf

Ignatov, N.Yu.: Multitasking and student performance. Open Distance Educ. **3**(55), 5–11 (2014)

Hembrooke, H., Gay, G.: The laptop and the lecture: the effects of multitasking in learning environments. J. Comput. High. Educ. **15**(1), 46–64 (2003). [Electronic resource] Access mode: http://www.ugr.es/~victorhs/gbd/docs/10.1.1.9.9018.pdf

Iqbal, S.T., Horvitz, E.: Disruption and Recovery of Computing Tasks (2007). [Electronic resource] Access mode: http://research.microsoft.com/en-us/um/people/horvitz/chi_2007_iqbal_horvitz.pdf

Jones, Q., Ravid, G., Rafaeli, S.: Information overload and the message dynamics of online interaction spaces: a theoretical model and empirical exploration. Inf. Syst. Res. **15**(2), 194–210 (2004). http://www.ravid.org/gilad/isr.pdf

Jones, C., Shao, B.: The NET Generation and Digital Natives: Implications for Higher Education. Higher Education Academy, York (2011). [Electronic resource] Access mode: http://oro.open.ac.uk/30014/

Jungo, R.: In-class multitasking and academic performance. Comput. Hum. Behav. **28**(6), 2236–2243 (2012)

Keen, A.: The Cult of the Amateur: How Today's Internet is Killing Our Culture. Nicholas Bradley Publishing, London (2007). [Electronic resource] Access mode: http://www.tc.umn.edu/~mill3239/home/presentations/panel3–1.pdf

Kirsch, D.: Few thoughts on cognitive overload. Intellectica **1**(30), 19–51 (2000). [Electronic resource] Access mode: http://adrenaline.ucsd.edu/kirsh/Articles/Overload/Cognitive_Overload.pdf

Levine, L.E., Waite, B.M., Bowman, L.L.: Can students really multitask? Comput. Educ. **54**(4), 927–931 (2010)

Lavie, N.: Distracted and confused? Selective attention under load. Trends Cogn. Sci. **9**(2), 75–82 (2005)

McPeak, D.: Multitasking vs Switch-tasking (2013). [Electronic resource] Access mode: http://inci-dent-prevention.com/ip-articles/multitasking-vs-switch-tasking-what-s-the-difference

Medina, J.: Brain Rules: 12 Principles for Surviving and Thriving at Work, Home, and School. Pear Press, Seatle (2008). [Electronic resource] Access mode: http://ext100.wsu.edu/pierce/wp-content/uploads/sites/9/2014/01/BrainRules-JohnMedina-MediaKit.pdf

Ophir, E., Nass, C., Wagner, A.D.: Cognitive control in media multitaskers. Proc. Nat. Acad. Sci. US Am. **106**(37), 15583–15587 (2009). [Electronic resource] Access mode: https://doi.org/10.1073/pnas.0903620106

Prensky, M.: From Digital Natives to Digital Wisdom: Hopeful Essays to 21st Century Learning. Corwin Press (2012). [Electronic resource] Access mode: http://marcprensky.com/from-digital-natives-to-digital-wisdom

Ravizza, S., Hambrick, D., Fenn, K.: Non-academic internet use in the classroom is negatively related to classroom learning regardless of intellectual ability. Comput. Educ. **10**(4), 109–114 (2014)

Rosen, L.: Teaching the iGeneration. Educ. Leadersh. **68**(5), 10–15 (2007). [Electronic resource] Access mode: http://www.ascd.org/publications/educational-leadership/feb11/vol68/num05/Teaching-the-iGeneration.aspx

Rosen, L.D., Carrier, L.M., Cheever, N.A.: Facebook and texting made me do it: media-induced task-switching while studying. Comput. Hum. Behav. **29**(3), 948–958 (2013)

Rubinstein, J., Meyer, D., Evans, J.: Executive control of cognitive processes in task switching. J. Exp. Psychol. Hum. Percept. Perform. **27**(4), 763–797 (2001)

Sanbonmatsu, D., Strayer, D., Medeiros-Ward, N., Watson, J.: Who multi-tasks and why? Multi-tasking ability, perceived multi-tasking ability, impulsivity, and sensation seeking. PLoS ONE **8**(1) (2013). [Electronic resource] Access mode: http://www.plosone.org/article/info:doi/10.1371/journal.pone.0054402

Savolainen, P., Mannering, F.: Effectiveness of motorcycle training and motorcyclists' risk-taking behavior. transportation research record. J. Transp. Res. Board, **2031**(2007), 52–58 (2008). [Electronic resource] Access mode: http://trb.metapress.com/content/3332n8q718k25830

Small, G., Vorgan, G.: Ibrain: Surviving the Technological Alteration of the Modern Mind. Collins Living, NewYork (2008). [Electronic resource] Access mode: http://www.barnesandnoble.com/sample/read/9780061340345

Study: Multitasking Hinders Learning. eSchool News, 26 July 2006. [Electronic resource] Access mode: http://www.eschoolnews.com/2006/07/26/study-multitasking-hinders-learning

Verhaeghen, P., Salthouse, T.: Meta-analyses of age-cognition relations in adulthood: estimates of linear and nonlinear age effects and structural models. Psychol. Bull. **122**(3), 231–249 (1997). [Electronic resource] Access mode: http://www.ncbi.nlm.nih.gov/pubmed/9354147 (1997)

Viadero, D.: Instant Messaging Found to Slow Students' Reading. Education Week, 15 Aug 2008. [Electronic resource] Access mode: www.edweek.org/ew/articles/2008/08/27/01im. h28.html

Willingham, D.: Have technology and multitasking rewired how students learn? Am. Educ. **34**(2), 23–28 (2009). [Electronic resource] Access mode: http://er-ic.ed.gov/?id=EJ889151

Health Management Informatics

Medical Care Safety - Problems and Perspectives

Yuriy Voskanyan[1] ⓘ, Irina Shikina[2,3(✉)] ⓘ, Fedor Kidalov[4] ⓘ,
and David Davidov[2] ⓘ

[1] Russian Medical Academy of Continuing Professional Education,
Ministry of Health of Russia, 2/1 Barrikadnaya Street, Building 1,
Moscow 125993, Russia
[2] Central Research Institute for Organization and Informatization of Medical
Care, Ministry of Health of Russia, 11 Dobrolyubova Street,
Moscow 127254, Russia
shikina@mednet.ru
[3] Central State Medical Academy Office of the President of the Russian
Federation, 19 Marshala Timoschenko Street, Building 1A,
Moscow 121359, Russia
[4] State Government Institution of Moscow City "Informational-Analytical
Center" for Medical Care, 10 Basmannaya Novaya Street, Building 1,
Moscow 107078, Russia

Abstract. The article presents modern views on the essence of the term "medical care safety", and describes the relevance of the topic, systemic approaches to adverse event risk management in medicine. The authors consider the term "medical care safety" as the ratio of the benefit for the patients and the risk of harm to the patient and medical staff, as well as the risk of unfavorable changes in the internal and external environment. Additional harm related to medical care (adverse events) is observed in 10.6% of hospitalized patients. Medical errors, as a cause of preventable harm, account for 45.5% of cases, other adverse events are classified as unpreventable. The cumulative probability of severe harm among patients with medical care-related complications is 11.8%, of unexpected death – 5.3%. Deaths due to adverse events account for 24.9% of hospital mortality and 9.7% of population deaths. High percentage of deaths unrelated to the progression of the underlying disease and comorbidities allows the authors to define medical care as a high-risk type of services. Systemic causes of medical care-related additional harm are latent threats that are constantly present and subsequently transformed into incidents and adverse events at the level of medical staff, patients and the environment in which medical care is provided. Identification of all latent threats and subsequent management of their transformation is the basis of the modern strategy of medical care safety management. It is impossible to ensure high level of safety in medical facilities without formation of a new culture not only at the level of an individual medical facility but also at the level of the government, the entire medical care system and the society.

Keywords: Medical care safety · Adverse events · Incident · Unexpected death

© Springer Nature Switzerland AG 2020
T. Antipova (Ed.): ICIS 2019, LNNS 78, pp. 291–304, 2020.
https://doi.org/10.1007/978-3-030-22493-6_26

1 Introduction

International studies in the field of medical care safety demonstrate the special role of medical errors and related adverse events, as well as unexpected deaths in the structure of hospital mortality and population mortality in developed countries. The key issues in additional harm risk management include definition of the essence of medical care safety, assessment of the problem severity, identification of systemic reasons for adverse events, formation of a new culture of safety, development, implementation and standardization of effective solutions taking into account the probability of incidents in medicine and the severity of their consequences [1, 2].

The authors have formulated a new approach to the definition of the term "medical care safety"; the epidemiology of medical care-related events have been studied and the basic principles of modern strategy of adverse event risk management in medicine have been formulated. Adverse events included unintentional physical or psychological trauma (additional harm), which was most likely related to medical care rather than the course of the main disease or concomitant diseases [3–6]. Information search was carried out by two researchers independently over a period of 1990–2017 using medical databases MEDLINE, Cochrane Collaboration; EMBASE, SCOPUS, ISI Web of Science. For the analysis, prospective and retrospective observational studies of high methodological quality were used, which presented data on the frequency and severity of adverse events in multispecialty short-stay hospitals. The frequency of new cases of adverse events (incidence) as well as the percentages of adverse events are shown together with confidence intervals at the 95% probability of accepting the null hypothesis. Published source data pooling was carried out in a meta-analysis, the mathematical model of which was determined depending on the statistical heterogeneity index I^2. In similar studies, the main principles of medical care management were evaluated. The authors necessarily took into account the relation between the principles used and the possibility of managing the systemic causes of additional harm.

2 Result and Discussion

Safety, along with clinical effectiveness and economic efficacy, is an important attribute of the medical care quality. However, it is necessary to strictly define the limits of this term. The majority of definitions (US Institute of Medicine, US Agency for Healthcare Research and Quality, World Health Organization etc.) connect the term of safety with the lack of unnecessary harm, minimization of its probability, or with the lack of unnecessary risk of adverse events [1]. All such definitions raise new questions related to the need to interpret such concepts as "unnecessary harm", "unnecessary risk", etc. However, most importantly, none of these definitions can quantitatively assess the degree of safety, which makes the management of adverse events difficult to implement. We believe that the medical care safety can be defined as the benefit/risk ratio when providing medical care. The mathematical expression of the harm will be its risk, which is the product of probability of an adverse event by the severity of its consequences. The mathematical expression of the benefit will be the probability of a planned favorable clinical outcome in the absence of the risk of additional harm. Without going into details of mathematical calculations using the values of the probability and rank coefficients of the assessment of the harm severity and the grade of the benefit, we will pay attention only to the fact that, from our point of view, it is not correct to eliminate reasonable risk of the concept "medical care safety". After all, is there any difference for the patient why he/she got hurt rather than benefited from medical interventions in which there was a reasonable risk? It was the disagreement of scientists with this "reasonable risk" that led to the rapid development of medical technologies, including surgery and drug therapy, which significantly reduced the amount of "reasonable risk" and significantly expanded the scope of their application in the target patient cohort. Next, it should be emphasized that the medical care safety is a complex concept that includes four components depending on the object of the potential risk of harm: patient's safety (unintentional physical and/or psychological trauma), staff safety (biological accidents, radiation exposure etc.), internal environment safety (high level of physical factors, pollution, state of the building and engineering systems etc.), safety of the environment (the impact of physical factors, chemical pollution etc.). This is what distinguishes safety in medicine from safety in other areas of economic activity where there are only three components (personnel, internal or external environment). Therefore, we suggest defining the term "medical care safety" as the ratio of the benefit for the patients and the risk of harm to the patient and medical staff, as well as the risk of unfavorable changes in the internal and external environment.

To assess the incidence and severity of additional harm (adverse events) when providing medical care, we selected 14 publications of high methodical quality from 9 countries: USA, Canada, UK, Denmark, New Zealand, Netherlands, Spain, Norway, Brazil. In these articles, the information about the results of inpatient treatment of 124,458 patients in 197 short-stay hospitals was analyzed. The index of statistical heterogeneity I^2 of the source data was equal to 0.37, so a fixed effects model based on binary data was chosen for the analysis. The cumulative probability of adverse events in inpatients was 10.6% of the total number of hospitalized patients [3–16] (Table 1).

Table 1. Probability of adverse events

Author, year of publication	Country	Number of hospitals	Number of observations	Frequency, % (95% CI)
Retrospective studies				
Brennan T. et al., 1991 [3]	USA (Harvard)	51	30,195	3.7 (3.5–3.9)
Wilson R. et al., 1995 [4]	Australia	28	14,210	16.6 (15.9–17.2)
Thomas E. et al., 2000 [5]	USA (Utah, Colorado)	28	14,565	5.4 (5.0–5.8)
Vincent C. et al., 2001 [6]	United Kingdom	2	1,014	108 (8.9–12.8)
Schioler T. et al., 2001 [7]	Denmark	17	1,097	10.4 (8.6–12.2)

(continued)

Table 1. (*continued*)

Author, year of publication	Country	Number of hospitals	Number of observations	Frequency, % (95% CI)
Davis P. et al., 2002 [8]	New Zealand	13	6,579	12.9 (12.1–13.7)
Baker G. et al., 2004 [9]	Canada	20	3,745	6.8 (6.0–7.6)
Zegers M. et al., 2009 [10]	Netherlands	21	7,926	8.4 (7.8–9.0)
Landrigan C. et al., 2010 [11]	USA (North Carolina)	10	2,341	18.1 (16.5–19.6)
Classen D. et al., 2011 [12]	USA (Massachusetts)	3	795	33.2 (29.9–36.5)
Deilkas E. et al., 2015 [13]	Norway	20	40,581	14.5 (14.3–15.0)
Prospective studies				
Andrews L. et al., 1997 [14]	Spain	3	1,047	17.7 (15.4–20.0)
Wanzel K. et al., 2000 [15]	Canada	1	192	39.1 (32.2–46.0)
Szlief C. et al., 2012 [16]	Brazil	1	171	55.0 (47.5–62.4)
Meta-analysis	–	197	124,458	10.6 (10.5–10.8)

Most of them (41.3%) were related to surgery, 14.4% - with drug therapy, 9.8% - with late or incorrect diagnosis, 9.1% - manipulations; 8.4% - incorrect treatment plan [3–5, 9, 10, 15] (Table 2).

Table 2. Types of medical interventions and percentages of adverse events

Types of medical interventions	Source						Meta analysis: percentage % (95% CI)
	Brennan T. et al., 1991 [3] (n = 1,117)	Wilson R. et al., 1995 [4] (n = 2,952)	Thomas E. et al., 2000 [5] (n = 787)	Wanzel M. et al., 2000 [15] (n = 32)	Baker G. et al., 2004 [9] (n = 360)	Zegers M. et al., 2009 [10] (n = 744)	
	Percentage %						
Surgery	47.7	39.3	44.9	31.2	34.2	54.2	41.3 (40.1–42.6)
Manipulation	7.0	6.7	13.5	9.4	7.2	17.0	9.1 (8.4–9.8)
Drug therapy	19.4	8.4	19.3	15.6	23.6	15.3	14.4 (13.5–15.3)
Incorrect treatment plan	7.5	9.3	4.3	15.6	11.9	5.1	8.4 (7.7–9.1)
Late or incorrect diagnosis	8.1	10.6	6.9	28.1	10.6	6.3	9.8 (9.0–10.5)
Other interventions	10.3	25.7	11.1	–	12.5	2.1	17.0 (16.1–18.0)
Total	100.0	100.0	100.0	100.0	100.0	100.0	–

Only 45.5% of adverse events were considered preventable [3, 4, 6, 10, 15, 17–20], and in 27.7% of cases additional harm was caused by negligence in the actions of the personnel [3, 5, 21] (Tables 3 and 4).

Table 3. Preventable adverse events

Author, year of publication	Total number of events	Number of preventable events	Percentage % (95% CI)
Hospital			
Brennan T. et al., 1991 [3]	1,117	308	27.6 (25.0–30.2)
McGuire H. et al., 1992 [17]	2,409	1,180	49.0 (47.0–51.0)
O'Neil A. et al., 1993 [18]	133	83	62.4 (54.2–70.6)
Wilson R. et al., 1995 [4]	2,302	1,178	51.2 (49.1–53.2)
Wanzel K. et al., 2000 [15]	192	88	45.8 (38.8–52.9)
Vincent C. et al., 2001 [6]	119	57	47.9 (38.9–56.9)
Schioler T. et al., 2001 [7]	114	46	40.3 (31.3–49.4)
Davis P. et al., 2002 [8]	850	315	37.1 (33.8–40.3)
Baker G. et al., 2004 [9]	289	106	36.7 (31.1–42.2)
Zegers M. et al., 2009 [10]	663	283	42.7 (38.9–46.4)
Outpatient clinic			
Singh H. et al., 2004 [19]	308	108	35.1 (29.7–40,4)
Woods D. et al., 2007 [20]	2,608	1,296	49.7 (47.8–51.6)
Meta-analysis	11,104	5,048	45.5 (44.5–46.4)

Table 4. Adverse events caused by negligence

Author, year of publication	Total number of events	Number of negligence-related events	Percentage % (95% CI)
Brennan T. et al., 1991 [3]	1,117	306	27.4 (24.8–30.0)
Leap L. et al., 1991 [21]	1,133	280	24.7 (22.2–27.2)
Thomas E. et al., 2000 [5]	787	236	30.0 (26.8–33.2)
Meta-analysis	3,037	822	27.1 (25.5–28.6)

Severe harm and disability in patients with adverse events were observed in 11.8% cases, and unexpected death - in 5.3% of cases [4–6, 8–11, 15, 22] (Table 5).

Table 5. Severe harm, disability and unexpected death in case of an adverse event

Source	Number of adverse events	Severity of harm			
		Severe harm and disability		Death	
		Absolute number	Percentage % (95% CI*)	Absolute number	Percentage % (95% CI)
Wilson R. et al., 1995 [4]	2,324	315	13.7 (12.3–15.1)	112	4.9 (4.0–5.8)
Thomas E. et al., 2000 [5]	787	130	16.6 (13.9–19.1)	52	6.6 (4.9–8.3)
Wanzel K. et al., 2000 [15]	144	10	6.9 (2.8–11.1)	2	1.4 (0.5–3.3)
Vincent C. et al., 2001 [6]	110	7	6.4 (1.8–10.9)	9	8.2 (3.1–13.3)
Davis P. et al., 2002 [38]	850	87	10.2 (8.2–12.3)	38	4.5 (3.1–5.9)
Baker G. et al., 2004 [9]	289	14	4.8 (2.4–7.3)	46	15.9(11.7–20.1)
Andrews J. et al., 2006 [22]	655	90	13.7 (11.1–16.4)	15	2.3 (1.1–3.4)
Zegers M. et al., 2009 [10]	663	33	5.0 (3.3–6.6)	52	7.8 (5.8–9.9)
Landrigan C. et al., 2010 [11]	588	67	11.4 (8.8–14.0)	14	2.4 (1.1–3.6)
Meta-analysis	6,388	753	11.8 (11.0–12.6)	340	5.3 (4.8–5.9)

The analysis of cumulative frequency of hospital mortality rate in a number of countries showed quite a large percentage of unexpected death caused by adverse events - 24.9% [4, 6, 9, 10, 22–29] (Table 6).

The authors from the Johns Hopkins Hospital [28] showed that adverse events form the third cause of mortality in the USA accounting for every tenth death in the country (Table 7).

Table 6. Hospital mortality rate and unexpected deaths

Source	Country	Percentage of unexpected deaths % (proportion)	Hospital mortality rate % (proportion)	Percentage of unexpected deaths of hospital mortality rate % (95% CI*)
Wilson R. et al., 1995 [4]	Australia	0.79 (112/14210)	1.90 (270/14210)	41.5 (35.6–47.4)
Vincent C. et al., 2001 [6]	United Kingdom	0.89 (9/1014)	–	26.6 (21.8–31.3)
Campbell M. et al., 2011 [23]		–	3.35 (1581358/47172030)	
Baker G. et al., 2004 [9]	Canada	1.23 (46/3745)	–	34.2 (30.0–38.3)
Canad. Inst. of Health Inf., 2005 [25]		–	3.60 (109989/3058901)	
Andrews J. et al., 2006 [22]	Spain	0.27 (15/5624)	–	20.8 (14.9–26.6)
Aiken L. et al., 2014 [26]		–	1.3 (283/21520)	
Zegers M. et al., 2009 [10]	Netherlands	0.66 (52/7926)	–	17.2 (14.0–20.4)
Jarman B. et al., 2010 [27]		–	3.84 (90873/2363332)	
Makary M. et al., 2016 [28]	USA	0.71 (251454/35416020)	–	34.8 (29.3–40.3)
Hall M. et al., 2013 [29]	USA	–	2.04 (715000/35049019)	
Meta-analysis	–	0.71 (251688/35448539)	2.85 (2497773/87679012)	24.9 (24.9–24.9)

Table 7. Mortality rate structure in the US population (Makary et al. [28])

Cause of death (2013)	Number of deaths	Percentage % (95% CI)
Cardiovascular disorders	614,348	23.6 (23.6–23.7)
Neoplasms	591,699	22.8 (22.7–22.8)
Harm related to provision of medical care	251,454	9.7 (9.7–9.7)
Chronic respiratory diseases	147,101	5.7 (5.6–5.7)
Unintentional damage	136,053	5.2 (5.2–5.3)
Cerebrovascular accident	133,103	5.1 (5.1–5.1)
Alzheimer's disease-related complications	93,541	3.6 (3.6–3.6)
Diabetes-related complications	76,488	2.9 (2.9–3.0)
Influenza and pneumonia	55,227	2.1 (2.0–2.0)
Kidney diseases	48,146	1.8 (1.8–1.9)
Suicide	42,773	1.6 (1.6–1.7)
Other causes	407,060	15.7 (15.6–15.7)
TOTAL	2,596,993	100.0

The causes and the mechanism of development of adverse events were studied in 20 publications of high methodological quality. The main reasons for additional harm were latent threats. These threats are not directly related to the source of the adverse event, are constant and do not carry any danger if they are inactive [21, 30, 31]. Under certain conditions, a latent threat is activated and turns into a dangerous situation (active threat 1), which, in its turn, lead to the development of dangerous processes (active threats 2) - unsafe actions (errors) of the personnel, unsafe patient behavior, unsafe processes in the environment in which medical care is provided. The result of a dangerous situation are dangerous events (incidents), which in the international literature are called incidents. The incident might cause harm to the patient (an incident without sequelae) or end with harm (an incident with sequelae) or might lead to the patient's death (a critical incident) [30–32].

Latent threats exist and transform into adverse events at three levels: at the level of the staff, at the level of the patient and at the level of the environment. At each of these levels, global latent threats (that are present regardless of the site of medical care provision and its profile) and specific latent threats (caused by the specific site of medical care provision and its profile) have been described. Quite often, a transformation of a latent threat moves from one level to the other [32–34]. In most cases, the incidents and resulting adverse events are the result of the transformation of several latent threats followed by a series of active threats that coincide in time and space [32].

So far, four groups of global latent threats have been studied at the staff level: related to personnel management (management system, procedural rules, communication (including identification and verification), team work); associated with personnel selection (staff positions, staff turnover, employment of part-time employees) associated with personnel competence (low baseline competence, freedom in the implementation of official duties, acquired competence deficiency) related to mental state and physiological condition of the personnel (personal problems and disease, a distrust in leadership and procedural rules, low level of commitment to procedural norms)

[30, 33–35]. These threats turn into a dangerous situation during the provision of medical care and lead to dangerous events – staff errors (blunder, miscalculation, omission, violation) [36, 37].

Global threats at the patient level can be divided into three groups: caused by mental state and physiological condition of the patient (pain, physical and mental disorders), caused by the personal characteristics of the patient (low general educational status, insufficient level of medical literacy, low level of motivation to fulfill medical prescriptions), caused by the personal data features (coincidence of personal data). The described threats turn into a dangerous situation at the time of the patient's movement, his/her communication with the staff, the implementation of medical prescriptions and in the process of self-monitoring of the patient's condition. As a result, dangerous events develop - incorrect actions (lack of actions) of the patient, or staff errors in the process of staff-patient interactions [33, 38–41].

At the level of the environment in which medical care is provided, two groups of global latent threats have been studied: those related to the social environment and those related to the technological environment. In the technological environment, there are threats associated with the workplace (tools and objects of labor, workspace) and threats associated with the building (constructions, engineering and logistics systems). The environmental threats turn into a dangerous situation during the provision of medical care [30–32, 35, 38]. As a result, unsafe processes occur in the environment itself (accidents, failures, failures in the building, equipment, engineering systems, direct harmful effects of physical, chemical or biological factors). Other variants of transformation of the latent threats of the environment include unsafe patient's behavior (e.g., stumbling) or a staff error that occurred in the process of its interaction with the tools (items) of labor and workspace [33, 38, 40, 41].

3 Conclusion

Successful experience of safety management in developed countries has shown that the solution to this problem can not be implemented only at the level of one medical organization. A comprehensive approach at the level of the public authorities, medical care system and society is needed. In the first case, it is necessary to make significant changes in the regulatory framework. These changes should define medical care as a high-risk service, admit the fact of existence of medical errors and adverse events, guarantee medical personnel the rights and freedoms in case of detection and recording of cases of additional harm to the patient, to formalize the powers and responsibilities of the patient, to provide additional funding to ensure the target level of the patient's safety, staff and the environment. The medical care system should adopt a new safety culture taking into account the existing high probability of adverse events and systemic causes of additional harm, provide management for the development and standardization of effective solutions to prevent the risk of additional harm, justify additional funding for activities and the necessary infrastructure to ensure the target level of safety in medical organizations. At the society level, it is necessary to arrange a discussion on the issues of medical care safety, the result of which should be an understanding of the need to form a single team of healthcare providers, patients and their relatives in the process of providing medical care - a team that will provide comprehensive management control of the feasibility, timeliness, occupancy and quality of implementation of the chosen treatment plan.

The modern strategy of medical care safety management in a medical organization includes the development of a new safety culture, arrangement of the incidents and threat accounting, incident stratification, the definition of the scope of incident management and the scope of activities, the introduction of the incident investigation algorithm, the standardization of the process of preventing the transformation of a latent threat into an adverse event. The introduction of this strategy usually includes two stages. At the first stage, the described events relate to latent threats, which ended in the development of incidents and adverse events. At the second stage, all latent threats at the clinic are identified and managed including those which did not end in the development of incidents [1, 32, 34, 35].

The new safety culture, which is part of the corporate culture, is based on the following provisions: incidents and adverse events are inevitable parts of medical care, they are based on permanent systemic causes - latent threats; safety is ensured by eliminating systemic causes and guaranteed not only by individual skills of the performer, but an integrated system that provides for the creation of conditions that prevent the transformation of a latent threat into an adverse event; prevention of harm associated with the provision of medical care is proactive, which implies the management of all latent threats identified in the clinic. The level of safety culture is assessed by means of special questionnaires, taking into account the orientation to the listed values of the top management, staff awareness of safety issues, staff perception of the general level of safety, staff commitment to procedural standards and their trust in management, as well as the quality of team management and the quality of communication [1, 30, 32, 34].

The accounting system includes obtaining information about the threats and incidents, their identification, registration, monitoring and measurement. For accounting, most countries use character encoding of threats and incidents suggested by the National Coordinating Council for Medication Error Reporting and Prevention (NCC MERP, 1998–2001) [1, 31, 33, 42, 43]. A big problem related to objective accounting is concealment and (or) masking of incidents and adverse events. In case of concealment, the incidents are simply not reported, in case of masking, they are interpreted as complications related to the course of the primary or concomitant diseases. The most common object of concealment and masking is an infection related to provision of medical care. Several effective solutions have been suggested in order to prevent concealment and masking of incidents: collection of information not only from medical records but also from colleagues, patients and independent auditors as well as accounting automation [30, 31, 33, 35, 38]. Another quite effective solution is a system of registration of procedure-related incidents and their consequences. An example is a system Global Trigger Tool used for this purpose in Europe and the USA. Procedure-related events include "rigid" (independent of a subjective factor) process indicators, non-typical complications, abnormal behavior and non-typical patient's condition as well as all cases of unexpected deaths. Information about a procedure-related event is the reason for a thorough audit, as a result of which it is usually possible to reveal medical care-related incidents [10, 44, 45].

Incidents' stratification determines the grade of their risk (a hazard class). For stratification, two main criteria measured in rank coefficients are used. They include the severity of the harm and repeatability (incidence) of the incident. The hazard class of the incident is determined in accordance with the classification proposed by experts of the UK National Healthcare System - NHS Commissioning Board Authority, which distinguishes four risk groups, each of which regulates the scope of activities, the scope of management and the possibility of further medical care provision. While investigating the incident, active and latent threats are identified and the adverse event route map is formed [32, 33, 35, 46, 47].

Prevention of transformation of the latent threat into an adverse event is achieved by the development and implementation of a management standard that envisages the elimination of preventable and minimization of the impact of unpreventable latent threats. For this purpose, a multilevel protection system envisaging the block of each stage of the latent threat transformation is normally created [21, 32, 33, 35, 48–50].

References

1. Roitberg, G.E., Kondratova, N.: Medical Organization According to the International Quality Standards: A Practical Guide to Implementation, 152 p. Moscow (2018) (In Russia)
2. World Health Organization: World Health Statistics: Monitoring Health for the SDGs, Sustainable Development Goals. Geneva. http://www.who.int/gho/publications/world_health_statistics/2018/en/ (2018)
3. Brennan, T.A., Leap, L.L., Larid, N.M., et al.: Incidence of adverse events and negligence in hospitalized patients: results of the Harvard Medical Practice Study I. N. Engl. J. Med. **324**, 370–376 (1991)
4. Wilson, R.M., Runciman, W.B., Gibberd, R.W., et al.: The quality in Australian medical care study. Med. J. Aust. **163**, 458–471 (1995)
5. Thomas, E.J., Studdert, D.M., Burstin, H.R., et al.: Incidence and types of adverse events and negligent care in Utah and Colorado. Med. Care **38**(3), 261–271 (2000)
6. Vincent, C., Neale, G., Woloshynowych, M.: Adverse events in British hospitals: preliminary retrospective record review. BMJ **322**(7285), 517–519 (2001)
7. Schioler, T., Lipczak, H., Pedersen, B.L., et al.: Incidence of adverse events in hospitals. A retrospective study of medical records. Ugeskr Laeger **163**(39), 5370–5378 (2001)
8. Davis, P., Lay-Yee, R., Briant, R., et al.: Adverse events in New Zealand public hospitals I: occurrence and impact. NZMJ **115**(1167), U271 (2002)
9. Baker, G.R., Norton, P.G., Flintoft, V., et al.: The Canadian adverse events study: the incidence of adverse events among hospital patients in Canada. CMAJ **170**(11), 1678–1686 (2004)
10. Zegers, M., Bruijne, M.C., Wagner, C., et al.: Adverse events and potentially preventable deaths in Dutch hospitals: results of a retrospective patient record review study. Qual. Saf. Med. Care. **18**, 297–302 (2009)
11. Landrigan, C.P., Parry, G.J., Bones, C.P., et al.: Temporal trends in rates of patient harm resulting from medical care. N. Engl. J. Med. **363**, 2124–2134 (2010)
12. Classen, D.C., Resar, R., Griffin, F., et al.: «Global trigger tool» shows that adverse events in hospitals may be ten times greater than previously measured. Health Aff. **30**, 4581–4589 (2011)

13. Deilkås, E.T., Bukholm, G., Lindstrøm, J.C., Haugen, M.: Monitoring adverse events in Norwegian hospitals from 2010 to 2013. BMJ Open 5, 1–6 (2015)
14. Andrews, L.B., Stocking, C., Krizek, T., et al.: An alternative strategy for studying adverse events in medical care. Lancet 349(9048), 309–313 (1997)
15. Wanzel, K.R., Jamieson, C.G., Bohnen, J.M.A.: Complications on a general surgery service: incidence and reporting. CJS. 43(2), 113–117 (2000)
16. Szlejf, C., Farfel, J.M., Curiati, J.A., et al.: Medical adverse events in elderly hospitalized patients: a prospective study. Clin. (Sao Paulo). 67(11), 1247–1252 (2012)
17. McGuire, H.H.J., Horsley, J.S., Salter, D.R., Sobel, M.: Measuring and managing quality of surgery. Statistical vs incidental approaches. Arch. Surg. 127(6), 733–737 (1992)
18. O'Neil, A.C., Petersen, L.A., Cook, E.F., et al.: Physician reporting compared with medical-record review to identify adverse medical events. Ann. Intern. Med. 119(5), 370–376 (1993)
19. Singh, H., Spitzmueller, C., Petersen, N.J., et al.: Primary care practitioners' views on test result management in EHR-enabled health systems: a national survey. J. Am. Med. Inform. Assoc. 0, 1–9 (2012)
20. Woods, D.M., Thomas, E.J., Holl, J.L., et al.: Ambulatory care adverse events and preventable adverse events leading to a hospital admission. Qual. Saf. Med. Care. 16(2), 127–131 (2007)
21. Leap, L.L., Brennan, T.A., Nan Laird, M.P.H., et al.: The nature of adverse events in hospitalized patients. Results of the Harvard medical practice study II. N. Engl. J. Med. 324, 377–384 (1991)
22. Andrews, J.M., Remon, C.A., Burillo, J.V., Lopez, P.R.: National Study on Hospitalisation-Related Adverse Events ENEAS 2005. Quality Plan of National Health System. Report. http://www.who.int/patientsafety/information_centre/reports/ENEAS-EnglishVersion-SPAIN.pdf, February 2006. Access 7 July 2018
23. Campbell, M.J., Jacques, R.M., Fotheringham, J., et al.: Developing a summary hospital mortality index: retrospective analysis in English hospitals over five years. BMJ 344, 1001–1012 (2012)
24. Wilson, L., Ferguson, C., Hider, Ph., et al.: Perioperative Mortality in New Zealand: Fourth Report of the Perioperative Mortality Review Committee. Report to the Health Quality & Safety Commission New Zealand. http://www.nzma.org.nz/journal/read-the-journal/all-issues/2010-2019/2015/vol-128-no-1424-30-october-2015/6712, June 2015. Access 7 July 2018
25. Hospital Trends in Canada. Canadian Institute for Health Information. National Health Expenditure Database. 2005 Ottawa, Ontario. https://secure.cihi.ca/free_products/Hospital_Trends_in_Canada_e.pdf. Access 7 July 2018
26. Aiken, L.H., Sloane, D.M., Bruyneel, L., et al.: Nurse staffing and education and hospital mortality in nine European countries: a retrospective observational study. Lancet 26, 1–7 (2014)
27. Jarman, B., Pieter, D., van der Veen, A.A., et al.: The hospital standardised mortality ratio: a powerful tool for Dutch hospitals to assess their quality of care? Qual. Saf. Med. Care. 19, 9–13 (2010)
28. Makary, M.A., Daniel, M.: Medical error—the third leading cause of death in the US. BMJ 353(3), 1–5 (2016)
29. Hall, M.J., Levant, S., De Frances, C.J.: Trends in inpatient hospital deaths: national hospital discharge survey, 2000–2010 (U.S. DHHS - Centers for Disease Control and Prevention), 118. NCHS Data Brief. March 2013
30. Beuzekom, M., Boer, F., Akerboom, S., Hudson, P.: Patient safety: latent risk factors. Br. J. Anaesth. 105(1), 52–59 (2010)

31. Lawton, R., Carruthers, S., Gardner, P., et al.: Identifying the latent failures underpinning medication administration errors: an exploratory study. Health Serv. Res. **47**(4), 1437–1459 (2012)
32. Reason, J.: Human error: models and management. Br. Med. J. **320**, 768–770 (2000)
33. Hoffmann, B., Rohe, J.: Patient safety and error management. Dtsch. Arztebl. Int. **107**(6), 92–99 (2010)
34. Mitchell, R., Williamson, A., Molesworth, B., Chung, A.: A review of the use of human factors classification frameworks that identify causal factors for adverse events in the hospital setting. Ergonomics **57**(10), 1443–1472 (2014)
35. Carayon, P., Schoofs Hundt, A., Karsh, B., et al.: Work system design for patient safety: the SEIPS model. Qual. Saf. Medical Care **15**(Suppl I), 150–158 (2006)
36. Clancy, C., Tornberg, D.: TeamSTEPPS: assuring optimal teamwork in clinical settings. Am. J. Med. Qual. **22**(3), 214–217 (2007)
37. Edmondson, A.: Learning from failure in medical care: frequent opportunities, pervasive barriers. Qual. Saf. Med. Care. **13**(Suppl II), 113–119 (2004)
38. Lyons, M.: Should patients have a role in patient safety? A safety engineering view. Qual. Saf. Med. Care **16**(2), 140–142 (2007)
39. Verstappen, W., Gaal, S., Esmail, A., Wensing, M.: Patient safety improvement programmes for primary care. Review of a Delphi procedure and pilot studies by the LINNEAUS collaboration on patient safety in primary care. Eur. J. Gen. Pract. **21**(Suppl 1), 50–55 (2015)
40. Molloy, G.J., O'Boyle, C.A.: The SHEL model: a useful tool for analyzing and teaching the contribution of human factors to medical error. Acad. Med. **80**(2), 152–155 (2005)
41. Takayanagi, K., Hagihara, Y.: Revised sunflower-SHELL model–an analysis tool to ensure adverse-events' factor analysis and followed by patient safety strategy. Jpn. Hosp. **25**, 11–18 (2007)
42. Pronovost, P., Weast, B., Holzmueller, C., et al.: Evaluation of the culture of safety: survey of clinicians and managers in an academic medical center. Qual. Saf. Med. Care **12**, 405–410 (2003)
43. Lilford, R., Mohammed, M., Braunholtz, D., Hofer, T.: The measurement of active errors: methodological issues. Qual. Saf. Med. Care **12**(Suppl II), 118–1112 (2003)
44. Hibbert, P., Williams, H.: The use of a global trigger tool to inform quality and safety in Australian general practice: a pilot study. Aust. Fam. Physician **43**(10), 723–726 (2014)
45. Michel, Ph, Quenon, J., de Sarasqueta, A., Scemama, O.: Comparison of three methods for estimating rates of adverse events and rates of preventable adverse events in acute care hospitals. BMJ **328**(24), 199–202 (2004)
46. Pietra, L., Calligaris, L., Molendini, L., et al.: Medical errors and clinical risk management: state of the art. Acta. Otorhinolaryngol. Ital. **25**, 339–346 (2005)
47. Shaw, R., Drever, F., Hughes, H., et al.: Adverse events and nearmiss reporting in the NHS. Qual. Saf. Med. Care. **14**, 279–283 (2005)
48. Shikina, I.B., Vardosanidze, S.L., Voskanyan, Yu.E., Sorokina, N.V.: Problems of patients' safety assurance in modern medical care, 336 p. Publishing House Glossarium LLC, Moscow (2006) (In Russian)

49. Joint Commission International Accreditation Standards for Hospitals 6th Edition. Effective. Including Standards for Academic Medical Center Hospitals, 1 July 2017, 37 p. https://www.jointcommissioninternational.org/assets/3/7/JCI_Hosp_Standards_6th_STANDARDS_ONLY_14Jan2018 pdf. Access 7 July 2018
50. Zadvornaya, O.L., Voskanyan, Y.E., Shikina, I.B., Borisov, K.N.: Socio-economic aspects of medical errors and their consequences in medical organizations. MIR (Modernization. Innovation. Research) **10**(1), 99–113 (2019). https://doi.org/10.18184/2079-4665.2019.10.1.99-113 (In Russia)

Materials Science in Digital Age

Stability of the Deep Neural Networks Learning Process in the Recognition Problems of the Material Microstructure

A. V. Klyuev, V. Yu. Stolbov, M. B. Gitman, and R. A. Klestov[✉]

Perm National Research Polytechnic University, Perm, Russia
klestovroman@gmail.com

Abstract. The paper investigates the algorithmic stability of learning a deep neural network in problems of recognition of the materials microstructure. It is shown that at 8% of quantitative deviation in the basic test set the algorithm trained network loses stability. This means that with such a quantitative or qualitative deviation in the training or test sets, the results obtained with such trained network can hardly be trusted.

Although the results of this study are applicable to the particular case, i.e. problems of recognition of the microstructure using ResNet-152, the authors propose a cheaper method for studying stability based on the analysis of the test, rather than the training set.

Keywords: Deep neural networks · Material microstructure · Image recognition · Deep learning · Algorithmic stability

1 Introduction

A well-known problem is the creation of the training and testing sets in machine learning problems, which also include deep learning tasks in order to predict the physicomechanical properties of functional materials [1, 2]. Many researchers in the field of applied calculations use the rule of dividing the entire available set into a learning and testing ratio of 80/20 or 70/30. Usually the reasons of such a division is not given. As for the recommendations on the size of the training set, they are completely absent, except for the statements that the power of the set must be 3–10 times the number of parameters of the neural network. Such a wide range of opinions says only that this task, for all its relevance, is poorly investigated.

In addition, there is the problem of assessing the quality of the training set. There is no clear answer to the question: how erroneously annotated images can affect the learning outcome? For example, quite often the work of the classifying neural network is compared with the human work process. Is such an assessment fair? After all, the training set contains annotated data that was prepared by people and which, in turn, contain the same errors. How can these errors distort the result?

It is known that the fundamental theory of uniform convergence of Vapnik and Chervonenkis [3] treats the fatal errors of the learning algorithm as a fail of algorithmic stability. Stable learning algorithms are understood as those that form only those

© Springer Nature Switzerland AG 2020
T. Antipova (Ed.): ICIS 2019, LNNS 78, pp. 307–313, 2020.
https://doi.org/10.1007/978-3-030-22493-6_27

hypotheses, the result of which changes only slightly with a small change in the training set [4, 5]. Within this theory, sets of such algorithms have been found, for example, regression, support vector machine, etc.

Learning deep neural networks can hardly be analyzed within the theory of uniform convergence, but numerical research on the stability of specific networks can be quite accessible. For example, in [6], the influence of the size of the training and testing sets on the accuracy and generalizing ability of the three-layer MLP neural network in the binary classification problem was studied. In [7], the accuracy and scatter of the GoogLeNet network in the problem of the classification of body parts by computed tomography images is investigated. The size of the training set is selected on the basis of the analysis of the accuracy of the training results. The disadvantages of this work include the fact that the dependence of the accuracy of training on the size of the training sample was calculated on the basis of 6 computational experiments (for each of the classes), and the testing set itself was obtained as a result of the separation of the training set in 75/25 proportion. However, if there is no sustainability of training, then these results can hardly be trusted.

2 Sustainability Research

The process of studying stability was to find such a neighborhood of a stable solution, in which there is a violation of the uniform nature of convergence to the mean. A stable solution was understood as the trained deep network ResNet-152. The network was trained to solve the problem of classifying microstructure images by the hardness of a metal iron-based alloy. Annotated images of the microstructure of the alloys were used as a training set, examples of which are shown in Fig. 1. In this case, the microstructure shown in Fig. 1 on the left corresponds to an alloy with a microhardness of 1900 MPa, and on the right – 7000 MPa.

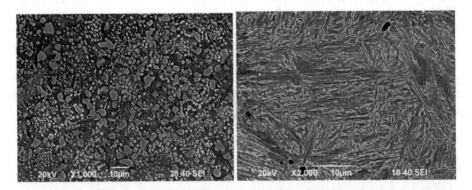

Fig. 1. Examples of images of microsection of material included in the training set

In the process of training the network, at the 200th epoch, an accuracy of 83.9% was achieved according to Top-3, after which the weights of the network were frozen.

Accuracy assessment was made on a set consisting of 1097 elements not presented to the network during its training. The process of convergence in accuracy calculated on the training set is depicted in Fig. 2.

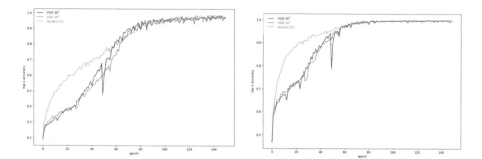

Fig. 2. Accuracy on the training set on the evaluation of Top-1 and Top-3

ResNet-152 network shows a high rate of convergence. For almost a few epochs, an accuracy of 80% on Top-3 was achieved (Fig. 2). The learning process is stable.

Table 1 shows the learning outcomes performed by different estimates on the training and validation sets.

Table 1. Network learning outcomes

Network	Training accuracy		Validation accuracy	
	Top-1	Top-3	Top-1	Top-3
VGG-16[1]	0.9751	0.9975	0.4704	0.8117
VGG 16[2]	0.9712	0.9970	0.5203	0.8142
ResNet-152	0.9593	0.9978	0.6243	**0.839**

A definite surprise was that, despite the instability of learning, the unexperienced VGG-16[2] network showed a much better Top-1 result compared to the pre-trained version of the same network. This can be explained by the fact that the first layers of the convolutional untreated network were able to better adapt to specific images of microstructure. The pre-trained network has previously been trained on a very large and diverse set of photos. This improved the universal properties of the VGG-16[1] network (in particular, its first layers), but worsened the degree of recognition of structures of a special type.

As expected, the ResNet-152 network showed the best result of 83.9% in the Top-3 rating. The shown accuracy allows the use of a trained network as the core of an intelligent system for the integrated assessment of the strength properties of functional and structural materials. For the above computational experiments, the training set was divided into 14 classes of metals and alloys microstructures. Each of the 14 classes corresponds to the range of values of microhardness and tensile strength. Thus, the classification problem within the framework of these parameters was solved [8, 9], but the accuracy indicated in Table 1 can be improved by changing the training set.

A significant improvement in ResNet-152 results was achieved while improving the quality of the training set. Samples of images with magnification of x40 were excluded from it. Firstly, their number was small, and secondly, images with such multiplicity were not present in all categories. It was suggested that these images impair learning outcomes. Table 2 presents the recognition accuracy by material class.

Table 2. Recognition accuracy by material class

Class	Validation accuracy	
	Top-1	Top-3
0	0.718	1.000
1	0.812	1.000
2	0.593	0.843
3	0.562	0.812
4	0.375	0.937
5	0.937	1.000
6	0.687	0.937
7	0.500	0.933
8	0.562	0.875
9	0.906	0.906
10	0.937	1.000
11	0.375	0.843
12	0.894	1.000
13	0.590	1.000

The best accuracy according to both estimates was achieved in class 10. The worst accuracy was estimated at Top-1 in classes 11 and 4. At the same time, according to Top-3, acceptable accuracy was reached at the same classes, which was 84.3%.

Determining the accuracy of the network on the test set showed that the density of the error distribution can be approximated by the lognormal distribution (Fig. 3), which once again confirms the correctness of the network. The optimization criterion in the learning process was cross-entropy, which is known to be proportional to the logarithm of the network error probability distribution.

To study the stability of a trained network, the test set was subjected to variations. The neighborhood of a stable solution was formed by excluding a certain number of elements from the test set. Elements were randomly selected on 100 implementations. Such a neighborhood is called Leave-one-out (LOO) [4]. Usually its depth is 1 element.

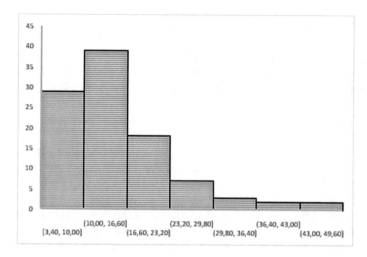

Fig. 3. The histogram of the distribution of accuracy in the test set

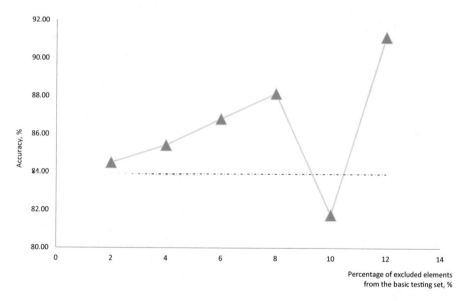

Fig. 4. Dependence of the accuracy of the trained neural network in the LOO neighborhood of the basic testing set

In this work, several neighborhoods were built: with a 2, 4, 6, 8, 10 and 12% deviation from the base number of elements in the test set, which was used to assess the accuracy of a stable solution. After running and determining the accuracy for all implementations in the neighborhood, the resulting accuracy was obtained by averaging. The results of the computational experiment are shown in Fig. 4.

3   Results and Discussion

From Fig. 4 it follows that the process of validation the trained network evenly con verges to the accuracy obtained on the base sample (indicated by the dotted line). When the number of elements in the test set decreases by more than 8%, there is a sharp oscillation of the accuracy estimate, which, in our opinion, is the loss of the algorithmic stability of the neural network in the classification problem under consideration, which it solves.

This statement may be applied to the learning process, since the test set is an independent random homogeneous sample. If the algorithm lost stability on the test sample during the control test of the accuracy of the network, then the network learning process will lose stability when the number of elements in the training sample decreases by more than 8%. Deviations of 8% can not be considered random, since the accuracy obtained by averaging over 100 computational experiments for each percent distribution where on each iteration random elements were removed from full elements set in according to that percent count.

It should be noted that the search for the loss of stability of the learning process in deep networks by the direct method is a difficult task. If the network learning process on a training set with a size of decades of thousands of elements can take several days, then the study of learning sustainability is difficult from a practical point of view. The authors of this paper propose a simplified approach — investigate uniform convergence on a test set, and then extend the findings to the entire training set.

The result obtained in this paper may impose not only quantitative but also qualitative limitations on the training set. The resulting limit may indicate that in the training set there can not be more than 8% of erroneously annotated images. Otherwise, the learning process will lose its stability and the learning outcomes cannot be trusted.

4   Conclusions

The paper studies the problem of uniform convergence of the learning process and the accuracy assessment of the deep ResNet-152 network in the problem of analysis of microsections of iron-based alloys. It is shown that with 8% quantitative deviation in the base set, the algorithm of the trained network loses stability. This means that with so many elements in the test set an adequate assessment of the accuracy of the network is impossible.

The methodology for assessing the stability of the deep network, applied in this work, can be extended to other networks and tasks. It does not require volumetric computations, since it allows one to estimate the sufficiency of the number of elements in the training set without performing network training on training sets having various capacities.

Acknowledgements. The reported study was funded by the Ministry of Science and Higher Education of the Russian Federation (the unique identifier RFMEFI58617X0055) and by the EC Horizon 2020 is MSCA-RISE-2016 FRAMED Fracture across Scales and Materials, Processes and Disciplines. The authors are grateful to the staff of the Institute of Nanosteels of MSTU named after G.I. Nosov, in particular to M. P. Baryshnikov, for the experimental data provided, which made it possible to train the constructed neural network with a given accuracy.

References

1. Gitman, M.B., Klyuev, A.V., Stolbov, V.Y., Gitman, I.M.: Complex estimation of strength properties of functional materials on the basis of the analysis of grain-phase structure parameters. Strength Mater. **49**(5), 710–717 (2017)
2. Kliuev, A., Klestov, R., Bartolomey, M., Rogozhnikov, A.: Recommendation system for material scientists based on deep learn neural network. In: Antipova, T., Rocha, A. (eds.) Digital Science. DSIC 2018. Advances in Intelligent Systems and Computing, vol. 850, pp. 216–223, Budva (2019)
3. Vapnik, V.N., Chervonenkis, A. Ya.: Teoriya raspoznavaniya obrazov, p. 416. M.: Nauka (1974)
4. Bousquet, O., Elisseeff, A.: Algorithmic stability and generalization performance. Adv. Neural Inf. Proc. Syst. **13**, 196–202 (2001)
5. Bousquet, O., Elisseeff, A.: Stability and generalization. J. Mach. Learn. Res. **2**, 499–526 (2002)
6. Brownlee, J. Impact of dataset size on deep learning model skill and performance estimates [Digital resource]. https://machinelearningmastery.com/impact-of-dataset-size-on-deep-learning-model-skill-and-performance-estimates (2019). Accessed 4 Feb 2019
7. Cho, J., Lee, K., Shin, E., Choy, G., Do, S.: How much data is needed to train a medical image deep learning system to achieve necessary high accuracy? arXiv:1511.06348 (2015) (preprint arXiv)
8. Klestov, R., Klyuev, A., Stolbov, V.: About some approaches to problem of metals and alloys microstructures classification based on neural network technologies. Adv. Eng. Res. (AER) **157**, 292–296 (2018)
9. Gitman, I.M., Klyuev, A.V., Gitman, M.B., Stolbov, V.Yu.: Multi-scale approach for strength properties estimation in functional materials. ZAMM Z. Angew. Math. Mechanik **98**(6), 945–953 (2018)

Deformation Behavior Numerical Analysis of the Flat Sliding Layer of the Spherical Bearing with the Lubrication Hole

A. A. Adamov[1], A. A. Kamenskih[2(✉)], and Yu. O. Nosov[2]

[1] Institute of Continuous Media Mechanics, Perm 614013, Russian Federation
[2] Perm National Research Polytechnic University,
Perm 614990, Russian Federation
anna_kamenskih@mail.ru

Abstract. The numerical simulation problem of the frictional contact interaction of the flat sliding layer periodicity cell of the spherical bearing is performed. The mathematical formulation of contact problems with a previously unknown contact area and all types of contact states (adhesion, sliding, no contact) is done. Three options for the sliding layer thickness of 4, 6 and 8 mm are considered. The deformation of the thin flat sliding layer of the spherical bearing is made on the example of an antifriction layer of modified PTFE. The deformation theory of elastoplasticity for the active loading case is chosen as the antifriction polymer behavior model. The thermomechanical and friction properties of the modified PTFE were obtained experimentally by a scientific group of Alfa-Tech LLC and IMSS of the Ural Branch of the Russian Academy of Sciences. The experiment to determine the frictional properties of the material was performed to a pressure level of 54 MPa. The analysis of the friction coefficient dependence on the level of pressure acting on the stamp is performed: approximating functions and for contact with $\mu(P) = 0.005 + 0.111/P + 0.623/P^2 - 3.57/P^3 + 3.335/P^4$ and $\mu(P) = -0.002 + 1.55/P - 17.166/P^2 + 64.979/P^3 - 55.745/P^4$ without lubricant on the mating surfaces are selected. The friction coefficient for a pressure level of more than 54 MPa is calculated from the obtained functions with an error of less than 1%. Simulation of the spherical bearing sliding layer deformation behavior is made taking into account the physicomechanical and friction properties of the polymer material using the example of a periodicity cell made of the modified PTFE with a hole for lubrication. Distribution fields of stress intensity and plastic strain intensity, as well as integral stiffness were obtained and analyzed. The relations of the maximum and minimum integral parameters values of the stress-strain state on the pressure level are established as part of the analysis. The influence of frictional properties and layer thickness on the contact zone parameters is considered. It was established that the 8 mm thickness layer enjoy a more favorable deformation behavior case than the other two variants of the layer thickness. The frictional properties have a slight effect on the stress-strain state parameters of the periodicity cell, their influence significantly on the pattern of the contact states zones distribution and contact tangential stress. It is established that the level of contact tangential stress when taking into account lubricant approximately is on lower 3 times than with contact without lubrication.

© Springer Nature Switzerland AG 2020
T. Antipova (Ed.): ICIS 2019, LNNS 78, pp. 314–325, 2020.
https://doi.org/10.1007/978-3-030-22493-6_28

Keywords: Modified PTFE · Contact · Friction · FEM · Polymer properties · Periodic cell · Lubrication hole · Spherical bearing

1   Introduction

A works number devoted to the analysis properties of the modified PTFE and composites based on it, including a comparison of its properties with classical antifriction materials, for example PTFE-4, can be noted [1–6 and etc.]. The possibility of using a modified PTFE in order to increase the reliability and service life of antifriction and sealing parts was noted in [1]. A works number are devoted to the analysis of mechanical [1–3 and etc.], antifriction [3, 6 and etc.], thermophysical and other operational properties [5] of the radiation-modified PTFE and composite materials based on it. The modified PTFE advantage in terms of wear resistance, radiation resistance, and elastic properties compared with PTFE-4 is noted in [1, 2]. The work [3] is devoted to the behavior mathematical description of the modified PTFE in the framework of the elastoplastic deformational theory. An experimental study of the material frictional properties in the specific operating pressures of the rotors supports range of 10–20 MPa was carried out in [6]. It was noted that the modified PTFE has an average friction coefficient of significantly less than that of ordinary PTFE-4, with a surface wear rate lower by a couple of orders. Modern antifriction materials are widely used in medicine [7, 8], aviation [9], bridge engineering [3, 5], industry and other branches of science and technology. Experimentally substantiated effective numerical models describing the polymers behavior, including within the framework of frictional contact, are necessary for their effective use as anti-friction coatings and layers. Effective numerical models will allow estimating the deformation behavior of the modified PTFE in the structural elements for various purposes.

The particular relevance of solving problems associated with transport and logistics systems elements and the problems associated with bridge construction is noted at the moment: geometric configuration and technology of deformation joints [10, 11], bearing [3, 5, 12], bridge spans [[13] and other elements. The study of the possibility of using modern antifriction polymer materials [12, 14 and etc.] in the bearing construction is one of the most important tasks of bridge building. The bearing of the bridges span structures is belong to the responsible unit, which are subject to increased requirements with respect to strength, wear resistance and durability. Such constructions are work within the framework of frictional contact interaction and belong to difficult-to-repair units. Of particular interest to the analysis of the carrying capacity, strength, wear resistance of the bearing elements in general and contact parameters in particular [3, 5, 15] including the materials from which their elements are made can be noted.

2   Problem Statement

Spherical and flat sliding layers are used as anti-friction interlayer's in the construction of bearing. Vertical and horizontal load from the bridge span perceive bearing. Analysis of the thin flat layer stress-strain state of the polymer material and contact zone parameters is an important task. The influence analysis of the sliding layer thickness and the properties of polymeric materials from which it is made on the stress-strain state and the contact zone parameters of the node in question is of particular interest. The deformation behavior of a thin flat antifriction sliding layer of modified PTFE with truncated spherical holes for lubrication (Fig. 1) is considered in the work. The periodicity cell is cut out from the antifriction material volume as shown in Fig. 1 with geometrical characteristics: maximum width 18 mm, maximum depth 15 mm, diameter of a spherical hole 8 mm, height of a spherical hole 2 mm. Periodic cell is deforms rigid steel plate thickness 3 mm. A comparative analysis of the deformation behavior of three variants of the anti-friction layer thickness of 4, 6 and 8 mm was performed. A fourth of the contact unit is considered. The discarded parts action is replaced by symmetry conditions. Lack of lubricant in the hole as an unfavorable case is considered. The periodic cell is deformed by a rigid steel plate with a constant pressure. The pressure level is from 5 to 90 MPa.

The friction properties influence of the modified PTFE on the stress-strain state of the periodicity cell is considered in the work. A experiments series to determine the frictional properties of modern antifriction materials suitable for use as a sliding layer in transport and logistics systems elements was made earlier by the research team of Alfa-Tech LLC and IMSS of the Ural Branch of the Russian Academy of Sciences. Original experimental equipment allowed determining the frictional properties of antifriction materials in the pressure range up to 54 MPa at the same time the operating pressure range of the spherical bearing can reach 90 MPa. The approximation of the

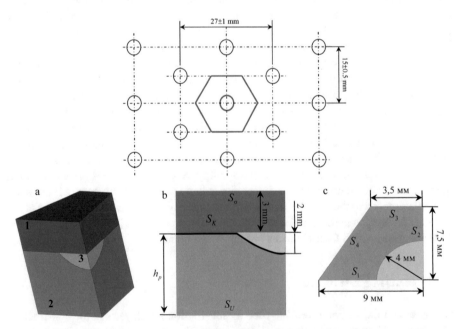

Fig. 1. Fragment of antifriction material layer with lubricant: a is 3D geometry, b is front view and c is top view with geometric characteristics of a periodicity cell.

experiments results was performed in the framework of the experimental data analysis. Selected functions were used to calculate friction coefficient values in the range of more than 54 MPa. The error of the friction coefficients obtained from the approximating functions on the experimental data does not exceed 1% for the modified PTFE. The results of experiments with and without lubrication as well as the functions approximating the experiments results are shown in the Fig. 2.

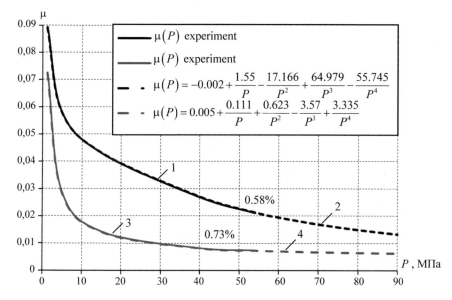

Fig. 2. Relation of friction coefficient on pressure: 1 is experimental data without lubrication; 2 is approximation of experimental data (1); 3 is experimental data with lubricant; 4 is approximation of experimental data (3).

The numerical simulation of the problem deforming of the periodicity cell of the spherical bearing thin flat sliding layer was performed using the finite element method in the ANSYS software package. The solution of the problem is conducted within the general mathematical statement of the problem of contact interaction of elastic bodies with elastoplasticity layer (Fig. 1) and includes:

– equilibrium relations

$$\text{div } \hat{\sigma} = 0, \ \vec{x} \in V; \tag{1}$$

– geometrical relations

$$\hat{\varepsilon} = \frac{1}{2}\left(\nabla \vec{u} + (\nabla \vec{u})^{\mathrm{T}}\right), \vec{x} \in V; \tag{2}$$

– physical relations

$$\hat{\sigma} = \lambda I_1(\hat{\varepsilon})\hat{I} + 2\mu\hat{\varepsilon}, \ \vec{x} \in V_1, \tag{3}$$

where λ and μ are Lame's parameters; $\hat{\sigma}$ is stress tensor; $\hat{\varepsilon}$ is strain tensor; $\vec{u}$ is vector of displacements; $\vec{x}$ is radius-vector of a random point; $I_1(\hat{\varepsilon})$ is the first invariant of strain tensor; $\hat{I}$ is the unit tensor; the area V consists of the subareas occupied by the plate of the press V_1 and the polymeric layer $V_2(V = V_1 \cup V_2)$.

To describe the behavior of the layer's material the deformation theory of elasto-plasticity is chosen, its physical relations are:

$$\hat{\sigma} = \frac{2\sigma_u}{3\varepsilon_u}\left(\hat{\varepsilon} - \frac{1}{3}I_1(\hat{\varepsilon})\hat{I}\right) + KI_1(\hat{\varepsilon})\hat{I}, \ \vec{x} \in V_2, \tag{4}$$

where $\sigma_u = \sqrt{3I_2(D_{\hat{\sigma}})}$ and $\varepsilon_u = \frac{2}{\sqrt{3}}\sqrt{I_2(D_{\hat{\varepsilon}})}$ are intensities of stress and strain tensors; $I_2(D_{\hat{\sigma}})$ and $I_2(D_{\hat{\varepsilon}})$ are the second invariants of deviators of stress tensor $D_{\hat{\sigma}}$ and strain tensor $D_{\hat{\varepsilon}}$; K is the volumetric elastic modulus; $\sigma_u = \Phi(\varepsilon_u)$ is the function determined by deformation diagram of the layer's material at uniaxial stress state.

The contact boundary conditions are applied to the surface S_K, where the two bodies 1 and 2 are in contact along S_K. The types of contact interaction that are implemented in the problem are given below:

– sliding with friction (friction of rest): $\vec{u}^1 = \vec{u}^2$, $\sigma_n^1 = \sigma_n^2$, $\sigma_{n\tau_1}^1 = \sigma_{n\tau_1}^2$, $\sigma_{n\tau_2}^1 = \sigma_{n\tau_2}^2$
   wherein $\sigma_n < 0$, $|\sigma_{n\tau}| < q(\sigma_n)|\sigma_n|$;
– sliding with friction (sliding friction): $u_n^1 = u_n^2$, $u_{\tau_1}^1 \neq u_{\tau_1}^2$, $u_{\tau_2}^1 \neq u_{\tau_2}^2$, $\sigma_n^1 = \sigma_n^2$,
   $\sigma_{n\tau_1}^1 = \sigma_{n\tau_1}^2$, $\sigma_{n\tau_2}^1 = \sigma_{n\tau_2}^2$, wherein $\sigma_n < 0$, $|\sigma_{n\tau}| = q(\sigma_n)|\sigma_n|$;
– no contact: $|u_n^1 - u_n^2| \geq 0$, $\sigma_{n\tau_1} = \sigma_{n\tau_2} = \sigma_n = 0$;
– adhesion: $\vec{u}^1 = \vec{u}^2$, $\sigma_n^1 = \sigma_n^2$, $\sigma_{n\tau_1}^1 = \sigma_{n\tau_1}^2$, $\sigma_{n\tau_2}^1 = \sigma_{n\tau_2}^2$,

where $q(\sigma_n)$ is the friction coefficient, τ_1, τ_2 are axes designation which lie in a plane tangent to the contact surface, u_n is displacement along a normal to a corresponding contact edge, u_{τ_1}, u_{τ_2} are displacement in a tangential plane, σ_n is stress along the normal to the contact boundary, $\sigma_{n\tau_1}$, $\sigma_{n\tau_2}$ are tangential stresses at the contact boundary, $\sigma_{n\tau}$ is the value of vector tangential contact stresses.

The mathematical statement (1)–(4) and contact conditions is supplemented by boundary conditions: the pressure is applied at the boundary S_σ, which varies from 5 to 90 MPa; the displacements along the y coordinate are forbidden on the boundary S_U; the symmetry conditions are valid on the boundarys $S_1 - S_4$.

3   Results and Discussion

A series of numerical calculations on a periodicity cell with a thickness of 4 mm was carried out as part of the convergence analysis of the contact problem numerical solution results. Four variants of the finite element mesh were considered: 15, 41, 169 and 443 thousand node unknowns. The numerical solution convergence assessment was performed by displacements of the contact boundary. The finite element mesh with gradient concentration of elements to the contact area was chosen for the main volume of the material according to the study results: the maximum element size is 0.5 mm, the minimum one is 0.125 mm (169 thousand node unknowns). The resolution about concentration the mesh near the edge of the hole, which was initially in contact with a rigid stamp, was taken to clarify the contact parameters near the hole for the lubricant. The element size in this area was 0.074 mm, increasing the number of node unknowns to 237 thousand. Discretization of a periodicity cell with different antifriction sliding layer thickness with a gradient decrease in the element size to the contact zone is performed with selected parameters of the finite element mesh: $h_p = 4$ mm is 237 thousand, $h_p = 6$ mm is 263 thousand and $h_p = 8$ mm is 277 thousand node unknowns respectively. A numerical experiments series aimed at analyzing the deformation behavior of the periodicity cell are performed as part of the work: analysis of changes in the hole profile for the lubricant, stress intensity analysis, plastic deformations level analysis, integral stiffness of the periodicity cell analysis

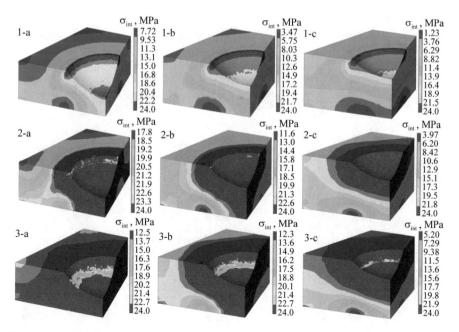

Fig. 3. Stress intensity in the periodicity cell (1 is $P = 30$ MPa; 2 is $P = 60$ MPa; 3 is $P = 90$ MPa; a is $h_p = 4$ mm; b is $h_p = 6$ mm; c is $h_p = 8$ mm).

320   A. A. Adamov et al.

The stress intensity distribution fields for the periodicity cell for the three considered variants of the sliding layer thickness of the spherical bearing for the three pressure
options acting on the rigid stamp of 30, 60 and 90 MPa are shown in Fig. 3. The
change in the pattern of the stress intensity distribution is shown for the case of contact
with lubricant under the mating surfaces.

The maximum level of stress intensity is observed near the lubricant hole for all
pressure levels. The volume of the material with the maximum stress intensity increases
nonlinearly. In the case of the periodicity cell $h_p = 4$ mm at maximum pressure
occupies more than 70% of the periodicity cell volume. An increase of the stress
intensity minimum level to the pressure of 70–80 MPa with a further decrease of the
minimum stress intensity level is observed for cases of 4 and 6 mm of the sliding layer
thickness, which is associated with a change in the area of contact interaction, redistribution of the integral stiffness and plastic deformations. A significant deformation of
the hole profile for the lubricant is observed in thicknesses of 4 and 6 mm with a load
of 70–80 MPa. Profile deformation is less significant for a sliding layer with a thickness of 8 mm, the minimum level of stress intensity increases non-linearly over the
entire pressure range.

The distribution fields of the integral stiffness for the periodicity cell are shown in
Fig. 4. The integral stiffness standard value for the modified PTFE is approximately
864 MPa. Integral stiffness periodicity cell is shown for the case of contact with the
lubricant on the mating surfaces similar to the stress intensity.

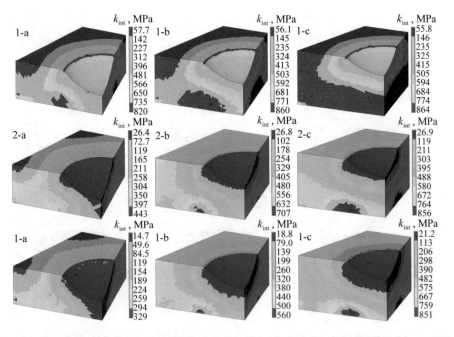

Fig. 4. Integral stiffness in the periodicity cell (1 is $P = 30$ MPa; 2 is $P = 60$ MPa; 3 is $P = 90$
MPa; a is $h_p = 4$ mm; b is $h_p = 6$ mm; c is $h_p = 8$ mm).

The integral stiffness of the periodicity cell decreased from the initial value by 53.59, 36.57 and 1.74% upon contact without lubrication and by 61.92, 35.19 and 1.5% upon contact with lubrication on the mating surfaces for 4, 6 and 8 mm thicknesses respectively. The maximum decrease integral stiffness of the periodicity cell with a thickness of 4 mm upon contact with lubricant on the mating surfaces is due to the highest level of plastic deformation.

Plastic deformation of the modified PTFE has a significant effect on the integral stiffness. The thickness of the sliding layer decreased by 8.7, 6.5 and 5.4% for the sliding layer with a thickness of 4, 6, 8 mm respectively due to the significant level of plastic deformation (Fig. 5).

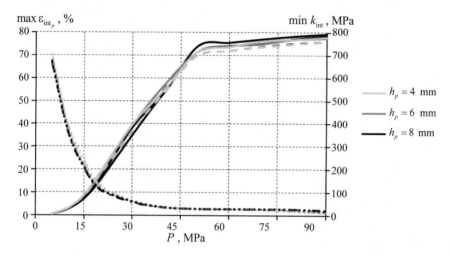

Fig. 5. Relation of the integral stiffness and the plastic deformations intensity to the pressure (max ε_{int_p} solid line is contact with lubrication; dotted line is contact without lubrication; min k_{int}: dash-dotted line is contact with lubricant; round points is contact without lubrication).

Accounting for lubrication at the mating surfaces of a periodicity cell with a rigid stamp has no significant effect on the stress-strain state of the thin flat sliding layer of a spherical bearing. Deformation of the hole profile on contact without lubricant on the mating surfaces by 2.24, 1.83 and 1.39% more than when contact with lubricant for a sliding layer 4, 6, 8 mm thickness respectively. The maximum stresses intensity in contact with the lubricant in the load range up to 15 MPa is more by 0.9% than in the case of contact without lubricant. The maximum plastic deformations intensity of the periodicity cell with a thickness of 6 and 8 mm is less than that of a periodicity cell with a thickness of 4 mm by 2.5 and 1.5% upon contact of the without and with lubricant on the mating surfaces respectively.

Accounting for lubrication over the mating surfaces has a significant effect on the parameters of the contact zone, especially the distribution of contact states zones and contact tangential stress. An example of the contact pressure distribution for the maximum pressure level of 90 MPa in contact with and without lubrication over the mating surfaces is shown in Fig. 6.

322   A. A. Adamov et al.

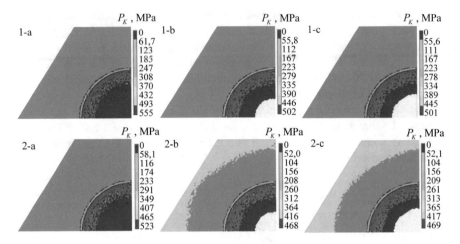

Fig. 6. Contact pressure at 90 MPa (1 is contact without lubrication; 2 is contact with lubrication; a is $h_p = 4$ mm; b is $h_p = 6$ mm; c is $h_p = 8$ mm).

The influence of the lubricant supply is observed on the mating surfaces on the distribution of the contact pressure of the slip layers with a thickness 6–8 mm: the decrease of the contact pressure level is observed when moving away from the lubricant hole. Accounting for lubrication when modeling contact has little effect on the contact pressure level: the maximum contact pressure level when taking into account lubricant on mating surfaces is lower by approximately 6% than when contacting without lubrication for all the options for the sliding layer thickness. The pattern of the contact tangential stresses distribution does not have differences from the pattern of the contact pressure distribution. The contact tangential stress level is more than 190 times lower than the contact pressure.

The dependences of the maximum level of contact pressure and contact tangential stress on the pressure acting on a rigid stamp are shown in Fig. 7. The significant effect

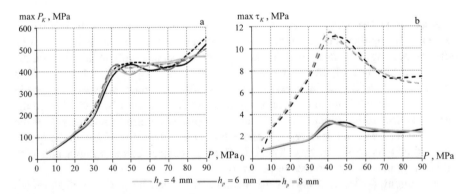

Fig. 7. Relation of the contact pressure (a) and the contact tangential stress (b): solid line is contact with lubricant; dotted line is contact without lubrication.

of taking lubrication over the mating surfaces on the contact tangential stress is observed as one would expect (Figs. 7 and 6).

The maximum contact pressure level at contact interaction, taking into account lubricant on mating surfaces, is on lower 5, 4 and 3.5% than at contact without lubrication for sliding layer thickness of 4, 6 and 8 mm respectively. The maximum contact tangential stress level during contact interaction, taking into account lubricant on mating surfaces, is on about 3 times lower than at contact without lubrication for all considered sliding layer thicknesses.

A significant change in the contact interaction area during plastic deformation of a periodic cell can be observed by the contact pressure distribution pattern (Fig. 7a). A decrease contact pressure level is observed with a significant increase in the contact interaction area: for a thickness of 4 mm at a pressure of 60 and 70 MPa; for a thickness of 6 mm 50 and 70 MPa. A decrease contact pressure level is not observed in the case of a sliding layer thickness of 8 mm. The maximum contact tangential stress level is increases to a pressure of 40 MPa, which is associated with a decrease in the contact surfaces adhesion area. The contact tangential stress level is decreases at a pressure of more than 40 MPa, since sliding is observed on most of the contact surface. A significant increase in the contact surfaces adhesion area is observed in the periodicity cells 4–6 mm thickness at contact without lubrication on mating surfaces, which leads to an increase in the P_K at a pressure of 90 MPa. This effect is associated with plastic deformations.

4 Conclusion

The deformation behavior analysis of the spherical bearing thin flat sliding layer with holes for lubrication was performed as part of the work. The periodicity cell with the lubrication hole is cut from the material volume. The task is modeled without lubrication in the hole, as an unfavorable case for a spherical bearing. Three options for the sliding layer thickness of 4, 6 and 8 mm are considered. Mathematical formulation of contact tasks is performed. The assessment of effect of the antifriction layer thickness and the layer material friction properties on the stress-strain state and on the contact zone parameters of the contact unit is in made.

A number of qualitative and quantitative patterns have been established as part of the results analysis of numerical simulation:

– The deformation behavior of the 8 mm thickness layer has a number of important features: the sliding layer thickness decrease is minimal and is 5.4% at the maximum load level; there is no significant contact pressure level reduction; the maximum contact pressure level is increases monotonically with increasing pressure; there are no pronounced zones of contact pressure redistribution when the changes contact area; deformation of the hole profile for the lubricant is minimal compared to the other two cases of the layer thickness in the operating pressure range; the maximum stress intensity volume is observed near the lubricant hole; the integral stiffness of the periodicity cell decreased from the initial value by 1.74 and 1.5% upon contact without and with lubricant on the mating surfaces respectively, which is minimal for all the considered sliding layer thicknesses.

– Anti-friction layer thickness of 4 mm is showed the most unfavorable case of deformation behavior: the maximum stress intensity level is observed on more than 70% of the periodicity cell volume; the maximum Integral stiffness decrease of the periodicity cell is observed due to the plastic deformation highest level; the integral stiffness of the periodicity cell decreased from the initial value by 53.59 and 61.92% upon contact without and with lubricant on the mating surfaces respectively; deformation of the geometrical configuration is greater than that of other variants of the layer thickness, inter the residual height of the hole is 20% of the initial height, the radius of the hole decreased by 46.3%, the sliding layer thickness decreased by 8.7%.

– Accounting for lubrication of mating surfaces has a slight effect on the stress-strain state of the contact node: the minimum integral stiffness on contact with lubricant is on average less by 2.5% than on contact without lubrication for all variants of the periodicity cell thickness; the maximum stresses intensity in contact with the lubricant in the load range up to 15 MPa is more by 0.9% than in the case of contact without lubricant; the deformation of the hole profile on contact without lubricant on the mating surfaces by 2.24, 1.83 and 1.39% more than when contact with lubricant for a sliding layer 4, 6, 8 mm thickness respectively; accounting of lubricant on the mating surfaces dose not significantly influence the contact pressure. Accounting of lubrication on the mating surfaces has the greatest impact on the contact tangential stress and the distribution of contact state zones: ehe maximum contact tangential stress level during contact interaction, taking into account lubricant on mating surfaces, is on about 3 times lower than at contact without lubrication for all considered sliding layer thicknesses; the occurrence of adhesion zones at loads of more than 30 MPa on a large contact surface area is not observed upon contact with lubricant; the significant increase in the contact surfaces adhesion area at a pressure of 90 MPa is observed for layer thicknesses of 4–6 mm, which leads to an increase in contact pressure.

Acknowledgements. The study supported by a grant of Russian Science Foundation (project No. 18-79-00147).

References

1. Khatipov, S.A., Serov, S.A., Sadovskaya, N.V., Konova, E.M.: Morphology of polytetrafluoroethylene before and after irradiation. Radiat. Phys. Chem. **81**, 256–263 (2012)
2. Syty, Y., Chursova, L.V., Khatipov, S.A., Sagomonova, V.A.: Properties and application of F-4RM radiation-modified fluoroplastic (polytetrafluoroethylene). Aviacionnye Mat. Tehnol. **4**(25), 48–55 (2012)
3. Kamenskih, A.A., Trufanov, N.A.: Numerical analysis of the stress state of a spherical contact system with an interlayer of antifriction material. Comput. Continuum Mech. **6**(1), 54–61 (2013)
4. Morgunov, A.P., Revina, I.V.: Radiation-modification impact on the filled polytetrafluoroethylene structure and mechanical strength. Key Eng. Mater. **736**, 29–34 (2017)

5. Kamenskih, A.A., Trufanov, N.A.: Regularities interaction of elements contact spherical unit with the antifrictional polymeric interlayer. J. Frict. Wear **36**(2), 170–176 (2015)
6. Balyakin, V.B., Pilla, C.K., Khatipov, S.A.: Experimental studies of tribotechnical characteristics of radiation-modified polytetrafluoroethylene to use in rotor supports. J. Frict. Wear **36**(4), 346–349 (2015)
7. Pinchuk, L.S., Nikolaev, V.I., Tsvetkova, E.A., Goldade, V.A.: Tribology and Biophysics of Artificial Joints. Elsevier, Amsterdam (2006)
8. Sitnikov, V.P., Kudaibergenova, S.F., Nugmanov, B.I., Shil'ko, S., El-Refai, H., Nadyrov, E.A.: Biocompatibility of prostheses-based prostheses with a diamond-like carbon nanocoating in ear surgery (an experimental study). Bull. Kaz. Natl. Med. Univ. **2**, 120–122 (2015)
9. Balyakin, V.B., Zhil'nikov, E.P., Pilla, K.K.: Calculating life spans of bearings taking into consideration the wear and failure of PTFE cages. J. Frict. Wear **39**(1), 19–23 (2018)
10. Anisimov, A.V., Bakhareva, V.E., Nikolaev, G.I.: Antifriction carbon plastics in machine building. J. Frict. Wear **28**(6), 541–545 (2007)
11. Yankovsky, L.V., Kochetkov, A.V., Ovsyannikov, S.V., Trofimenko, Y.: Deformation seams of small structures of small movements: device, repairability, texture. Tech. Regul. Transp. Constr. **3**(7), 6–12 (2014)
12. Choi, E., Lee, J.S., Jeon, H.Kw., Park, T., Kim, H.-T.: Static and dynamic behavior of disk bearings for OSPG railway bridges under railway vehicle loading. Nonlinear Dyn. **62**, 73–93 (2010)
13. Ivanov, B.G.: Diagnostics of Damage to the Span of Metal Bridges: A Monograph. Marshrut, Moscow (2006)
14. Peel, H., Luo, S., Cohn, A.G., Fuentes, R.: Localisation of a mobile robot for bridge bearing inspection. Autom. Constr. **94**, 244–256 (2018)
15. Wu, Y., Wang, H., Li, A., Feng, D., Sha, B., Zhang, Yu.: Explicit finite element analysis and experimental verification of a sliding lead rubber bearing. J. Zhejiang Univ.-Sci. A **18**(5), 363–376 (2017)

Information Management System

The Basic Processes of Creating a "Megascience" Project

Nurzhan Nurakhov[(⊠)]

National Research Center "Kurchatov Institute", 1, Akademika Kurchatova Pl.,
Moscow 123182, Russia
n.nurakhov@gmail.com

Abstract. This paper is devoted to the study of the basic processes for the implementation of "megascience" projects and to ensure the possibility of their effective implementation using management information systems. All stages of the "megascience" project are inherent in all sorts of risks that may arise from the lack of accounting for resources and the limitations of the project. The work showed that the "megascience" project is a large and complex system, each of the stages of the project life cycle is an independent project and it can be considered as a business project, the result of which can be a commercially viable installation that has no analogues in the World, or new knowledge and technology. There is also showed the need to create an appropriate integral information management system of the "megascience" project. An approach to the creation of such a system was also proposed, and 12 sets of tasks of information systems were considered, requiring solutions for its creation.

Keywords: Megascience · Project management · Information system · Risk

1 Introduction

In the process of creating research facilities and project implementation, it is necessary to be able to take into account available resources and project limitations in order to minimize risks arising at each stage of project implementation. These can be economic, legal, technological, construction, management, environmental and cyber risks.

For effective risk management, it is necessary to consider any megascience project as a business project whose goal is to create a commercially viable research facility or the emergence of new knowledge and technologies. As a result of using this approach, the need to create an integrated project management information system becomes obvious. The creation of such a system is then an integral step of the megascience project.

In this paper, the main stages of the project to create a megascience class research facility are considered. The necessity of considering such a project as a business project and a large and complex system with various risks is shown. The necessity of creating an appropriate project management information system is shown and an approach to creating such a system is proposed.

The reported study was funded by Russian Foundation for Basic Research (RFBR) according to the research project № 18-29-15015.

© Springer Nature Switzerland AG 2020
T. Antipova (Ed.): ICIS 2019, LNNS 78, pp. 329–339, 2020.
https://doi.org/10.1007/978-3-030-22493-6_29

2 Basic Processes of the "Megascience" Projects Creating

To create and ensure the effective functioning and operation of the "megascience" project we must have an integral view of the concept of "megascience" and about the basic processes (stages) of its creation and functioning.

The "megascience" project is "a project for the creation of research facilities, the financing of the creation and operation of which is beyond the capabilities of individual States" [2].

We can also define "megascience" project as a "supranational" organization with the "independent representations" [12]. For example, the ITER project unites four main research forces in the world – the USA, the European Union, Russia and Japan [10]. It is also an organizational and managerial innovation [13].

In accordance with the works [6, 14] "megascience" project must meet the following criteria:

- research-intensive, than they can afford for their implementation one country;
- research that is more cost-effective to conduct on a multilateral basis;
- research, by its nature, can bring more results if they are carried out in a wide range, including geographical;
- research contributing to the creation of a single EU market and the unification of European science and technology;
- research promoting social and economic cohesion;
- research that encourages the mobility of scientific and technical personnel and the coordination of scientific and technical policies of member countries.

Analyzing the definitions proposed by different authors, we can come to the conclusion that it is necessary to consider the project of "megascience" class as a large and complex system that requires an appropriate administrative apparatus, legal and information support.

Summing up the experience of implementation of "megascience" project [3, 5–8] it is possible to allocate following basic stages of the life cycle of a project of creating a "megascience" research facility:

1. Development of the concept and scientific program of the project:

   - Consideration of the concept and scientific program by potential participants at the national and international level. In particular, the analysis of conformity of the proposed project the criteria for inclusion of the project to the projects class "magicians" and used the scientific expertise of the project.
   - Economic and other types of project expertise;

2. Formation of international collaboration and conditions of participation by signing a Memorandum of understanding and cooperation (letters of intent).
3. Development of the project roadmap and technical design of the project;
4. Creation of an appropriate organization, ensuring the interaction of all members of the collaboration and the actual implementation of the project. At this stage, the legal form of the organization is selected, taking into account regulatory restrictions and the possible format of participation of the members of the collaboration.

5. Implementation of the direct construction of the installation and the creation of the necessary infrastructure.
6. Implementation of control and monitoring of construction and infrastructure. Control and monitoring is carried out in accordance with the normative indicators of the road map, in accordance with the requirements of TDR. Control is also exercised by various Supervisory bodies.
7. Project completion. At this stage, acceptance and verification of compliance is carried out both by the collaboration itself and by various Supervisory bodies.

All the listed stages are broadly in line with the methodology of the Project Management Institute [1]. Here the life cycle of a project is defined as a series of phases that represent the evolution of a product, from concept through delivery, maturity, and to retirement.

Since the project of the "megascience" class is a large and complex system, each of the stages of the project life cycle is an independent project, each of which has its own life cycle consisting of the following stages: initiation, planning, implementation, monitoring and control, completion.

Thus, the project of the "megascience" class can be considered as a business project, the result of which can be a commercially viable installation that has no analogues in the World, or new knowledge and technology.

An integrated information system (IS) is required to ensure the effective implementation of such a project. Such information system should take into account the available resources and limitations and minimize the risks of the project.

3 Creating of a Complete "Megascience" Project Management Information System

We will use the following definitions of system and consistency, corresponding to the integral method of complementarity [15].

The system is a complete set of methods and/or means of ensuring the interaction of the internal environment of the elements (parts) of the system with the external environment of the system. The external environment of the system is presented, as a rule, structured from the position of the system in the form of sources of resources and consumers of products of the system;

Consistency – is the integrity of the element (part) of the system in relation to the system; the integral element (part) of the system is intended for activities in the interests of the system.

There is a high cost of errors that can be made in the design and operation of the IS, if not sufficiently reliable models of the user and the IS are used. One of the reasons for the errors is the lack of an integral model of IS tasks that clearly reflects the problems, tasks and goals of the user, as well as incomplete consideration of the phenomenon of information system complexity.

Parts of the information system – databases, information, information technologies, technical instruments, as well as produced information products, used models of business processes, as well as the IS itself have the integrity of the first type [18].

When considering information system as a complex system [4, 11], a complete model of IS tasks should contain at least two sets of tasks.

For this reason, we pay special attention to the methodology of structuring the problems, tasks, goals of the user and drawing up a model of an integrated set of IS tasks necessary throughout the life cycle of complex IS.

3.1   The Integrity Principle of the Information System

The IS, the IS management system, the IS product is an activity triad "subject – object – result" [17, 18], which functions in the interests of the user of the IS – IS triad.

The components of the IS triad are defined as follows.

The object of the triad, the object of management is actually the IS itself, which carries out activities for the production of information products for the user.

The subject of the triad is an IS management system aimed at providing the necessary qualitative and quantitative indicators of information products of the IS.

The result of the triad is the information products produced by the IS for the user under the control of the subject of the triad.

In accordance with the general Integrity Principle [15, 16], the Integrity Principle of the IS can be formulated as the following condition: to ensure the integrity of an IS triad, it is necessary to comply with an IS triad common to similar IS triads of an IS triad model in the form of an integral complete-triad corresponding to a set of tasks of user support information.

3.2   The Integral Complete - Approach to the Formation of the Tasks of IS

Let us review the process of forming an integral set of 12 sets of IS tasks from the standpoint of the integral complete approach [16].

We proceed from the fact that in the activity of the IS user there are constantly problems that require resolution using the IS. Solving problems is accomplished by achieving certain goals and solving relevant tasks (Fig. 1).

Let us proceed to the structuring of problems, tasks, goals of the user and the compilation of an integral model of complexes of IS tasks that are necessary throughout the entire life cycle of any IS.

The solution of all the sets of IS tasks described in this section can be combined with the help of the data bank of the considered IS intended for the centralized storage and sharing of data. The data bank includes all the databases listed here, a database directory, DBMS, as well as query and application libraries.

Let us accept, in the general case, that there is some universal, possibly, an environment that we define as containing parts (parts of the environment), as well as the potentials of the activities of the parts of the environment.

Part of the environment using the potential of the environment as a resource for the solution to the problem of its own survival, conservation, development. The potential of the environment includes human (social), information, natural, energy potentials, management potential, and others.

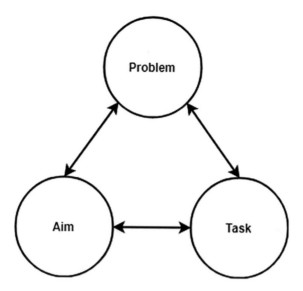

Fig. 1. "Problem-Aim-Task" interaction

Each part of the environment can be considered the carrier of the problems of their own survival, conservation and development. The parts of the environment of the user's IS activity are employees, production and management technologies, services, departments and other parts of his internal environment. Part of the environment of an IS user's activities are also its customers, suppliers, tax authorities and other parts of its external environment. User problems can arise in both its internal and external environments.

The problem of a part of the environment (an IS user, for example) in a general sense is defined as a stable contradiction between the desired and actual states of a part of the environment in the sense of survival, preservation and development. Problems, as it known, cannot be solved directly, immediately. To solve problems, tasks are solved, goals are achieved.

A goal is a description of the completion of an activity, a part of an activity (stage, phase, stage, step, etc.). Achieving the goal is accomplished by solving the problem: "the process of solving the problem … is the process of achieving the goal …", "the task implies the need to consciously search for the appropriate means to achieve a clearly visible, but directly unattainable goal" [9].

We detail the definition of the problem as follows. A task is the process of finding a method to achieve a goal by using available resources with given constraints on goals, methods, resources. The solution of the problem is the application to achieve the goal of the chosen variant of the method with the use of the corresponding set of available resources when defining certain restrictions.

The first set of IS objectives is to support the formation and implementation of the problematic triad of the IS user "a problem is a carrier of a problem is a result" (Fig. 2).

We are interested in parts of the environment - potential users of IS operating in an economic environment, for example. As an IS user, as already defined, are considered

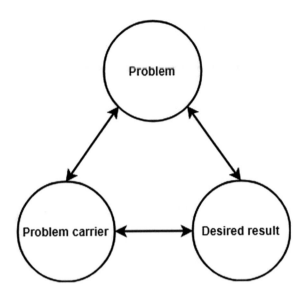

Fig. 2. Triad "problem – carrier of problem – result"

as individual individuals (employee, manager, etc.), legal entities (enterprise, institution, organization, etc.) and their parts (divisions, services, departments, etc.), and associations and groups of persons (holding, social group, society, etc.).

As an example of an IS user problem, one can cite the problems of budgeting, strategizing, environmental, social and other problems, problems of conflicts in collaborations, etc.

IS user problems may be up-to-date (up-to-date) or in a state of satisfactory solution (resolution).

As we consider the user of the IS, we proceed from the fact that the contribution to the solution of the user's problems is impossible without the information support of his activity, without "satisfying the information hunger". It is also possible to permanently resolve the user's problems locally - providing informational support as the problems become actual. Infrastructure solutions are also possible to provide information support for solving user problems.

The first complex includes 18 tasks [18], which require an information infrastructure. These are problems of problem modeling, as well as business processes of problem carriers and problem solving results. This includes the task of forming the appropriate database, building the necessary information technology.

For the complete solution of the first set of tasks, it is necessary to apply the described conditions of the Integrity Principle.

The first set of IS tasks should be addressed both for the user of the IS and for all its parts. For example, for an enterprise (organization, institution) the first set of tasks should be solved both for the enterprise as a whole and for its production departments, economic services and management structures, individual employees, managers, managers, the enterprise team and its departments, services.

The second set of IS objectives is to support the formation and implementation of the production-technological triad of the IS user "subject-object-result" of production (Fig. 3).

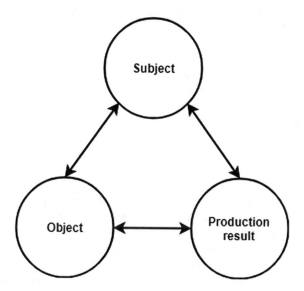

Fig. 3. Triad "subject – object – result" of production

The second complex includes 12 tasks [18], for the solution of which the information infrastructure is necessary. These are the tasks of modeling the business processes of an IS user. This also includes the task of forming the appropriate database, building the necessary information technology.

For the complete solution of the second set of tasks, it is also necessary, as for the first set, to apply the described conditions of the Integrity Principle.

The second set of IS tasks, as well as in the first case, should be solved for all parts of the IS user.

The third set of IS objectives is to support the formation and implementation of the triad of delivery of the result (product) of the IS user "delivery management - result delivery technology - getting the result by the carrier of the problem".

The fourth set of IS objectives is to support the formation and implementation of the triad of consumption of the user's IS result "consumption management - technology of consumption of the result - solution of the problem by its carrier".

The fifth set of IS objectives is to support the formation and implementation of the triad of the missionary goal of the IS user "managing the achievement of the missionary goal — technology for achieving the missionary goal — assessing the achievement of the missionary goal".

From the standpoint of an integral method of action in the direction of achieving the missionary goal of an IS user, they are an integral factor [15], i.e. form and strengthen its focus on the development of the functioning environment.

The sixth set of IS objectives is to support the formation and implementation of the triad of its own and group goals of the IS user "managing the achievement of own and group goals — technology for achieving one's own and group goals — assessing the achievement of one's own and group goals".

The user's own goal of IS is the survival, preservation and development in a globalizing market environment. This goal is realized, for example, by obtaining benefits (material, financial, spiritual, informational, other) from the realization of the result produced.

The group goal of an IS user is to promote the development of a group of one class of users operating in a common economic environment, competing in a common "market niche". The activity of a group of enterprises (organizations, institutions), social groups and other users of IS is regulated by the voluntary establishment of general rules for activities in a common market environment. The group goal of an IS user is to create a policy of survival, preservation and development of a certain type from the production and realization of the results (products) that are uniform with similar users. The group goal can be interpreted as the goal of maintaining the environment of one's own kind, the goal of the group, the "goal of the community".

Following its own and group goals, the user (for example, an enterprise, organization, institution) partially demonstrates the integrity of the third type, directing its activities to the interests of users of its own IS level, including its own interests.

As a result, the carrier of a problem, as a consumer of an IS user's products, always has the opportunity to assess how consistently a certain product manufacturer is guided by the rules established for a given producer group.

The seventh set of IS objectives is to support the formation and implementation of the triad of its own, group and missionary goals of the IS user;

Here, the IS is aimed at forming an integrated system of agreed weights of goals (own, group and missionary) in the activities of the user of the IS, which creates a balanced influence of the goal-forming coherent factors in the functioning of the user IS, enhance its role in the environment.

Eighth complex IS tasks - support for the unity of formation and implementation of the complex of problem-oriented triads IS user activity in the form of a triad "subject - object - the result."

To solve each set of tasks, a number of the above triads of activities are needed. All these triads can be represented as a common complete, triad of the integral activity "subject-object-result". For their effective construction in the IS it is necessary to apply the definitions, postulates and models of the integral complete -method; The principle of integrity, the postulate of the need for integrity of the thinking and practice of a specialist, the law of integrity, the law of integral development, the principles of development of the whole integral complete approach, and the method of system complete technology [15, 16].

In addition, as a triad, in fact, one can consider the actual IS as the unity of production technologies of information products (object of the IS-triad), technologies of production management of information products (the subject of the IS-triad), a set of information products (the result of the IS-triad).

The ninth set of IS objectives - support the formation and implementation of each process in the activity triads of an IS user "subject - object - result" in the form of an integral technological process that meets the requirements of the integrity of the third type.

On the basis of type of user activity that IS should maintain, as part of the IS can be identified analytical, expert, research, design, production, management, archive, licensing, control of information technology.

The tenth set of IS objectives - support the formation and implementation of each process in the activity triads of an IS user "subject - object - result" in the form of an integral process of achieving a goal, solving problems.

Complete- model of the process of achieving the goal (Complete-model of PAG) is an integral combination of the task model and the task solution model [15–17]. The use of a unified model of a PAG enables the IS to create a single integral set of methodological support for all targeted processes of the user of the IS.

The eleventh set of tasks for IS is to support, in the early stages of designing an IS, systematizing and bringing the entire set of concepts of building models, information technologies and databases into conformity with the concepts of integrity and integrity.

The eleventh set of tasks of the IS supports the implementation of the integral complete approach at all stages of the design of the IS.

The twelfth set of IS objectives - support for managing the development of IS.

The basis for the formation and development of the IS system structure is a growing set of activity triads for all user problems. The triad, as already noted, is the IS itself, as the unity of production technologies of information products (object of the IS-triad), technologies of production management of information products (the subject of the IS-triad), complex of information products (the result of the IS-triad). The triads are parts of the IS and elements of the IS.

Therefore, the formation of IS subsystems as triads is reduced to the formation of sets of triads aimed at solving certain problems of the IS user. The basis for the formation of the IS system process is a typical set of conditions for the construction of information technologies, databases, business process models [18] for all the problems of the IS user.

This, as it has been established, is analytical, research, design, production, managerial, expert, permissive, control, archival information technologies of IS. From this set, processes of elements, parts of the IS and the IS itself are formed.

For all technologies, uniform conditions for implementation and a general complete process model are also established.

The twelfth set of tasks of the IS supports the implementation of the integral complete approach at such design stages of an IS as technical design, working documentation, commissioning, maintenance of IS.

All sets of IS tasks are necessary for the IS user at all stages of the IS life cycle. These sets of tasks of the IS allow the user of the IS to carry out a complete interaction with the designer and complete control over the course of the design of the IS. The IS designer, with such interaction with the IS user, is aimed at creating and developing an integral and integral problem-oriented IS.

In turn, using the twelve sets of IS tasks formed here, the IS user will be able to fully influence the formation of the information environment of his activity.

The principle of integrity and twelve complexes of IS tasks enable the IS and the IS user to form a complete interaction between the IS user and the information system.

4   Conclusions

This paper is devoted to the study of the basic processes for the implementation of "megascience" projects and to ensure the possibility of their effective implementation using management information systems.

During the research, the following main stages (processes) of the "megascience" project were identified:

1. Development of the concept and scientific program of the project;
2. Formation of international collaboration;
3. Development of the project roadmap and technical design of the project;
4. Creation of an appropriate organization;
5. Installation of the necessary infrastructure.
6. Implementation of control and monitoring of construction and infrastructure;
7. Project completion.

All the listed stages of the "megascience" project are inherent in all sorts of risks that may arise from the lack of accounting for resources and the limitations of the project.

The work showed the need to create an appropriate holistic information management system « megascience » project. An approach to the creation of such a system was also proposed, and 12 sets of tasks of information systems were considered, requiring solutions for its creation.

The proposed approach can also be applied to build information systems for managing activities in all areas of the economics.

The reported study was funded by Russian Foundation for Basic Research (RFBR) according to the research project № 18-29-15015.

References

1. A Guide to the Project Management Body of knowledge (PMBOK® Guide). Fifth edn. Project Management Institute, Inc. (2013)
2. A Meeting on Megasiens was Held at the Ministry of Education and Science of the Russian Federation. Ministry of Education and Science of the Russian Federation. http://www.xcels. iapras.ru/img/MON%20press-reliz.pdf, June 2011 (Press releases)
3. Soldatov, A.V.: Megascience installations as the most important tool for integrating world-class science and education, No. 8–9, pp. 94–98. Higher education in Russia (2015)
4. Berg, A.I.: Questions of Cybernetics, VK-72. In: Suslova, R.M., Reutov, A.P. (eds.) Scientific Council of the Academy of Sciences of the USSR "Cybernetics", p. 3 (1980)
5. Kuhn, M., Remoe, S.O.: Building the European Research Area: Socio-economic Research in Practice. Peter Lang Publishing, New York (2005). ISBN 0-8204-7471-1
6. EU Framework Programs for Research and Innovation. ISBN 978-92-846-1687-9. DOI: https://doi.org/10.2861/60724

7. European XFEL Annual Report 2016. European XFEL GmbH. https://www.xfel.eu/sites/sites_custom/site_xfel/content/e35178/e56171/e56388/xfel_file56396/European_XFEL_Annual_Report_16_eng.pdf
8. Quevedo, F.: The importance of international research institutions for science diplomacy. Sci. Diplomacy. http://www.sciencediplomacy.org/files/the_importance_of_international_research_institutions_for_science_diplomacy_science__diplomacy.pdf, September 2013
9. Polya, J.: Mathematical discovery (translated from English), 448 p. Science (1976)
10. Ratchford, J.T., Colombo, U.: Megascience. Reprinted from Unesco World Science Report (1996)
11. Large Systems and Control (p. Unit V. VI Chernetsky), 206 p. Ed. LVWIKA them. A.F. Mozhaisky, Leningrad (1969)
12. Phys. Perspect 18, 355. https://doi.org/10.1007/s00016-016-0193-0 (2016)
13. Platonov, V.: Conceptual framework for the study of mega science as an organizational and managerial innovation. Innovation **228**, 11–16 (2017)
14. Priority-Setting in the European Research Framework Programs. Dan Andrée - The Swedish Ministry for Education and Research. Vinnova Analysis VA 2009: 17. ISBN 978-91-85959-69-3. ISSN 1651-355X. VINNOVA –Swedish Governmental Agency for Innovation Systems, July 2009
15. Telemtaev, M.M.: Complement or philosophy, theory and practice of integral solutions, 234 p. Irisbuk (2012)
16. Telemtaev, M.M.: From scattered ideas and knowledge to a complete system. Completion: from theory to implementation, 312 p. M.: Book House "LIBROKOM" (2013)
17. Telemtaev, M.M.: Organization of large computer systems, 186 p. Almaty, KazSU them. CM. Kirov (1989)
18. Telemtaev, M.M., Nurakhov, N.N.: Information systems in economics, educational edn., 102 p. REA them. G.V. Pleha New (2010)

On Markov Chains and Some Matrices and Metrics for Undirected Graphs

Victor A. Rusakov$^{(\boxtimes)}$ (iD)

National Research Nuclear University MEPhI
(Moscow Engineering Physics Institute), Moscow 115409, Russia
VARusakov@mephi.ru

Abstract. Metric tasks often arise as a simplification of complex and practically important problems on graphs. The correspondence between the search algorithms of the usual shortest paths and Markov chains is shown. From this starting point a sequence of matrix descriptions of undirected graphs is established. The sequence ends with the description of the explicit form of the Moore-Penrose pseudo inversed incidence matrix. Such a matrix is a powerful analytical and computational tool for working with edge flows with conditionally minimal Euclidian norms. The metrics of a graph are represented as its characteristics generated by the norms of linear spaces of edge and vertex flows. The Euclidian metric demonstrates the advantages of the practice of solving problems on graphs in comparison with traditional metrics based on the shortest paths or minimal cuts.

Keywords: Markov chains ·
The Moore-Penrose pseudo inverse of the incidence matrix · Shortest paths ·
Minimal cuts · The Euclidian metric on graphs

1 Introduction

Graphs are widely used as a model of objects in the surrounding world. In their turn, the various ways of presenting these graphs are dependent on a whole series of factors. The discrete nature of objects in the real world modeled in combination with the simplicity of their representation through a set of points connected by lines is among such factors. The introduction and use of various understood distances between such points, i.e. different metrics, by using the characteristics of the aggregate of such lines, is just as natural.

The matrix tools of linear algebra are well developed and well suited to the representation of many features of graphs. It is natural to begin with an adjacency matrix. In this matrix the correspondence between its entries, their organization and the absence or presence of lines connecting the points in the visual representation of the graph is certainly the most simple. Likewise the usual shortest paths serve as the traditional basis of measurement in the graph [1, 2]. Different connectivity evaluations of the graph are used as well as the spectral characteristics of its adjacency matrix with the quantitative evaluations of some of the features of the graph [2]. Here the spectral features of some graph matrices will also be used.

© Springer Nature Switzerland AG 2020
T. Antipova (Ed.): ICIS 2019, LNNS 78, pp. 340–348, 2020.
https://doi.org/10.1007/978-3-030-22493-6_30

From now on a graph will be understood as a finite undirected graph without loops and multiple edges having k vertices and h edges. If there are no stipulations to the opposite, then the graph is supposed to be connected.

2   Adjacency Matrix. Normalizations. Shortest Paths

The traditional adjacency matrix $C = \|c_{ij}\|$ has entries $c_{ij} = c_{ji} = 0$ where there is no edge (i,j) in the graph and $c_{ij} = c_{ji} = 1$ when the edge is present, i.e. when there is adjacency of the vertices i and j. A zero main diagonal symbolizes the absence of loops. The adjacency matrix is often used in studying and using spectral features of graphs. Though the unit in the latter equality is traditional, in essence for many of the problems solved only the non-zero value of the quantity in its place is important. Let's use this obvious arbitrariness to go over to the class of stochastic matrices. Their spectral features are well studied [3] and will be used. To go over from C to the matrix $P = \|p_{ij}\|$ divide each line i of the matrix C into d_i – the degree of the vertex i of the graph. Due to the graph connectivity, the matrix P is irreducible. Note that C normalization gives the matrix P a zero main diagonal and that for the irregular graph there will be no symmetry. If you take away the line t and the column t of the matrix P, you turn it into the matrix Q. The entries of Q will from now on, without fear of confusion, be indicated just as elements P. The eigenvalues of the matrix Q are such that the maximum absolute value of them is strictly less than 1 [3]. So the sum $I + Q + Q^2 + \ldots$ of the members of the matrix power series converges absolutely to $(I - Q)^{-1} = N$, which in form corresponds to the usual numerical geometric progression. Matrix I, here and from now on, designates an identity matrix whose size is clear from the context. When necessary the number of the line and column removed will be designated by a lower index for Q and N. The designation of both matrices taken from [4] where the matrix Q is the transition probabilities matrix of the finite absorbing Markov chain with absorbing state t, while $\forall$ entry i,j of N gives the mean number of times the chain is in state j before being absorbed into t when its start has been in state i. It is clear that the entries of N are non-negative and equal to 0 if and only if t is the articulation vertex, while the state of the start of the chain and the state where we are observing the mean number of times it is being fallen into belong to the blocks of the graph on different sides of t.

The features of the Markov chain will from now on be used in comparing, on one hand, the known algorithms for determining the shortest paths in a graph and, on the other hand, the sets of realizations of the conduct of the Markov chain at the start in some transient state s and its absorption into t. The results with this correspondence are too great to be accidental. It leads at the end of the constructions to another metric [5] that is not traditional for tasks on graphs.

1. In both cases the first step is to give the beginning vertex (state) s a definite beginning value (weight). To get the shortest paths it is usually 0 [6–8]. For each separate realization of the absorbing chain this is 1 – a unit corresponding to the contribution of the starting state to the mean number of times the chain is in the given state. All other vertices (states) are also given some value at the beginning. For the shortest paths it is a quite big (positive) number. For each realization of the absorption chain it is 0.

2. In both cases the set of vertices (states) $\{j\}$, the adjacencies of the already viewed set of vertices (states) $\{i\}$, are viewed in some way. $\{i\} = s$ at the first step. To the adjacent vertex (state) $j \in \{j\}$ a weight is ascribed according to a certain law on the basis of some value which relates to the edge (i,j), $i \in \{i\}$. Such a value for the algorithm for finding the shortest paths is the weight of the edge. The weight of the vertex j is replaced by the sum of the weight of vertex i and the weight of the edge (i,j), if that sum is less than the old weight of vertex j [6–8]. When realizing the absorbing chain, the current weight of the adjacent state j – the number of previous visits to this state – increases by 1 in correspondence with the probability p_{ij} of the chain going from i to j in one step. The value p_{ij} depends on the weight of the edge connecting the vertices (states) i and j. The less the weight of the edge (i,j), the less the probability of the current weight of the state j increasing by 1.

3. The set of vertices (states) $\{i\}$ is quite arbitrary. When searching for the shortest paths it can be the current tree [6] or just any vertex [7, 8]. It is only important that there be *uniformity* of the rule according to which the weight of the vertex j is modified. When being realized the absorbing chain, $\{i\}$ is the single state which the chain is in at the given moment. Any time the chain is in state i, increasing the current weights of the adjacent states $j \in \{j\}$ by 1 depends only on the probability of the transition in one step from i to $j \in \{j\}$ which will occur due to the definition of the Markov chain.

4. In both cases, in spite of the arbitrariness of the sequence of viewing the vertices (states), the weights of the latter always converge to certain values. When it comes to algorithms for searching for the shortest paths see, for example, [7, 8]. For the set of realizations of the absorbing chain, the convergence of the mean number of times of being in the transient state to the values of the corresponding entries of N and the uniqueness of the latter is guaranteed by the fundamental theorem about absorbing chains [4], also see the comment above about the $I + Q + Q^2 + \ldots$ convergence to N.

5. In both cases the question about whether the arbitrary edge (i,j) belongs to the extreme paths is decided on the basis of the correlations between the weight assigned to that edge and the difference of the final weights of the vertices (states) i and j. When searching for the shortest paths, the edge (i,j) is part of the shortest path if and only if the difference of the final weights of the vertices i and j is equal to the weight of the edge (i,j) [6–8]. Here note that the loops (with a non-negative weight) obviously can't be part of the shortest paths. That means that in metric tasks which involve using the usual shortest paths, the presence or absence of loops in the graph doesn't matter. When the absorbing chain edge (i,j) is part of the shortest (as we will see later, in another metric) path with a weight which is determined also through the difference of final weights of the states i and j, i.e. n_{si} and n_{sj}. The value of the edge is gotten through the additional weighing of each of the mentioned two weights of the states with the weight of their edge (i,j) – remember there is a lack of symmetry in general when you have the matrix P. For the state i it is p_{ij}, for state j it is p_{ji}. In other words, the weight of entry of the edge (i,j) is $n_{si}^{(t)} p_{ij} - n_{sj}^{(t)} p_{ji}$.

Whereas the weight of any edge is one and the same for both of its directions for the initial undirected graph. The set of values $\{n_{si}^{(t)} p_{ij} - n_{sj}^{(t)} p_{ji}\}$ for all edges (i,j) and any vertices $s \neq t$ of the graph describes the shortest s,t-path in a metric that is nontraditional for graphs [5]. The following Lemma allows us to further use the valuable property of the restored symmetry of the matrices P, Q_t, N_t. Recall that for irregular graphs these matrices with zero main diagonals are asymmetric in general. The Lemma is an analogue of the obvious property for ordinary shortest paths (minimal cuts), namely, the absence of the influence of loops with nonnegative weights on these paths (cuts) and their lengths (values).

Let **U** be a set of transient states of the absorbing Markov chain.

Lemma 1   [5]. For $s,i \in \mathbf{U}, j = 1, \ldots k$, $i \neq j$, when p_{ii} changes in the interval $[0,1)$ and there is proportional renormalization of p_{ij}: $\sum_{j=1}^{k} p_{ij} = 1 = \text{const}$ then the $n_{si}^{(t)} p_{ij}$ remain the same.

Proof. Simply compare the changes of the corresponding non-diagonal elements of the line i of Q_t and the elements of the column i of N_t when changing the diagonal entry i of Q_t in the indicated interval [5].

Further designated

$$d_{\max} = \max_{1 \leq i \leq k} d_i, p = (d_{\max} + c_n)^{-1}, 0 \leq c_n < \infty,$$

$$\text{and let } p_{ij} = \begin{cases} 0, & \text{there's no edge } (i,j),\ i \neq j, \\ p, & \text{there's edge } (i,j),\ i \neq j, \\ 1 - p d_i, & i = j. \end{cases} \tag{1}$$

The matrix P determined according to (1) is symmetrical as are any of its sub matrices Q and the corresponding $N = (I-Q)^{-1}$. It is possible that such P (and Q) have non zero elements on the main diagonal that corresponds to the loops of the graph. But according to Lemma 1, for any $s \neq t$ and edge (i, j) the value $p(n_{si}^{(t)} - n_{sj}^{(t)})$ is the same as the value of that weight calculated by using $n_{si}^{(t)} p_{ij} - n_{sj}^{(t)} p_{ji}$ for P (and Q) with a zero main diagonal and even without their (and matrices N) being symmetrical. Further the matrix P determined according to (1) and the Q and N gotten from it will be used everywhere.

In view of the already mentioned convergence of $I + Q + Q^2 + \ldots$ to N you get the product $N_t Q_t = N_t - I$, $\forall t$. From this matrix equality for $\forall s \neq t$ line we get [5] a set of $p(n_{si}^{(t)} - n_{sj}^{(t)}) \geq 0$ for all edges (i,j) of the graph that can be looked on as an edge flow transferring the flow of 1 from vertex s to vertex t. That circumstance initiates the transfer to yet another matrix of the representation of an undirected graph using numbering not only of the vertices, but also the edges.

3   Incidence Matrices. Traditional Metrics. The Pseudoinverse Lemma. The Euclidian Metric

The incidence matrix $A = \|a_{ij}\|$ has k lines and h columns. $a_{ij} \neq 0$ if and only if the vertex with the number i is incident to the edge with the number j. Since in a graph without loops each edge is incident to two different vertices, each column contains exactly two non-zero elements. The further interpretation of $a_{ij} \neq 0$ in their correspondence to the graph depends on the object of the following construction. For undirected graphs both non-zero elements of each column often [7, 9–14] are assumed to be equal to 1. For directed graphs, as a rule [7, 10, 11, 14], but not always [15], one of these units is replaced by −1. On the other hand, when working with undirected graphs with the help of such a replacement for each edge they sometimes give the orientation [10, 16, 17] and it is used for certain goals. Of course [16, 17], with this one can then use only those results that are invariants to the arbitrariness in the course of such an orientation.

The traditional edge flow is determined as a set of non-negative numbers having some additional features. Each of these numbers corresponds to a concrete edge of the graph and is called the magnitude of the flow along that edge [6]. The requirement of non-negativity simply means that the *direction* of the flow is by default based on the representation which contains the *magnitude* of the flow. For example, through the order of indicating the vertices incidental to a given edge, as this was above for $p(n_{si}^{(t)} - n_{sj}^{(t)}) \geq 0$ for all edges (i,j) of the graph. Such an approach is not always acceptable. When simultaneously viewing two or more flows of one and the same color, for some edges the flows can go in opposite directions. In this case the magnitudes of such flows as those having opposite signs are subtracted from one another [6]. Therefore it makes sense in such cases to choose, arbitrarily and invariably for further work, one of the directions as positive ahead of time for each edge of the graph. After that it is always possible to treat the negative numbers corresponding to any edge of the graph as the magnitude of the flows for these edges going in the opposite direction. It is an analogous situation with the directions of the vertex flows, i.e. the flows from the vertices outside the graph and flowing into them from outside.

Let X and Y be vector spaces with dimensions h and k correspondingly over the field of real numbers. The represented orientations in the absence of limits on the capacity of the edges and vertices turns the incidence matrix A of an undirected graph having ± 1 as non-zero elements in each column into a useful instrument for the *linear* transformation $Ax = y$ of the edge flows $x \in X$ into the vertex flows $y \in Y$. From now on the incidence matrix A will be understood as just such a matrix.

The use of linear transformations usually involves using those or other norms of elements of linear space. The norms for transforming these spaces are also formed based on them. Moreover the calculation of the additional conditions is always possible. For example when studying graphs one can look for norms $\|A\|^{s,t}$ corresponding to a certain pair of vertices s,t. Let $\{x_{st} \in X \mid Ax_{st} = y_{st}\}$, where y_{st} has an entry s equal to $+1$, an entry t equal to -1, while the remaining entries are equal to 0. Any of the vectors x_{st} represent the transfer of the flow of 1 from s to t along the edges of the graph. Then, without additional designations one should search for the extreme norms along all such x_{st} with the clear possibility of changing sup to max.

We introduce a cubic norm $\|\bullet\|_c$ into Y, while into X – the octahedral $\|\bullet\|_o$, and let each edge of the graph be given the length 1. Then $\|A\|_{c/o}^{s,t} = sup\|Ax_{st}\|_c(\|x_{st}\|_o)^{-1} = max\|y_{st}\|_c(\|x_{st}\|_o)^{-1} = max(\|x_{st}\|_o)^{-1} = (min\|x_{st}\|_o)^{-1} = l_{st}^{-1}$, where l_{st} is the length of the shortest s,t-path.

Now we also introduce a cubic norm into X, and let m_{st} be a maximum number of the paths that have no common edges between s and t. Then $\|A\|_{c/c}^{s,t} = sup\|Ax_{st}\|_c(\|x_{st}\|_c)^{-1} = max(\|x_{st}\|_c)^{-1} = (min\|x_{st}\|_c)^{-1} = m_{st}$, as if the edges were given the traditionally understood capacity equal m_{st}^{-1}, and the maximal flow from s to t of 1 was found with the help, for example, of the known [6] Ford-Fulkerson algorithm.

As we see, the norms of X used give entirely traditional metrics of a graph. Note that in each of these two cases x_{st} giving a maximum ratio of $\|y_{st}\|$ to $\|x_{st}\|$, won't, generally speaking, be unique.

The next step is possible by using the flows described above in the form of a multitude of magnitudes of the form $p(n_{si}^{(t)} - n_{sj}^{(t)})$, $\{s \mid s \neq t\}$. However it is simpler [5] to again use the power series, this time of the matrix P–F, where the matrix $F = \|f_{ij}\|$, $f_{ij} = k^{-1}$, $\forall i,j$, – the limiting matrix of the ergodic Markov chain. That is $I + (P-F) + (P-F)^2 + \ldots = (I-P+F)^{-1} = Z$, which gives $ZP = Z-I + F$, where $Z = \|z_{ij}\|$ is the fundamental matrix of the ergodic Markov chain [4]. Similarly with the product NQ from the last equality we get [5] the set of values $p(z_{si}-z_{sj})$, $\forall s$, for all edges (i,j) of the graph. They can be considered as the edge flow transferring the flow of $1-k^{-1}$ from the vertex s to all remaining $k-1$ vertices equally, i.e. k^{-1} units of the flow each. From this set we form the h-dimension vector in correspondence with the numbering of the edges. According to *indirect* analysis [18, 19], the entries of the column s, $\forall s$, of the matrix representation of the graph that we need should possess exactly this feature. That same analysis suggests that the second feature which must be there is the independence of the result of the special summation (sign-adjusted sum) of the entries of the column s, $\forall s$, from the routes of the graph it was done from. Once there is this feature too, then one can immediately formulate and prove [5] the following.

Lemma 2. Let A be the incidence matrix of an undirected graph, $Z = (I-P + F)^{-1}$, where p and the matrix P are determined according to (1), through A^+ the designated Moore-Penrose pseudo inversed matrix A, while « ′ » symbolizes the transposition. Then $A^+ = pA'Z$.

Proof. By checking using the conditions of Penrose: for the matrices A and A^+ one can see that four equalities have been carried out $AA^+A = A$, $A^+AA^+ = A^+$, $AA^+ = (AA^+)'$, and $A^+A = (A^+A)'$.

According to a well-known [20, 21] property of the Moore-Penrose pseudo inverse, the Euclidian (spheral) norm $\|\bullet\|_e$ of the vector $x_{st} = A^+y_{st}$ will be minimal among $\{x_{st} \in X \mid Ax_{st} = y_{st}\}$. Leave for Y a cubic norm, but for X we will use a Euclidian norm. Then we get $\|A\|_{c/e}^{s,t} = sup\|Ax_{st}\|_c(\|x_{st}\|_e)^{-1} = (min\|x_{st}\|_e)^{-1} = (p(z_{ss} + z_{tt} - 2z_{st}))^{-1/2}$.

For an arbitrary edge (i,j) of a graph let its column in the matrix A have +1 in the i-th row, and –1 in the j-th row. For an arbitrary graph $A^+ = pA'Z$, so for this edge the component of the flow A^+y_{st} is equal to $p(z_{si} - z_{sj} + z_{tj} - z_{ti})$. Then we use the well-known [4] matrix equality $Z = F + (I - F)N^*(I - F)$. Here the $k \times k$ matrix N^* is the matrix N_t, $\forall t$, with the zero row and column inserted into the t-th places. We multiply the matrices of this equality on the left by the row-vector y_{st}', and on the right – by the vector y_{ij}. The last vector, like the aforementioned column of the (i,j) edge in the matrix A, has an i-th entry equal to +1, a j-th entry equal to −1, and the other entries are equal to 0. We obtain the equality $z_{si} - z_{sj} + z_{tj} - z_{ti} = n_{si}^{(t)} - n_{sj}^{(t)}$. Next multiply both sides by p.

We obtain the component equality of the flows along the edges described, on the one hand, in terms of the entries of the matrix Z, i.e. A^+, and in terms of the entries of the matrix N^* on the other hand. This equality demonstrates the validity of the assumption that the set of values of the form $p(n_{si}^{(t)} - n_{sj}^{(t)}) = n_{si}^{(t)} p_{ij} - n_{sj}^{(t)} p_{ji}$ can be used as a description of the shortest s,t-path in some specific (quadratic, Euclidian) metric.

Finally, unlike the cases of traditional metrics of a graph, the use of the Euclidian metric always gives uniqueness of the vector x_{st} of the maximizing ratio $\|y_{st}\|$ to $\|x_{st}\|$ that is the vector $A^+ y_{st}$.

When there is a disconnected graph the discontinuous nature of the pseudoinverse make the variant of lemma 2 trivial. The creation of the Euclidian metric of a disconnected graph on the basis of the components metric when restoring graph connectivity considered in [22] is the reverse situation.

4   Examples of Use

The distribution of flows in the analysis of the throughput of the computer network [5, 23, 24], the graphs throughput evaluations [5, 23, 24], the analysis of variability of throughput in conditions of non-stationary traffic [25, 26], the analysis and evaluations of the graphs reliability [5, 23, 24], the synthesis of high reliability graphs [23, 24].

5   Results

1. The correspondence between search algorithms of the usual shortest paths and implementations of the Markov absorbing chain is described. From this starting point, a sequence of matrix descriptions of undirected graphs is established.
2. Restoring the valuable symmetry of matrix descriptions of irregular graphs is justified (Lemma 1). The core of Lemma 1 is an analogue of the property for ordinary shortest paths (minimal cuts), namely, the absence of the influence of loops with nonnegative weights on these paths (cuts) and their lengths (values).
3. The need to make a representation of each flow component $\{n_{si}^{(t)} p_{ij} - n_{sj}^{(t)} p_{ji}\}$ in the same form for t as for any other vertex leads to the ergodic Markov chain. The explicit form of the Moore-Penrose pseudo inversion of the incidence matrix is described on the basis of two special properties of the edge and vertex flows. These properties follow from the matrix equality which is characteristic of Markov ergodic chains. The symmetry of the matrices describing these chains is used. This symmetry can always be restored even for irregular graphs using Lemma 1.
4. The way of proving the truth of an explicit form of the Moore-Penrose pseudo inversion of the incidence matrix is indicated (and verified, Lemma 2). Lemma 2 is a metric analogue of the descriptions of the well-known search algorithms for ordinary shortest paths and minimal cuts. The core of this metric analogue, i.e. the explicit form of the Moore-Penrose pseudo inversion of the incidence matrix, has powerful new (analytical) capabilities.

5. Applied problems on the graphs are listed, where the Euclidian metric demonstrates its advantage over traditional metrics, which are based on the usual shortest paths and minimal cuts.

6 Conclusion

The sequence of matrix representations of undirected graphs starting from the adjacency matrix and ending with the explicit form of the Moore-Penrose pseudoinverse of the incidence matrix has been established. Matrix spectral properties are used. The family of algorithms for finding the usual shortest paths has been compared to the realizations of the finite absorbing Markov chain. From the matrix power series connected with the Markov chains one gets a representation of the special combination of edge and vertex flows. Such a representation makes it possible to make conjectures about the explicit form of the pseudo inversed incidence matrix. Then confirmation of the hypothesis was made by checking the carrying out of the conditions of Penrose. The use of Euclidian metrics ensures better results and more possibilities in solving a number of applied tasks on graphs in comparison to the practice of using traditional metrics.

References

1. Kozyrev, V.P., Yushmanov, S.V.: Graph theory (algorithmic, algebraic and metric problems). The results of science and technology. Probability theory. Math. Stat. Theor. Cybern. Ser. **23**, 68–117 (1985)
2. Hernandez, J.M., van Mieghem, P.: Classification of graph metrics, p. 20. Delft https://www.nas.ewi.tudelft.nl/people/Piet/papers/TUDreport20111111_MetricList.pdf (2011). Accessed 27 March 2019
3. Gantmakher, F.R.: Matrix Theory, 3rd edn, p. 576. Nauka, Moscow (1967)
4. Kemeny, J., Snell, J.: Finite Markov Chains. University Series in Undergraduate Mathematics, p. 210. Van Nostrand, Princeton (1960)
5. Rusakov, V.A.: Analysis and Synthesis of Computer Network Structures. Part 1. Analysis, p 122. Moscow Engg. Phys. Inst. Report: All-Union Sci. Tech. Inform. Center No. Б796153 (1979)
6. Hu, T.C.: Integer Programming and Network Flows, p. 452. Addison-Wesley, Menlo Park-London (1970)
7. Ore, O.: Theory of Graphs, p. 270. American Mathematical Society, Providence US (1962)
8. Davies, D., Barber, D.: Communication Networks for Computers, p. 575. Wiley, Hoboken (1973)
9. Diestel, R.: Graph Theory Graduate Texts in Mathematics, vol. 173, p. 322. Springer-Verlag, New York (2000)
10. Van Dooren, P.: Graph Theory and Applications, p. 110. Dublin. https://perso.uclouvain.be/paul.vandooren/DublinCourse.pdf (2009). Accessed 27 March 2019
11. Ruohonen, K.: Graph Theory, p. 114. Tampere University of Technology. http://math.tut.fi/~ruohonen/GT_English.pdf (2013). Accessed 27 March 2019
12. Zykov, A.A.: Fundamentals of Graph Theory, p. 382. Nauka, Moscow (1987)

13. Wilson, R.J.: Introduction to Graph Theory, 4th edn, p. 171. Edinburgh Gate, Harlow and Essex, Addison Wesley Longman Limited (1996)
14. Harary, F.: Graph Theory, p. 274. Addison-Wesley, Reading (1969)
15. Anderson, J.A.: Discrete Mathematics with Combinatorics, 1st edn, p. 799. Prentice Hall, Upper Saddle River (2000)
16. Biggs, N.: Algebraic Graph Theory, 2nd edn, p. 211. Cambridge University Press, Cambridge-New York-Melbourne (1993)
17. Godsil, C., Royle, G.: Algebraic graph theory. In: Graduate Texts in Mathematics, vol. 207, p. 443. Springer-Verlag, New York (2001)
18. Idjiry, Y.: On the generalized inverse of an incidence matrix. J. SIAM **13**(3), 827–836 (1965)
19. Boullion, T.L., Odell, P.L.: Generalized Inverse Matrices, p. 107. Wiley Interscience, New York-London-Sydney-Toronto (1971)
20. Albert, A.E.: Regression and the Moore-Penrose Pseudoinverse, p. 180. Academic Press, New York (1972)
21. Beklemishev, D.V.: Additional Chapters of Linear Algebra, p. 336. Nauka, Moscow (1983)
22. Rusakov, V.A.: Reconstruction of the Euclidian Metric of an Undirected Graph by Metrics of Components. Nat. Tech. Sci. **2**(52), 22–24 (2011)
23. Rusakov, V.A.: A Technique for Analyzing and Synthesizing the Structures of Computer Networks Using Markov Chains. Computer Networks and Data Transmission Systems, pp. 62–68. Znaniye, Moscow (1977)
24. Rusakov, V.A.: Implementation of the methodology for analysis and synthesis of computer network structures using Markov chains. Engineering-mathematical methods in physics and cybernetics. Issue 7, pp. 41–45. Atomizdat, Moscow (1978)
25. Rusakov, V.A.: Synthesis of computer network structures and the problem of small certainty of initial values. USSR AS's Scientific Council on Cybernetics. In: Proceedings of 5th All-Union School-Seminar on Computing Networks, 1, pp. 112–116. VINITI, Moscow-Vladivostok (1980)
26. Rusakov, V.A.: On the regularity of the displacement of the mean estimate for the throughput with non-stationary traffic. USSR AS's Scientific Council on Cybernetics. In: Proceedings of 9th All-Union School-Seminar on Computing Networks, 1.2, pp. 48–52. VINITI, Moscow-Pushchino (1984)

Social Science in Digital Age

The Digital Technologies for Improving the Operational Efficiency: Case of Russian Industry of Ferrous and Non-ferrous Metals Scrap

Efim Popov[1,2] and Zhanna Mingaleva[1,3(✉)]

[1] Perm National Research Polytechnic University,
Perm 614990, Russian Federation
[2] RosMetTrade Company, 2, Plekhanov St., Perm 614065, Russia
[3] Perm State University, Perm 614990, Russian Federation
mingall@pstu.ru

Abstract. The paper discloses the experience of increase in efficiency of operating activities in the industry of ferrous and non-ferrous metals scrap on the example of the Russian company "RosMetTrade". The article discusses the ways of business processes automation in industry of ferrous and non-ferrous metals scrap using the modern digital technologies. The authors reveal the sources and factors for increasing the operational efficiency due to digitization of basic operations within business processes. Also, the information technologies applicable to business processes automation in industry of ferrous and non-ferrous metals scrap are defined. Each part of business processes which can be improved by the means of information technologies are described. There are three factors for increasing the operational efficiency using information technologies: (1) decreasing the time of implementing the internal transactions, (2) reducing the production cycle, (3) lowering the cost of operations.

Keywords: Operational efficiency · Information technologies ·
Digitization of business processes operations

1 Introduction

Digitization is going into a depth of modern society. It is widely recognized that the effectiveness of operation activities in a company can be significantly increased by the business processes automation. According to foreign researchers, the most developed in the application of digital technologies are the banking and credit and financial sectors, health care, transportation [1]. However, there are great number of Russian companies which did not realize the opportunity mentioned above. The high potential of automation and increase in efficiency of business processes is available for companies operating in the Russian industry of ferrous and non-ferrous metals scrap. Its great importance is noted at the Strategy of development of ferrous metallurgy of Russia for 2014–2020 and on prospect till 2030 and the Strategy of development of nonferrous metallurgy of Russia for 2014–2020 and on prospect till 2030 [2].

T. Antipova (Ed.): ICIS 2019, LNNS 78, pp. 351–363, 2020.
https://doi.org/10.1007/978-3-030-22493-6_31

The importance of the secondary metal industry becomes especially relevant in the context of limited primary raw materials. One side, in 2017, the cost of ore reached $ 80 per ton. On the other side, according to the Ruslom agency [3], in Russia annually 17 million tons of ferrous metal scrap goes to landfills and is not used in steel production. This volume of ferrous scrap is equivalent to 1/3 of the annual iron smelting volume in Russia (in 2017 according to Federal State Statistics Service in Russia 52.1 million tons of cast iron were produced).

Such landfills pollute the environment, harm the environment, do not allow efficient use of land resources. It is necessary to emphasize that companies carrying operations in this industry provide not only the supply of valuable metallurgical raw materials, but, first of all, collect and utilize waste. Also, they are engaged in cleaning the lands polluted by metal scrap, thereby providing environment protection that, eventually, allows to preserve mineral resources and to save the clean environment for future generations.

From economic view the industry of ferrous and non-ferrous metals scrap in any country has rising impact both on internal and international economic relations. For Russian economy the following facts can testify to the scale of operations in this sphere. According to the Translom company [4], one of the leading operators of scrap metal market, in 2017 the volume of ferrous metals scrap shipment in Russia was 28.2 million tons. At current moment more than 5 million tons of ferrous metals scrap is exporting from Russia (see Fig. 1). It is important to notice that Russian industry of ferrous and non-ferrous metals scrap has a high potential for export operations.

Thus, increasing the efficiency of operators of the secondary metal market in Russia is an important scientific and practical task. In the digital economy, this task is even more relevant for several reasons.

The first reason is the considerable remoteness of places of formation of scrap metal from consumers (metallurgical enterprises). Under these conditions, the digitization of business and production processes eliminates many bottlenecks and solves some of the logistical problems.

The second reason is the rising cost of resources - fuel, materials. Also, the industry of ferrous and non-ferrous metals scrap is characterized by a high capital intensity of the business of collecting and recycling scrap metal, because it requires expensive special transport and equipment to conduct operations. Any opportunities to reduce the cost of these elements through the introduction of digital technologies (for example, optimizing scrap transportation routes for fuel economy) can significantly affect the efficiency of the entire industry of ferrous and non-ferrous metals scrap.

The third reason is the difficulty of organizing and managing the network of branches for the collection, transportation, processing and shipment of secondary raw materials to consumers. And here the introduction of information technologies, remote access operations and other digital methods of communication can significantly increase the efficiency of activity.

The fourth reason is related to such a factor as the pricing policy of metallurgical enterprises. The latter, in turn, depends on the level of domestic demand, the balance of supply and demand in the world market, prices for ore and many other circumstances. Companies engaged in secondary metallurgical raw materials cannot influence these factors, but they can predict them and take them into account in their activities [5].

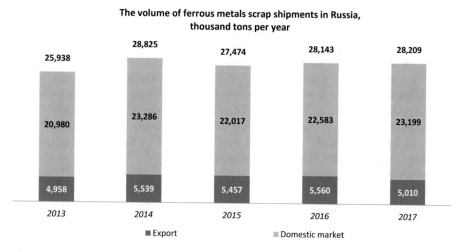

Fig. 1. The volume of ferrous metals scrap shipments in Russia (*Source: own processing*)

It is necessary to note the complexity of the organization and management process in Russian companies operating in industry of ferrous and non-ferrous metals scrap. All these circumstances demand application of modern methods of management, first of all, the information systems and technologies, digitization of all operation processes.

1.1 Review of Scientific Literature

At the first time the problem of effectiveness of operation activities in a company got to the center of attention of academicians in post-war years when reduction of military orders became the reason of crisis in many North American companies, in particular, in Ford Motor corporation. The era of fighting for the consumer had come, the attention of top managers and scientists was focused on factors of competitive success. The first works in this direction were made in Harvard business schools by professors Bain and Andrews [6]. In particular, the experience curve and economy on production scale were called one of such factors. Later, in 1988, at the conference in Academy of management (Anaheim, California, the USA) trends of information technology development and their influence on industries and separate firms were considered [7]. It was noted that, first of all, the information technologies are capable to change extent of the competition by formation of new entrance barriers, to impact on relationship between suppliers and consumers and creation of new goods substitutes. Secondly, appropriate applying the information technologies are able to give competitive advantages for a company. And at last, use of advanced information systems can lower operational expenses, increase the speed of transactions and offer clients new products. The present article is on the third direction of a scientific thought – reduction of operating expenses and increase in operational efficiency due to use of the modern digital technologies.

At this moment the set of scientific and practical works on pages of the leading foreign magazines, such as *Management Decision, Journal of Strategic Information Systems, Performance Measurement and Metrics* is devoted to the problem of operational efficiency and use of information technologies and systems. The digital infrastructure literature provides important perspectives on the intrinsic relations between information systems in today's organizations [8, 51]. Nowadays both the automation and digitization are the key idea for improving the efficacy of operations and building superior business model in a company. For example, D.C. Pantaleo is considered the business process automation as one of the most important directions of innovative activity in a company [9]. Researchers at "The Practical Real-Time Enterprise" have the same opinion [10]. "Cloud technologies" is also considered by the academicians and managers as the most important component of the business processes connected with customer relationship management and entrance into the new markets [11].

Much attention in foreign literature is paid to creation of management accounting systems and internal control on the basis of modern information technologies and as well as supply chain management, with is important for the studied industry and the analyzed company. "Sustainable manufacturing techniques, such closed-loop supply chains, is another trend in this area" [12]. There are already several researches in the literature on the study that connects business, information, engineering and analytics perspectives on digitization and the supply chain risks [13, 14]. The studies on the creation of a digital architecture for time management [15], the formation of an archive of software files as part of industrial management [16], the management of digital infrastructures [17], the expansion of digital infrastructure to various business operations [12] are increasingly being conducted in recent years.

The issues of the capacity and readiness of organizations for the introduction of digital technologies are being actively investigated by scientists and consultants of MIT [18, 19].

The Russian scientific literature and periodicals have insufficient attention to a problem of operational efficiency in the industry of ferrous and non-ferrous metals scrap. Only the general-theoretical questions were raised [20–24]. In Russia at the government level considerable attention is paid to a problem of automation and digitization of business processes in companies [25, 26]. But in spite of this fact the problem of scientific evidence and application of an information systems for increase in operational efficiency in all branches of the national economy is developed insufficiently; it is noted the shortage of conceptual approaches and methodical recommendations.

2   The Methodology of Research and Data Base

The methodology of the current research is as follows. First of all, characteristics of the Russian market of scrap of ferrous and non-ferrous metals are given. These characteristics of this market define a value chain and its part in which a company operates. Then the scheme of business processes for the certain participant of the industry – the company which conducts operations on collecting, processing and transportation of scrap of ferrous and non-ferrous metals is built. To this end, we use the basic principles of the theory of network formation in Russian organizations and management of network interaction [27]. Further the attention on those elements of operational model which are carried out insufficiently and which improvement can give operational and economic effect for the company in general is fixed.

Scrap metal is the metal products which served the term, the tool, the equipment, machines or their metal parts, a container and packing from ferrous and non-ferrous metals (from cast iron, steel, aluminum and tin), metal working waste (shaving, sawdust, dust), scrap, production wastes of metals, metal cutting waste, the fulfilled accumulators which lost consumer properties of a wire (steel, aluminum, copper).

In Table 1 comparative data on use of scrap metal are provided in world steel industry [28].

According to Deloitte consulting firm, world consumption of steel in 2017 grew by 5.1%, in 2018 growth is expected for 2.8%, in 2019 the increase of world steel production can reach 1.3% [29].

The industry of scrap of ferrous and non-ferrous metals in Russia, according to the Ruslom agency [3] represents 60 thousand employees engaged inside 5,000 enterprises. Consolidated revenues exceed 1 trillion rubles per year, the volume of procurement of ferrous and non-ferrous metals scrap is more than 45 million tons per year. In terms of physical volume, the scrap metal market is consisting of ferrous scrap metal – 96% of the total amount of scrap metal formation in Russia.

Sources of scrap of ferrous metals are: depreciation scrap (56%), waste of rolling production (23%), waste of steel-smelting production (10%) and other (11%). The largest suppliers of scrap in Russia are the oil and gas extraction companies, engineering enterprises, railroad companies, the enterprises of energy industry. The largest consumers of scrap metal (more than 1 million tons per year) are PJSC "Magnitogorsk Iron and Steel Works", CJSC "Nizhneserginsky hardware-metallurgical plant", JSC "West Siberian Metallurgical Plant", NLMK Group[1], JSC "Belarusian Metallurgical Plant" (the management company of the holding BMK), CJSC "NLMK Kaluga", JSC "Severstal" [28]. The largest market players are CJSC "Oris Prom", CJSC "Vtormetall"/CJSC "Met-Profit", CJSC "Uralvtormet", CJSC "Regionalnye gruzoperevozki", JSC "Troyka-Met", CJSC "Stroymeteko" [28].

The value-added chain of industry of ferrous and non-ferrous metals scrap in Russia consist of three key stages:

1. Suppliers: the organizations which use metal products in operating activities: oil and gas extraction companies, engineering and metalworking enterprises, rail carriers, civil engineering firms, etc.
2. Scrap metal traders: scrap processing enterprises, other traders, small scrap metal collectors.
3. Consumers: Russian and foreign metallurgical enterprises.

---

[1] NLMK Group (NLMK) is an international steel company with assets in Russia, the USA and European countries. The main asset of the Group is the Novolipetsk Metallurgical Plant, the first largest metallurgical plant in Russia.

Table 1. Consumption of scrap metal in world steel industry, million tons

Country	Period				
	2013	2014	2015	2016	2017
China	85,6	87,5	83,3	90,1	147,9
EC-28	90,3	91,6	90,6	88,4	93,4
USA	59,0	62,0	56,5	56,7	58,8
Japan	36,7	36,9	33,5	33,6	35,8
South Korea	32,7	32,6	29,9	27,4	30,5
Turkey	30,4	28,2	24,1	25,9	30,3
Russian Federation	25,9	30,7	27,2	27,8	28,5

Source: own processing

3   Research

3.1   Features of Scrap Metal Trading's Business Model

RosMetTrade company as a market player is scrap metal trader. Its operational activities have a set of features:

1. **Network of scrap metal processing and storage workshops**. Each workshop is situated at a great distance from headquarters and other workshops. Procurement of scrap is performing in places of supplier's operating activities. In particular, the RosMetTrade gather scrap metal on oil & gas fields of PAO "LUKOIL" in Western Siberia, the Komi Republic – Russian regions that are far removed from steel-making plants consuming scrap metal in cast iron production. Each division have to organize and the perform production cycle which consists of operations related with procurement, logistics, processing and shipment of scrap metal. Meanwhile, this circumstance defines the high level of overheads.
2. **Economy on production scale**. Inbound scrap metal logistics is based on lorries, trucks, and other motor vehicles. As said above supplier's storage places are located far from RosMetTrade's scrap metal processing and storage workshops. Therefore, the efficiency of transport procurement activities is defined by extent of loading of trucking facilities. It is obligatory to have obtained the weight of the transported freights as much as possible. So, transport costs would lower to minimal level. It is notable the business performance depends on logistics' efficacy on all stage of value added chain in industry of ferrous and non-ferrous metals scrap.
3. **High-polluted scrap metal**. The most profitable type of scrap – pipes of small diameter – have the largest specific weight and in smaller degree are polluted. However, under the terms of contracts the RosMetTrade company is obliged to gather all formed scrap metal – parts of constructions, metal boxes, wires, metal shavings and so forth, and in certain cases has to clean up lands polluted with scrap metal stored. Scrap of ferrous metals, as a rule, contains the remains of nonmetallic designs, an upholstery, isolation, etc. and also is subject to corrosion owing to what there are considerable non-productive losses.

4. **Large number of participants of operating activities**. RosMetTrade's scrap metal processing and storage workshops attract the local companies as suppliers and contractors. Control of an agreement discipline and activity of contractors from headquarters becomes a difficult task. In the same time the problem of constant control and internal audit of operating activities is an important factor of increase in efficiency of operating activities.

These factors define efficiency of operating activities of the of scrap metal trader, in particular, the RosMetTrade company.

3.2   Problems of Increase in Efficiency of RosMetTrade Company's Operating Activities

Efficiency analysis of the core business processes allowed to reveal a number of problems which the scrap metal trader, including the RosMetTrade company face.

1. **Planning and implementation of operations on scrap metal procurement**. Scrap metal is forming during productive activity – as the sorted metalwork, pipelines, half-ruined buildings taken out of service of a construction, lifting mechanisms, automobiles, tanks, platforms and so forth. For collecting, preparation for export and transportation of scrap metal from the storage location it is necessary develop the precise operating plan: to involve the required number of workers, the necessary lifting mechanisms and trucking facilities. It is necessary to remember that operations on gathering of scrap metal are performing under conditions of 4–6 categories of complexity of labor activity. The longer the crew works at the place of scrap metal gathering, the more production costs. Under Far North conditions it is important to gather scrap metal as soon as possible because it can be done only in winter time when there are blizzards, snowfalls, hard frosts or, opposite to thaw. Scrap metal needs to be examined, defined in category, foreman needs to plan and to quickly carry out work on its gathering and transporting.
2. **Cost efficiency of operations on procurement and processing of scrap**. Marginal income of the scrap metal trader, as a rule, does not exceed 2,500 rubs per one ton of scrap of ferrous metals. The cost of operations on scrap metal gathering and its transporting from the storage location can be from 1,000 to 10,000 rubs per ton. Expenses on processing of scrap metal can make from 500 to 1,500 rubs per ton. Hence, the efficiency of operations on gathering and processing of scrap metal before transporting to the consumer becomes the critical success factor. Considering what scrap metal appears on sale, as a rule, in the conditions of hard price struggle on stock and non-stock exchanges, it is become obvious that cost efficiency of transactions is the critical success factor in market competition.
3. **The control over loading and unloading works**. Steel-making plants and railroad companies are imposing very strict requirements to scrap metal's loading in railroad cars and its suitability for production of cast iron. Therefore, the quality of the scrap metal prepared for shipment and also its radiation purity, lack of pollution and unrelated inclusions, is a problem of the organization of operating activities too. Therefore, constant control over activity of scrap metal processing and storage workshops' network is needed.

4. **Accounting and internal audit**. The problem of prevention of plunders, proper accounting and control of appliances, safety of fixed assets, respect for financial discipline also is important for the company having a network on scrap metal trading. It is difficult to trace the accuracy of measurements and indications of scales, quality of the arriving scrap metal, volume of shipments, etc. Lots won at scrap metal tenders have low profitability therefore the advanced accounting systems and internal audit are the critical success factor too.

3.3 Digitization and Automation of the Core and Managerial Business Processes

1. **Introduction of decision-making procedures in an electronic form**. In "Bitrix" software application the following electronic procedures are developed and realized:

   - planning and control of execution of directives, tasks, plans, project etc.;
   - planning and coordination of holidays;
   - coordination of contracts and agreements;
   - coordination of invoices for payment;
   - planning and accounting of working hours;
   - electronic archive of documents.

   Thus, time for commission of internal transactions, storage and document retrieval is reduced, the performing discipline increases, the transparency and high speed of document flow are provided.

2. **Planning the operating activities of scrap metal processing and storage workshops' network**. For mobile devices in foremen's hands were developed software applications for planning and control of operations on gathering and transporting scrap metal:

   - survey and photography of scrap metal in storage locations;
   - determination of quantity and category of scrap metal;
   - list of the main works;
   - the report on the performed works.

   The foreman and its crew carry out works, make in electronic form the report on the performed works, including photos which remain in an information system and can be checked by the division leader in the remote mode. So, the production discipline increases, effective use of all types of human and material resources is provided.

3. **The standardized accounting procedures**. At "1C" software application is developed a number of standard procedures for primary information arriving into core accounting procedures. In particular, accounting and processing of applications for scrap metal transporting, accounting of scrap metal processing, storage and shipment to consumers. It allowed to organize carrying out accounting transactions in territorially remote divisions, to reduce document handling time, to provide complete control of all transactions.

4. **Internal control system of the movement of material resources inside scrap metal processing and storage workshops**. By means of a software and hardware the problem of control of cargo weighing on automobile scales is solved. Now a

freight in a body of a car is putting on a photo along with the operation of weighing a car with scrap metal's freight. Also, there is photo- and video- fixing of the movement of trucks across the storage place. Thus, in real time it is possible to control results of freights weighing, to check the movement of trucks on the scrap metal processing and storage workshops. Besides, vehicle's traffic between storage locations and processing workshop is monitored by navigation's means, while results of loading and unloading works are under constant control.

4   Discussion

4.1   The Factors of Operational Efficiency of Scrap Metal Trader

The RosMetTrade company is engaged in operations at the second stage of a chain of formation of cost: prepares scrap in places of its storage, performs its delivery to production bases, carries out its preparation for sale and makes shipment on steel works. The map of business processes inside the RosMetTrade company is submitted in Fig. 2.

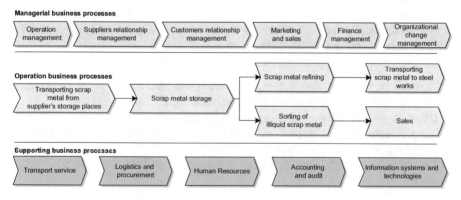

Fig. 2. The map of business processes inside the RosMetTrade company (*Source: own processing*)

From the analysis of the business processes map in Fig. 2 it becomes obvious that increase in operational efficiency can be achieved within business processes "Transporting scrap metal from supplier's storage places", "Scrap metal storage", "Scrap metal refining". Regarding operational business processes there is a set of opportunities to increase the operational efficiency by means of informatization and automation:

(1) **Transporting scrap metal from supplier's storage places**: drawing up work assignments for foremen, preparation of reports with photo- and video- writing. Decision: development of the mobile application with cloudy data storage and an integrating into corporate information system;

(2) **Scrap metal storage, transporting scrap metal to steel works:** reporting about the movement of scrap metal cargoes based on: photo- and video- writing of loading and unloading works and weighing, the control over trucks moving through the scrap metal workshop and storage place. Decision: additional modules for corporate information system, the organization of remote jobs on workshops, development of the standardized accounting procedures;

(3) **Scrap metal refining, sorting illiquid scrap metal:** drawing up work assignments for foremen, preparation of reports of implemented works with photo- and video- writing and its integrating into corporate information system, development of the standardized accounting procedures;

Further we will consider the ways for increasing of operational efficiency of RosMetTrade company's.

4.2 The Ways for Increasing the Efficiency of RosMetTrade Company's Operating Activities

The main direction of increase in efficiency of operating activities of the RosMetTrade company's operating activities as scrap metal trader is digitization and automation of the core and managerial business processes. Nowadays the company's successful performance is impossible without information systems and technologies. But the few companies in Russian industry of ferrous and non-ferrous metals widely apply them as a source of outstanding operating efficiency.

Efficiency factors, respectively, are:

1. **Reduction of time for internal transactions** (coordination of documents, accounting entries, calculations, planning, drawing up reports, etc.).
2. **Decrease of time for preparation and carrying out working activities** (survey of scrap, preparation for export, transportation on production bases, loading and unloading works and so forth).
3. **Lowering of specific cost of separate transactions** (transportation, processing, etc.).

5 Conclusion

On the basis of the business processes analysis and production operations of the RosMetTrade company conducting activity in the Russian industry of scrap of ferrous and non-ferrous metals the main problems were revealed and described. These problems were solved by digitization and automation of the core and managerial business processes.

Application of means of digitization and automation of business process at RosMetTrade company allowed to receive the following operating rooms and economic results:

1. Increase in efficiency of operations on gathering and transporting of scrap metal on processing and storage workshop. Digitization promoted that the requirements of suppliers are fulfilled completely at the scheduled time. Specific production costs do not exceed the established standards.
2. Reduction of time for internal transactions. Introduction of decision-making procedures in electronic form, standardization of accounting transactions, electronic coordination of documents allowed to achieve high operational efficiency for managerial business processes, to provide openness and transparency of transactions on scrap metal processing and storage workshops.
3. Control over activity of contractors and suppliers of motor transportation services, control of the movement of material resources on processing and storage workshops, allowed to reduce significantly scrap metal storage losses.

The economic effect reached as a result of use of modern information technologies consists in accomplishment of standards for business profitability and return on capital employed, providing planned targets on net profit and return on investment in RosMetTrade company.

In article it is shown how tasks relating to electronic procedures, standardization of means of accounting, planning of operating activities, internal control can be solved by means of applied information technologies. It is revealed that as a result of digitization of a number of processes and transactions essential operational and economic effect due to which company competitiveness at the conclusion of the stock exchange transactions on electronic trading platforms is provided can be reached. The decisions applied in the RosMetTrade company in many respects are universal, can be used by other operators of the market of scrap of ferrous and non-ferrous metals.

Further researches in the field in relation to improvement of activity of the RosMetTrade company are going to be conducted concerning digitization and automation of the core and managerial business processes. In particular, use of "cloud technologies" for data exchange in the course of operational planning is discussed.

References

1. Koutsikouri, D., Lindgren, R., Henfridsson, O., Rudmark, D.: Extending digital infrastructures: a typology of growth tactics. J. Assoc. Inf. Syst. 19(10), 1001–1019 (2018)
2. Strategy of Development of Ferrous Metallurgy of Russia for 2014–2020 and on Prospect Till 2030 and the Strategy of Development of Nonferrous Metallurgy of Russia for 2014–2020 and on Prospect Till 2030. http://legalacts.ru/doc/prikaz-minpromtorga-rossii-ot-05052014-n-839. Accessed 17 Feb 2019
3. Ruslom Agency: https://ruslom.com/wp-content/uploads/2019/02/spravka-po-rynku-loma-2018.pdf. Accessed 10 Feb 2019
4. Translom Company: http://www.translom.ru/uploads/annual_report2017.pdf. Accessed 19 Feb 2019
5. Kane, G.C., Palmer, D., Phillips, A.N., Kiron, D., Buckley, N.: Strategy, not technology, drives digital transformation. MIT Sloan Manag. Rev. Deloitte Univ. Press 14, 1–25 (2015)
6. Bain, J.S.: Relation of profit rate to industry concentration: American manufacturing, 1936–1940. Q. J. Econ. 65, 293–324 (1951)

7. Reimann, B.C.: Getting value from strategic planning. The conference board's 1988. strategic planning conference. Plann. Rev. **16**, 42–48 (1988)
8. Augustsson, N.-P., Nilsson, A., Holmström, J., Mathiassen, L.: Managing digital infrastructures: negotiating control and drift in service provisioning. Int. J. Bus. Inf. Syst. **30**(1), 51–78 (2019)
9. Pantaleo, D.C.: The Agile Enterprise: Reinventing Your Organization for Success in an Demand World. Springer, New York (2005)
10. Kuhlin, B., Thielmann, H.: The Practical Real-Time Enterprise. Springer, Berlin (2005)
11. Berman, S.J., Kesterson-Townes, L., Marshall, A., Srivathsa, R.: How cloud computing enables process and business model innovation. Strategy Leadersh. **40**(4), 27–35 (2012)
12. Panetto, H., Iung, B., Ivanov, D., Weichhart, G., Wang, X.: Challenges for the cyber-physical manufacturing enterprises of the future. Annu. Rev. Control **47** (2019)
13. Ivanov, D., Dolgui, A., Sokolov, B.: The impact of digital technology and industry 4.0 on the ripple effect and supply chain risk analytics. Int. J. Prod. Res. **57**(3), 829–846 (2019)
14. Ivanov, D., Sethi, S., Dolgui, A., Sokolov, B.: A survey on the control theory applications to operational systems supply chain management and industry 4.0. Annu. Rev. Control **46**, 134–147 (2018)
15. Wajcman, J.: The digital architecture of time management. Sci. Technol. Hum. Values **44**(2), 315–337 (2019)
16. Paliy, R.V., Spiridonov, E.S., Telyatnikova, N.A., Cerniauskaite, L.: The logical structure of the software file archive formation as a part of industrial management. In: Proceedings of the 2019 IEEE Conference of Russian Young Researchers in Electrical and Electronic Engineering, ElConRus, pp. 1435–1439 (2019)
17. Koutsikouri, D., Lindgren, R., Henfridsson, O., Rudmark, D.: Extending digital infrastructures: a typology of growth tactics. J. Assoc. Inf. Syst. **19**(10), 1001–1019 (2018)
18. Kane, G.C., Palmer, D., Phillips, A.N., Kiron, D., Buckley, N: Aligning the organization for its digital future. MIT Sloan Management Review and Deloitte University Press (2016)
19. Kane, G.C., Palmer, D., Phillips, A.N., Kiron, D.: Is your business ready for a digital future? MIT Sloan Manag. Rev. **56**(4), 37–44 (2015)
20. Belousov, V.V., Ivanova, T.A., Trofimova, V.: The statistical overview of the market of scrap of ferrous metals in the Russian Federation. Theor. Technol. Metall. Prod. **1**(18), 76–80 (2016)
21. Chumak, I.A.: Introduction of a new system of digital control in management by development of the sixth and origin of the seventh of technology ways. Bull. Plekhanov Russ. Acad. Econ. **3**(23), 61–67 (2018)
22. Ivanova, T.A., Trofimova, V.Sh., Kalitaev, A.N., Stepanov, D.G.: Regional logistics of purchase of scrap of ferrous metals by the metallurgical enterprises in the Russian Federation. Reg. Econ. **13**(1), 170–178 (2017)
23. Mistyukova, S.V., Kuznetsova, E.D., Fedulova, I.Yu.: Methods of information support and optimization of the choice of business solutions in the unstable markets. Econ. Entrepreneurship **5**(94), 666–671 (2018)
24. Vasileva, Z.A., Likhacheva, T.P., Globa, S.B.: The industrial development strategy of Krasnoyarsk Krai: challenges and actors of development. J. Siberian Fed. Univ. Humanit. Soc. Sci. **11**(9), 2582–2591 (2016)
25. The Strategy of Development for an Informational Society in the Russian Federation for 2017–2030. It is approved as the Presidential decree of the Russian Federation of 9 May 2017, No. 203
26. Digital Economy of the Russian Federation Program. It is approved as the order of the Government of the Russian Federation of 28 July 2017, No. 1632-r

27. Mingaleva, Z., Bykova, E., Plotnikova, E.: Potential of the network concept for an assessment of organizational structure. Proc. Soc. Behav. Sci. **81**, 126–130 (2013)
28. MetalPlace.ru: http://metallplace.ru/news040918_6. Accessed 9 Mar 2019
29. Deloitte: Overview of the market of ferrous metallurgy. https://www2.deloitte.com/ru/ru/pages/research-center/articles/overview-of-steel-and-iron-market.html. Accessed 1 Mar 2019

Big Data: Nil Novo Sub Luna

Artem A. Balyakin[1], Andrey S. Malyshev[1], Marına V. Nurbina[1]([✉]),
and Mikhail A. Titov[2]

[1] National Research Centre Kurchatov Institute, Moscow 143968, Russia
Nurbina_MV@nrcki.ru
[2] Lomonosov Moscow State University, Moscow 119991, Russia

Abstract. Big data technology belongs to the one of high-tech solutions that will form the forthcoming digital epoch. We present short overview of the current approach to big data, highlighting the main problems it gives rise to. We point out that big data seems to be revolutionary new technology, but in fact should be treated just a new tool to produce knowledge, thus it produces the same risks and challenges as other breakthroughs the humanity endured in the past. We also conclude that cultural aspects of big data implementation should be considered the most crucial one, and the experts community should form the balanced attitude towards this technology. We show the impossibility to control big data with attempts to prohibit some peculiar features it possesses, and propose to focus on such practical steps as terminology improvements, and the evaluation of societal outcomes of the new technology.

Keywords: Big data · Scientific infrastructure · Megascience ·
Research and innovation policy · Socio-Economic challenges

1 Introduction

Digitization is considered to be one of the promising directions of development of the society. In the near future, it is expected the construction of a global information infrastructure and the development of digital technologies worldwide. Subsequently, according to experts, this will serve as the basis for the transition to the "Internet of Things", where an important element is the correctly collected, structured, processed information, and enabling to make optimal management decisions [1, 2]. Thus, in the EU, the proper organization of scientific research, where the first step is to build a digital infrastructure that ensures effective interaction between researchers and infrastructure elements, is the main idea for the development of high technologies [3]. In the Russian Federation, digital technologies are included in the number of breakthrough technologies, and their use in the future of 10–15 years with a high probability can ensure Russia's global technological competitiveness [4].

At the same time, a new industry will be highly individual: it is expected to bring a flexible network approach to the industry (i.e., network-centric approach), when each user (i.e., consumer) becomes a manufacturer, constructing the necessary goods. Secondly, most of the actions will have to be done online, starting from the design of the sample and its numerical processing, and to the exchange of data. Thirdly, it is

T. Antipova (Ed.): ICIS 2019, LNNS 78, pp. 364–373, 2020.
https://doi.org/10.1007/978-3-030-22493-6_32

assumed the geographical dispersed of various elements of production (i.e., production of the constituent parts of the goods). In optimistic scenarios, the innovation cycle is also expected to be reduced, which, however, seems highly questionable due to the strengthening regulatory function of the state in the field of high technologies.

With the development of electronic computer technologies and the Internet, the role of the management of large amounts of data is becoming increasingly prominent. The issues of transferring, processing and storing information from purely applied tasks of building hardware and software are transformed into a problem of infrastructure organization, and the problem of data management is moving into the field of economics, sociology and public administration. In fact, the point is that the proper organization of large data sets contributes to the adoption of the most effective management decisions, and the "big data" technology (and related ones) is one of the likely ways of organizing science in the coming Digital Age.

The most common approach is to consider the Internet as a source of "big data", and, more narrowly, - social networks [5]. At the same time, such areas as science (unique scientific facilities), retail, and medicine play an important role [1, 6–8]. The apparent diversity of "big data" sources is being neglected by similar approaches in collection, storage and analysis of the gathered information. Hereby, the algorithms used in the scientific field can be transferred to other areas of knowledge. Moreover, the growths of "big data" technologies gives an impetus to the development of a number of specialized scientific areas, including a dual purpose ones. In particular, we are talking about the creation of a system of highly specialized artificial intelligence (i.e., intelligent big data processing systems), the development of mechanisms for optimizing data selection using a statistical-probabilistic approach, and the creation of new methods for in-depth analysis of large amounts of data, methods for solving a multidimensional incorrect problems.

A feature of the development of digital technologies and related areas is a serious heterogeneity in the introduction and implementation into everyday life; at the same time, the dynamics of the spread of information technologies in world regions does not meet optimistic expectations [9], which ultimately leads to the emergence of new socio-economic and political challenges, the response to which requires increased attention of society, with the involvement of experts in various industry areas. Thus, one of the possible consequences is the information (digital) spatial inequality, which is irremovable in the short term [1].

In particular, one of the problems requiring a scientific approach is the legal implementation of processes of the "big data" circulation and the development of a common (cultural) approach to "big data" that takes into account the socio-economic and moral-ethical dimensions of new technologies.

In this paper, we consider the current situation of perception of "big data": in the first section a general description of the problem of "big data" is given, with corresponding definitions and concepts. The second section describes some features of the legal regulation of the "big data" circulation and existing practices, and shows the important role of the cultural aspect of the considered problem. In the third section, we postulate the idea of non-uniqueity of "big data", we provide parallels with already existing technological innovations in human history. Finally, in conclusion, a number of practical steps are proposed in the field of "big data" regulation.

Our main position is that the novelty of the practical use of "big data", as well as the problems produced by them, is alleged: most of the difficulties have already arisen in human history throughout scientific and technological progress, thereby, we are able to assume (predict) the main direction of development and try to make the most optimal management decisions in this area.

Based on the proposed approach, the conclusion contains brief summary and recommendations on the need and the possibility of regulating "big data".

2   Main Approaches to Big Data. The Current State-of-the-Art

The term "Big Data" does not have a common definition which is generally accepted. Moreover, a number of researchers exclude the technology of "big data" as an independent direction, considering it as "...a title that includes a large number of technologies that are actively used in everyday life, related to the various areas of activity and do not have signs of innovation..." [1].

Previously, the main criterion for referring to "big data" was the amount of information processed, "the size of which exceeds the capabilities of typical databases for writing, storing, managing and analyzing information" [10], and the "big data" themselves were determined by specifying the following main characteristics of the operated data: (1) large volume (Volume), (2) diversity of data (Variety) and (3) high rate of their change (Velocity) [2].

In a broader sense, "big data" was understood as a socio-economic phenomenon associated with the emergence of technological capabilities to analyze huge amounts of data, in some problem areas - the entire volume of worldwide data, and the resulting transformational consequences [11].

Gradually, the limitation of computing capacity for information processing began to wash out of this definition, and the main emphasis was placed on the methods and approaches to the processing of initial/raw data. Thus, the main software methods used for the processing of "big data" were the means of mass-parallel processing of vaguely structured data, first of all, database management systems of the NoSQL category, MapReduce algorithms and software frameworks with corresponding libraries implementing the Hadoop project [12]. A variety of information technology solutions further began to be attributed to a series of big data technologies, to a greater or lesser extent, providing similar characteristics to the possibility of processing extra-large amounts of data.

Since "big data" involves a combination of information, methods of its processing and obtained results (new knowledge), the question of the source of information arose. As one of the possible definitions, the term "data lake" is used, which means either all the initial unstructured information in general, or the infrastructure that makes it possible to operate the "big data".

In the practical aspect the most important point in regard to "big data" is their commercialization. Accordingly, when considering the "big data", it is necessary to represent their nature, where the source of information comes from:

- virtual data (e.g., Internet: social networks, forums, electronic media, etc.);
- corporate data (e.g., banks, advertisement, retailers - it is, usually, customer data and currency transactions);
- physical data (e.g., data from sensors, detectors, measuring tools and other devices);
- technogenic data (information exchanged between devices during operation; auxiliary information).

The resulting data set relates to various areas of human society: social (e.g., electronic data), economic (e.g., data of corporations and banks), medical (e.g., collection of personal data and a number of corporate data), scientific (e.g., physical data from scientific facilities) - which, in turn, leads to different prioritization. Thus, for electronic data, the size (volume) aspect is important, for corporate data - the emphasis is on variability (dynamics), while the data itself is often initially structured. For medical data, such parameters as volume and dynamism are important, but there is also a clearly postulated parameter "value of human life".

Working with the "big data" implies the structuring information into categories (tags), the corresponding accompanying information (what, where, how, by whom, etc.) is often comparable in volume to the original one. The rapid growth of the data archive leads to problems with scalability, efficiency of the monitoring system and delays in receiving a response to external requests. At the physical level, this leads to the fact that working with "big data" means [partial] segmentation of data on tasks (in accordance with the metadata model adapted to the requirements of the system: monitoring of terrorist activity, updates in the Facebook, analysis of scientific data, etc.). The requirement for computational efficiency (e.g., minimizing the amount of data processed in memory while generating a report) leads to the need to generate reports at different levels of detail [8]: in fact, "big data" is a data model with stepwise aggregation (in the limit of infinite recursion, fractal).

In the future, we will consider "big data" as a kind of entity with the following properties and meeting the following criteria:

- The initial data set is considered to be some "data lake": unstructured, dynamically changing, heterogeneous and loosely coupled.
- "Big data" is the result of applying some methods (models) of selecting data from the all information available in the "data lake", and includes both the data itself with corresponding categories, and the mechanism (algorithm) of their selection and analysis.
- "Big data" is a dynamic, non-stationary system that is in constant process of filling, updating and adjustment[1].

---

[1] Any arbitrarily large structured data is just a large database.

- Inside the "big data" is embedded the algorithm of their processing, issuing a response upon some request. In this case, a feedback mechanism is implemented[2]: the data obtained, in turn, modify the "big data" and/or the mechanism of their selection from the data lake[3].
- Any "big data" is scientific in the sense that it is measurable, selected by the chosen model, processed in advance by given methods and algorithms.

Towards the creation of a digital infrastructure that uses large amounts of data, several problems are highlighted, both of technical and institutional nature, in particular [13]:

- Data encoding;
- Elimination of "garbage", data "noise" management (e.g., filtering out unnecessary information, extracting useful information, evaluation of data adequacy);
- Addressing issues of the long-term content preservation, the development of new storage devices, backup technology;
- Compatibility of data from different periods of time (methods of data writing and encoding are different today and 10 years ago). Compatibility of data from different fields of science;
- Numerous duplications and repeatability of data, information redundancy;
- The need for continuous data verification and its "repackaging" (saving in a more compact form and/or more accessible). Data reduction (for writing) and their recovery (for adequate performance on request);
- Data presentation (visualization);
- The problem related to the creation of metadata: the transition from simple records to complex ones, having external and internal references for navigation (e.g., the analogy to the Internet); metadata structure development;
- Organization of data search and retrieval: formalization of search queries, caching of search data, allocation of servers for storage depending on the tasks and stored information. Organization of multi-level data access (e.g., the analogy with the library);
- Compliance with the legal issues of storing information, which is related to different countries (jurisdictions);
- Resolving the issue of territorial distribution of stored information.

In our opinion, the most important problem to date is not technical difficulties, but issues of regulating "big data" circulation, providing for both following national interests and ensuring the protection of human rights (in case of using and processing personal information). In that question, science provides examples of both the appearance of the first difficulties of the legal regulation of "big data" and the emergence of moral and ethical problems, as well as possible ways of addressing them.

---

[2] Data should be continuously updated or modified (including its structure).

[3] In practice, it can be said that the data at the time of its retrieval from the "data lake" is ALREADY outdated.

3 Big Data. Cultural Approach as a Main Issue

The fact that modern science and technology are inseparable from the socio-economic and political life: all areas of activity are so intertwined that it is impossible to separate technological progress from changes in social norms. In the formal language of title deeds, this led to the fact that, for example, in the European Union, the basis of decisions made is the need to solve social and humanitarian challenges in all their manifestations [3, 13]. With respect to the "big data" in the EU there is a discussion about the ways of their management and regulation; Biomedicine (Artificial Intelligence for Decision Making) was selected as the first field of application: metadata is now being collected and ethical principles relating to the regulation are being developed.

In the United States, the first concerns are not about controlling the circulation of "big data", but about the technical access control: the so-called "neutral" Internet involves changing the physical parameters of access to information and its processing according to its content; in practice, data and users of these data are ranked according to their status. In the United States, issues with the protection of private information and its transfer to third parties for analysis and processing led to a number of serious scandals with social network companies, but it was not possible to formulate clear definitions of permissible information disclosure (a discussion of existing legislative initiatives is given in [14]).

In Russia, the topic of "big data" has not yet gone beyond the highly specialized approach: thus, it raises the question of the need to create a public operator, allowing private sector participation, who will manage information about users' social data (e.g., user preferences on the Internet, social connections, the circle of communication, etc.) [15]. At the same time, legal issues are mainly limited to the regulation of access to personal user data by third parties [5, 16] or the use of collected information to solve legal problems (e.g., user localization on the basis of geo-tracking of his mobile phone) [17].

The first problem that arises in the field of "big data" circulation is connected with personal data protection. Main issue is the necessity of personalized consent (confidence agreement) that includes the information to which consent is given, and the procedure for its use. Practically, this leads to the emergence of new a new segment of "big data" (the volume of regulatory legal documents, with the details of the regulated base, is comparable with the initial volume, i.e., it is also "big data"). On the other hand, the set of all available data is so large (the size of data lakes is significant) that even without their personalization it becomes possible to identify the individual without any doubt, i.e., "data protection" function becomes useless. A way to break this vicious circle could be classifying "data lakes" as "natural" (i.e., data with no possessor, belonging to anyone), as suggested by the McKinsey report [18].

It is also important to note that the existing legal restrictions on the processing of personal data solely in accordance with the originally stated processing objectives, as well as the inadmissibility of combining different databases with originally stated and incompatible processing objectives, contradicts the existing technology and business practices, since it eliminates the advantages provided by "big data" technologies [19].

Note that a superficial way on the issue of regulating "big data" circulation leads to the thesis about the prohibition of social networks or their artificial restriction by introducing forbidden words, topics, etc. However, this approach is doomed to failure according to the above concept of "data lakes" and "big data": the initial data per se can be any, and their selection and analysis play the most important role. As practice shows, while maintaining and enhancing the existing attitude to the information regulation, "big data" technologies will increasingly migrate towards DarkNet, thus, leaving from the legal field. Thereafter, it should be about regulation at a level higher than the formal prohibition of certain words and/or pictures.

For example, in the EU, the concerns expressed about the recently expanding requirements for the protection of personal data will lead to the suspension of work in the field of "big data". As an example, the General Data Protection Regulation (EU) is provided (2016/679, EU GDPR) [20]. This block of laws is aimed at giving citizens the control over their own personal data, at the same time suggesting the simplification of the regulatory framework for international economic relations by unifying regulation within the EU. The law expands the concept of personal data, introduces the concepts of "cross-border data transfer", "pseudo-anonymization", establishes the "right to oblivion", introduces the role of a security officer. The continuation of this policy is the recently adopted directive on the protection of copyright (EU Copyright Directive).

Another problem is the task of storing and accessing "big data". In general, a more capacious data warehouse, an improved search system, maximum complementarity and coherence of information are offered. In order to obtain maximum results and implement the right of equal access, the EU actively implements the principles of open science: thus, 779 organizational declarations governing open access were noted in the ROARMAP report for 2016. Of these, 133 were prepared by investors, 636 were formulated by scientific organizations. The "open science" movement in the Russian Federation has not yet become one of critical technology, but the creation and promotion of "open" principles for building digital infrastructure is included in the number of recommendations for adjustments to the scientific and technological priorities of research and development. Thus, in the foreseeable future, it is planned to implement a system of distributed remote access to both unique scientific facilities and databases, the development of systems for cloud computing and information storage.

This, in turn, leads to the fact that "big data" creates the illusion of knowledge, when the quantity replaces the quality. Easily accessible information leads, for example, to the desire to launch an "unconditional digital income", which contradicts to the fact that most of the information is useless and is never used (e.g., the CISCO report indicates that the "data lakes" themselves are growing at an enormous rate and according to expert estimates, at present, up to 90% of lakes are useless, since they are overfilled with information collected for some unknown purposes [1]).

4 Big Data. Perspectives from a Retrospective Viewpoint

Such concerns, as can be seen, were already in human history: it is enough to indicate the invention of typography and the change in the status of monasteries at the end of the middle ages, when they lost their functions of knowledge accumulation and preservation. We can see the same tendency with respect to "big data". It is observed when the volume of information puts, first of all, the question of its preservation and transference, rather than the extraction of new knowledge. In fact, we are witnessing the return to the present of the difficulties and problems of the middle ages: how to properly organize the functioning of libraries? Storage and preservation of manuscripts? How we can sure that their rewriting by literate monks?

The amount of information created today generates the same scholastic problems as before: does new knowledge exist or does the overall objective only narrow down to the correct codification of the existing one? To date, the answer is most likely negative: quantity (amount of data) does not yet turn into quality. However, there is a tendency of the dilution of the notion of "big data", when society's demand shifts from quantity (size) to quality (algorithms and results of the technology use).

An obstacle to the development of the technology of "big data" is, in our opinion, a bias against it: the strong term is shifted from the field of scientific analysis to the field of hype. The main problem now in the field of using "big data" is the lack of a culture of handling a new tool of scientific and technological progress. Therefore, many problems and difficulties arising in the application of the "big data" technology do not have a conscious cause and malice: they result from ignorance (both the mechanisms and algorithms of the "big data" and the interaction of high technologies with the society).

The lack of a culture of handling "big data" means no practices in the relevant field. Accordingly, at this stage there is no object for legislative work. In this regard, the expert community can and should form these practices.

In this case, the holistic approach is important. Currently, the most common attitude involves answers to ongoing challenges, the regulation of particulars: restrictions and prohibitions are created, which in fact are very easy to get around - this is not about lacunae, but about holes. The whole experience of mankind shows that such an approach is doomed [to failure]: an attempt to manage "big data" "piece by piece" creates new "big data" that governs the original "big data": it turns out a vicious circle.

It is required to perceive "big data" as a new tool of cognition of the world, carrying both positive and negative sides. This leads us to the conclusion that it is necessary to manage the social dimension of high technology.

In addition to this "from above" approach, the induction method is also possible. For example, one of the options for resolving such problems is the case law, which fixes the established tradition. Practically, society is waiting for some event to happen (e.g., the Cambridge Analytica data scandal) in order to begin to regulate this field. Until a certain moment, there is a fear of the new and unknown, to which unique features are attributed. It does not take into account the fact that similar problems have already arisen earlier, and it is just required their adaptation, "translation" from the old language and terminology into a contemporary perspective. As an example, here is the dispute about the responsibility of artificial intelligence, driving cars. It is proposed, for example, to use the Roman approach and to consider AI as an analogy for a slave (servile) in Roman law, i.e., not a subject, but an object of law. In such interpretation, there is no need to expand the concepts and introduction, as some researchers suggest, of a new - digital personality (which AI would be endowed with).

As the authors see it, the development of "big data" technology still raises more questions than answers, which is why it is now important to promote the development of new technologies with taking into account their social consequences. It is, in our opinion, is equivalent to the formation of the scientific culture of using digital infrastructure. It is people who write the rules and set the language of the future. Its symbols are the digital infrastructure, but the logic of the organization of communication will be set by human. This is a hermeneutic approach: a well-formed language structure solves half of the problems.

This defines the existence in the coming Digital Age, as well as a format of Social Science in the Digital Age.

5   Conclusion

In our opinion, the problem of "big data" represents an important task of modern social science: the versatility of methods and approaches in the technology of "big data", the "transborder nature" of their consequences, the level of impact on society - all this makes us consider new technologies both as an opportunity and as a serious challenge at the same time.

In doing so, there is a serious overestimation of the "big data" technology, when it appears to be something revolutionary that drastically changes human nature. This is an overstatement, to our mind, that is associated with the topic's being "trendy" and its direct connection with daily life. In practice, much of the moral and ethical issues have already arisen to humanity in the past, and the most important way to solve them was not partial prohibitions, but an understanding of the problem as a whole, developing a culture of dealing with a new phenomenon.

As for the practical steps, at the initial stage it is required:

First, the coordination and promotion of a single glossary in the field of "big data", taking into account the experience from various fields of activity;

Second, the rejection of attempts to control the information used by "big data", and move the emphasis on the result of their use.

Third, bridging the gap between the technologies themselves and the consequences of their use, which requires taking into account social and economic effects of high technologies.

Acknowledgments. This work was supported by RFBR grant № 18-29-16130 MK.

References

1. Hype Cycle for Emerging Technologies: https://www.gartner.com/doc/3100227/ (2015). Accessed 31 Mar 2019
2. Lynch, C.: How do your data grow? Nature **455**, 28–29 (2008)
3. Florio, M., Sirtori, E.: Social benefits and costs of large scale research infrastructures. Technol. Forecast. Soc. Chang. **112**, 65–78 (2016)

4. The Decree of the President of the Russian Federation: About the strategy for the development of the information society in the Russian Federation for 2017–2030. No. 203. 9 May 2017
5. Almyrzaeva, A., Kostyuk, V., Nevredinov, A.: The role of Big data in modern society. J. Econ. Entrepreneurship **9**(3), 580–582 (2017)
6. Yuchinson, K.: Big data and legislation on competition law. J. High. Sch. Econ. **1**, 216–245 (2017)
7. Grigorieva, M., Golosova, M., Ryabinkin, E., Klimentov, A.: Exascale store for scientific data. Open Syst. DBMS **23**(4), 14–17 (2015)
8. The Global Information Technology Report: The World Economic Forum. http://www3. weforum.org/docs/GITR2016/WEF_GITR_Full_Report.pdf (2016). Accessed 31 Mar 2019
9. Zobova, L., Shcherbakova, L., Evdokimova, E.: Digital spatial competition in the global information space. Fundam. Res. **5**, 64–68 (2018)
10. Manuka, J., Chui, M., Brown, B., Bughin, J., Hobbs, R., Roxburgh, C., Byers, A.: Big Data: The Next Frontier for Innovation, Competition, and Productivity. McKinsey Global Institute, New York (2011)
11. Mayer-Schönberger, V., Cukier, K.: Big Data: A Revolution That Will Transform How We Live, Work, and Think. Houghton Mifflin Harcourt, Boston (2013)
12. Making Sense of Big Data. PwC Tech Forecast, issue 3 (2010)
13. Balyakin, A., Mun, D.: Formation of an open science system in the European Union. Information and Innovations. In: Proceedings of the Conference "Scientometrics and Bibliometrics", pp. 33–37 (2017)
14. Net Neutrality 2019 Legislation: National Conference of State Legislatures. http://www.ncsl. org/research/telecommunications-and-information-technology/net-neutrality-2019-legislation.aspx (2019). Accessed 31 Mar 2019
15. Hu, J., Zhang, Y.: Discovering the interdisciplinary nature of big data research through social network analysis and visualization. Scientometrics **112**(1), 91–109 (2017)
16. Sokolova, A.: The impact of big data analysis technologies (big data) on personal data legislation. In: In the Collection: Jurisprudence 2.0: A New Look at the Right Materials of the Interuniversity Scientific-Practical Conference with International Participation, pp. 282–285. Russian University of Peoples' Friendship, Moscow (2017)
17. Bulgakova, E.V.: Methods for analyzing big data in solving legal problems. In: In the Collection: Law and Information: Questions of Theory and Practice, a Collection of Materials of the International Scientific-Practical Conference, pp. 90–96. Ser. "Electronic legislation" of the FSBI Presidential Library Named After B.N. Yeltsin (2017)
18. Digital Russia: A New Reality. http://www.tadviser.ru/images/c/c2/Digital-Russia-report.pdf (2017). Accessed 31 Mar 2019
19. Savelyev, A.: The Issues of Implementing Legislation on Personal Data in the Era of Big Data. Law. J. High. Sch. Econ. **1**, 43–66 (2015)
20. The EU General Data Protection Regulation (GDPR): https://gdpr-info.eu. Accessed 31 Mar 2019

State and Development Prospects of the Marine Transport Infrastructure of Russia

Igor Shevchenko[1], Vyacheslav Ponomarev[2], Marina Ponomareva[2],
and Nikolay Kryuchenko[2(✉)]

[1] Department of Economics, Kuban State University, 149 Stavropolskaya Street,
Krasnodar 350040, Russia
[2] Department of Humanitarian Economics, Novorossiysk Institute, Moscow
Humanitarian and Economic University, 36/37 Communisticheskaya/Sovetov
Street, Novorossiysk, Russia
gazetagel@mail.ru

Abstract. The article is devoted to the state and development prospects of the marine transport infrastructure of Russia. The authoring team analysis the current situation in the transport industry in general and in the marine transport complex in particular. The search of casual relationship between changes of opportunities of port infrastructure and results available to the transport industry was carried out. Most attention has been paid to the dynamics of the general trade and commercial goods turnover of the Russian Federation, to the changes in export and import flow and also to the terms of benefit relative to the previous periods of the transport complex activity. The analysis of dynamics of a change in the infrastructure capacities and existence of favorable logistic routes of ports was carried out. There was revealed domination in cargo handlings at the ports of the Far East Basin on such nomenclature freights as is revealed: the wood, coal, liquefied gas and ports of the Azovo-Chernomorsky pool that process most part of metals and grains. The article proposes a model allowing to adapt modern infrastructure to the technical and economic requirements of the enterprises of the marine transport of Russia with prospects for the further development in order to increase in opportunities of the transport sector of economy and improving competitiveness of the enterprises of the international marine transport.

Keywords: Infrastructure · Transport · Prospects of development ·
Foreign trade turnover · Goods turnover · Seaports

1 Introduction

The subject "State and development prospects of the marine transport infrastructure of Russia" offered to a consideration is important today because in the context of the modern international relations Russian transport is one of the key industries of the national economy. Infrastructure of the marine transport promotes creation of the technological conditions for more effective forwarding implementation. Infrastructure of the international marine transport enterprises helps to develop a logistic component as a complex of interdependent objects (machines and mechanisms, buildings and

© Springer Nature Switzerland AG 2020
T. Antipova (Ed.): ICIS 2019, LNNS 78, pp. 374–381, 2020.
https://doi.org/10.1007/978-3-030-22493-6_33

constructions, human resource – workers and personnel, information program complexes, means of communication, financial resources – as a source of ensuring this process, document flow) as well. Methodological and applied part of the logistic performance and the use of the developed infrastructure have to allow maritime transportation companies to accelerate delivery of goods from the producer to the end user with minimal risks of losses.

This article allows the reader to understand the state and development prospects of marine transport infrastructure of Russia taking into account the analysis of macroeconomic indicators and change impacts on them that happen in the marine transport industry.

The importance of a problem of the Russian marine transport infrastructure is evidenced in the risks increase connected with losses related to the breakages of goods, change of their packing, delays of delivery and many others.

The goal of research is the analysis of the existing state and prospects of the marine transport of Russia in view of improvement of organizational and economic bases of the transport infrastructure formation and the strategic directions of its development in the system of economic territory of the country.

It should be noted that formation of the modern infrastructure of the marine transport is caused by the following prerequisites characterizing improvement in the investment climate of the Russian Federation. The first one includes a large number of companies throughout the whole country such as opening of new organizations, for example, the project with China: construction of new enterprises, both in oil and metallurgical sectors.

There is a need for the further development of the system of the sea transport infrastructure of the Russian Federation due to the fact that China plans to process products made in our country. The advantageous geographical location of our country is the second prerequisite for development of the modern infrastructure of the Russian marine transport. Russia connects Europe and Asia and is the entry point for the CIS countries.

The third prerequisite includes enterprises that provide the current effective infrastructure base for further development of logistics such as: JSC "Russian Railways" (JSC «RZhD»), JSC "Aeroflot", JSC "Transaero", PJSC "NCSP Group" and etc.

Thus, the companies are ready to changes, there is a good base, but the best practices will be needed. Cooperation will be mutually fruitful for both Russia and foreign partners. According to the aforementioned it is necessary to emphasize the possibility of developing projects with a short payback time that is extremely relevant in post-crisis conditions.

It is also possible to claim about high profitability of projects (the expected traffic flows). Political stability in Russia (both in national and provincial levels) and attractive tax system (as an example: companies pay 20% tax on profits) are very important as well as the large number of positive changes for the last period in the development of legislative framework.

The Foreign Investments Act also identifies such concepts as the direct foreign investments and the priority investment project.

2 Prior Literature Review

Issues relating to the state and development prospects of the marine transport infrastructure of Russia and also economic infrastructure that affected the author's point of view were reflected in the works of domestic and foreign authors: Bennet M., Zhamin V., Zotova T., Iokhimsen R., Marx K., Muravyev D., Oreshin V., Rakhmangulov A., Fedko V., Chernoka A., Sharipov A., Engelya E., etc. The market-orientation of the infrastructure transformations was analyzed by Zadvorny Yu., Karnaukhov S., Morozova I., Popova L., Fedko V., Ford R.'s works, etc.

3 Statemen of Basic Materials

Transport in the Russian Federation promotes the formation of a sectoral basis of the national and world systems that include technical and economic, political, foreign trade, customs and standard legal relations and allow to increase the tempo of structural growth of the latter.

Russian transports role as a major mechanism is seen not only in its natural need, but also in diversity of solvable tasks for satisfaction of requirements of the market, including sellers and buyers, intermediaries and insurers, national institutions and public institutes for consumer protection of goods, works, services [1].

The tools of the corresponding infrastructure from the sphere of referring motivation and movement of people apart from application of effective distribution and thus of transport processes production is used through the prism of innovative technologies that contribute the growth of transport availability to any natural and legal entity. This promotes development of various industries and social sphere by means of multiplicative effect of the investments and innovations in the transport complex that are directed at qualitative improving the process of rendering transportation services, updating objects of transport infrastructure and vehicles 6 [5].

Transport and all necessary infrastructure still play the main role in supporting the growth of economy despite the essential negative conditions promoting slowdown in the social and economic development of the country based on the dependence of export and raw model of national economy and significant capital outflow abroad.

Transport gives Russia an opportunity to provide unity of economic space, territorial integrity of the state. The expanded range of transport infrastructure strengthenes and contributes to the development of correlation communications of all regions of the country that promotes accumulation of material resources for ensuring requirements of foreign economic relations of Russia in being integrated into the global economy [7].

At the same time Russia creates conditions to improve competitiveness of its goods and services in the world market by forming basic infrastructure of transport hubs and setting out the preconditions of increase the quality of life of its population and creating demand for hi-tech products of various spheres of national economy.

Figure 1 presents the dynamics of the foreign trade turnover of the Russian Federation (one billion US dollars) from 2013 to 2017.

As it is seen at the Fig. 1 the significant recession of the foreign trade turnover of Russia took place in 2015/Then there was a stabilization of national currency rate –

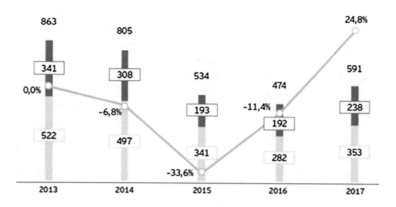

values (522, 497, 341, 282, 353) are export operations;
values (341, 308, 193, 192, 238) are import operations;
values (863, 805, 534, 474) are export and import operations during the entire period;
circle represents the growth rate compared to the previous year, %.

Fig. 1. Dynamics of the foreign trade turnover of the Russian Federation (one billion US dollars) [8, 12]

ruble, at the background of price increase of oil and a considerable economic breakthrough and exit from a crisis situation. In 2017 the foreign trade turnover of Russia increased by 25% against to 2016 according to the Central Bank of Russia and made 591 billion US dollars. Export volume growth of such goods as wheat, ore, mineral fertilizers and coal significantly influenced the growth of transport goods turnover as well.

Due to the positive dynamics of trade in domestic market for the last three years increase of volumes of the wholesale was 5.7% and in retail 1.3%.

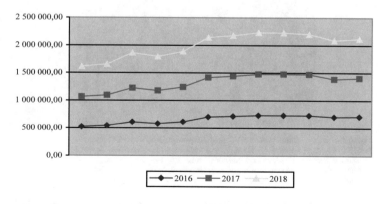

vertical axis – units of measure in thousands of tons,
horizontal axis – the periods marked by points in months (from January to December of the corresponding year)

Fig. 2. Dynamics of commercial goods turnover of Russian transport (for all means of transport) [11]

Figure 2 shows the dynamics of commercial goods turnover of Russian transport from 2016 to 2018 which includes the following means of transport: railway, automobile, sea, internal water, air (transport aircraft), pipeline.

Data of Fig. 2 focuses attention on the most significant growth in the volume of transportation of goods including all means of transport. In August 2018 the value of goods turnover resulted in 754583.5 thousand tons in comparison with the values of the same period in 2017-745865.6 thousand tons. and in 2016-731512.7 thousand tons.

In general change of goods turnover by all means of transport in Russia during the period from January 2016 to December 2018 is positive due to the dynamics growth on the rate of goods turnover.

Today the Government and the Ministry of Transport of the Russian Federation set the task for its transport industry to strengthen the position of our country in the global market of transport and logistic services. Besides it is noted that it is impossible to use reserves of transit potential effectively without development of sea port infrastructure.

According to the Office of National Statistics in 2018 year the number of the organizations by types of economic activity which are engaged in transportation and storage of goods is 256.5 thousand units.

In 2017 year according to the information of Association of the Sea Trade Ports the cargo handling volume in seaports was 786.9 million tons that exceeds a similar indicator of last period for 9%. According to the volume of goods in structure of trans-shipment in seaports the next types of goods dominate: oil products – 142 million tons, coal – 154 million tons, crude oil – 253 million tons. At the same time the volume of trans-shipment of bulk goods grew by 7.2% and was 414 million tons (that is 53% of the general goods turnover), and the volume of trans-shipment of dry goods was 372.9 million tons (47% of the general goods turnover).

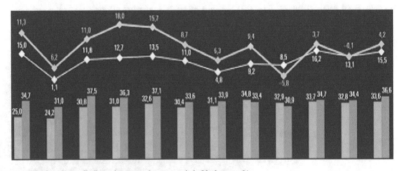

top diagram – changes of transfer volume of bulk goods compared to same period of the last year, %;
lower diagram – changes of transfer volume of dry goods compared to same period of the last year, %;
histogram (is closer to the left edge of the fig) transfer volume of dry goods, millions of tons;
histogram (is closer to the right edge of the fig) transfer volume of bulk goods, millions of tons;
periods marked by points in months (from January to December of the corresponding year).

Fig. 3. Dynamics cargo handling volumes of Russia seaports in 2017 [9]

Figure 3 represents the volumes of cargo handling in seaports of Russia in 2017 on the basis of the available and modernized capacities of the existing transport infrastructure.

According to the experts changes and fluctuations in dynamics of cargo handling volumes in Russian seaports in 2017 are mainly connected with the geographical disproportion in trans-shipment between Basins.

By virtue of existing and developed infrastructure capacities for the above mentioned period and availability of the logistic routes the port farms of the Far East Basin dominate in cargo handlings of such nomenclature goods as: wood, coal, liquefied gas. Whereas, the Azovo-Chernomorsky Basin process the most part of metals and grain [2].

According to the Russian Railway News Agency partner.ru the goods turnover of the Russian sea trade ports in 7 months of 2018 grew by 3.1% in comparison with the same period of last year, container turnover – for 11.5% [10].

The largest volume of cargo handling operations is held by the ports of the Azovo-Chernomorsky Basin. Since the beginning of year they handled 157.1 million tons that is 6.2% higher compared to the level of the previous one. Growth of indicators is due to the volume increase of dry goods trans-shipment by 14% in comparison with the same period of the first half of the 2017 year [3].

Instead, the quantity of the processed bulk goods increased more significantly than dry ones. In the ports of the basin there were handled 68.8 million tons of dry goods and 88.4 million tons of bulk ones within 7 months.

Trade experts that study issues of transport, logistics, technologies of goods delivery with work performance and rendering the corresponding services, come to a conclusion that one of the key directions to overcome the problems constraining development of the transport sector is the creation of conditions that can attract the private investments necessary for the rapid rates of the transport infrastructure development. This is pointed out by the authors of article "Development of sea port infrastructure of the region on the basis of "Dry ports"" Rakhmangulov and Muravyev [4].

4   Results

Growth of throughput and overworking ability of seaports can only be achieved if the modern complex of mechanisms that provides reduction of unevenness of internal goods traffic in ports or designing of "the rear cargo terminal" – the "dry port" is available. An important point in this offer is the fact that despite the minimum technical and technological expenses the overworking port possibilities dramatically increase by modernization and reconstruction.

Methodical tools of the experimental imitating and models that are partially realized in practice helps to create the scheme of identification and approbation of the most effective criteria for evaluation of fundamental parameters of "dry ports".

Thereafter we are to reveal and consider phases of implementation of the "dry port" project. The first stage is to include assessment of all the technical and economic indicators of efficiency of the master plan of seaport, including specifications and reports on results of the performed works and developments, functional features of port economy within 5 – 10 years with due regard to the port waters, port highways, specialized railway access roads and stations.

The second stage is to assume accumulation of information about time periods for which handling and warehousing services will be carried out by all means of transport, logistic schemes and information, commodity, documentary flows.

It is important to note the necessity of creation of a program complex to calculate the actual indicators of technical and economic activity with due regard to the statistical data confirmed by the financial statements of the enterprise [6]. The third stage is to develop the imitating paradigm of the modern transport enterprise in view of an opportunity of diversification given by the "dry port".

For the purpose of determination of the maximum overworking port ability the fourth stage is to perform an experimental component having in consideration the imitating modeling of planned and expected values of intensity and correlation dependence of export and import operations.

The fifth stage represents modernization of the existing capacities with account of implementation of the "dry port" into the technological scheme or creation of the new complex of constructions based on the principle of cost minimization and increase of operational loading and thereby the volume characteristics. The sixth stage is to develop the operating plan according to geographical and technological features of the territory, its strategic potential for concentration of production capacities of the "dry port".

The seventh stage is a variable part for further direct production of the "dry port", but having in consideration the following elements: length and width between the main port and "dry" one, quantitative opportunities (platforms, territory), port capacities, cost of all construction works.

The eighth stage is the diversification and efficiency increase of interaction in modeling joint activity between the sea and "dry" ports by means of implementation of the specialized sections imitating work of the "dry port". The ninth stage is to simulate work of the "dry port" including various technical and economic risks direction by means of specialized programs.

The previous three stages are insulated by the final one that represents the experimental realization of the system "seaport – "dry port" when changing the values of intensity and correlation of incoming and outgoing traffic flows in order to definite the optimum values of key parameters of the "dry port" for each scenario of the port infrastructure development.

5   Conclusion

The following conclusion has been made by summarizing the above. Russian transport is an important sector of the national economy that is necessary to ensure the effective functioning of domestic and foreign trade relations, and that contributes to the economic development of the country.

The multiplicative effect of transport consists in use of infrastructure to meet the goals and objectives of the industry that are to ensure timely delivery of goods to the addressee at the minimum expenses.

Transport infrastructure performs both economic and public function providing employment in the port farms at the same time.

Nevertheless there are problems in transport including maritime one, that couldn't provide the further development of the industry without effective solution.

Public-private partnership remains in priority. It allows to minimize risks of losses of time and funds for business and provides control for improvement of trade relations.

It should be noted that the central problem of logistic infrastructure functioning in the seaports area is the insufficient capacity of port facilities and massive unequality of incoming car traffic and shipment flows. It results in the deterioration of indicators of promptness of freight process flows in ports. The efficient way to handle this problem is to create the rear terminal potentially capable to increase the capacity of nearby seaport and provide costs reduction of handling service by way of shorten waiting time of vehicles and reduction of stocks in the "seaport-dry port" system.

Group of authors offered the system of "dry port" key parameters. It is aimed to determinate the optimum combination of parameters in the course of strategic planning and port infrastructure development and is based on using scenario approach in combination with the method of simulation.

The methodology can be used to justify the investment decisions on development state of marine transport infrastructure of Russia and its prospects.

References

1. Nikulina, O.V., Kryuchenko, N.N.: Financial justification of the strategy of implementation of innovative project of the international transport company. Mag. Econ. Bus. **4–1**(45), 371–375 (2014)
2. Ponomarev, V.V.: Organizational-legal bases of formation and development of the Black Sea province. In: The monograph, p. 75. Moscow humanitarian and economic institute Novorossiysk branch, Center of scientific knowledge of "Logos", Stavropol (2016)
3. Ponomareva, M.Yu.: Function of usefulness as the indicator of development of commodity production in society. In: Ponomaryova, V.V., Kutkovich, T.A. (eds.) Section of the Monograph: Resource Potential as Basis of Competitiveness of the Region, pp. 78–85, Center of scientific knowledge of "Logos", Stavropol (2018)
4. Rakhmangulov, A.N., Ants, D.S.: Development of the regional sea port infrastructure on the basis of "dry ports". Mag.: Reg. Econ. **12**(3), 924–936 (2016)
5. Williamson, O.E.: Managerial economics: structure and consequences. In: Langlois, Ad.R. N. (ed.) Economy as a Process: The essay in the New Established Economy, pp. 171–202. Cambridge University Press, Cambridge (1986)
6. Ford, R.: Infrastructure and Productivity of the Private Sector, Paris (1991)
7. Shevchenko, I.V., Korobeynikova, M.S.: Russia's role in the domestic and world economic system diversification. In: Collected Works of the Conference: Economic Development of Russia: Restructuring and Diversification of a World Ecosystem, pp. 195–198, Krasnodar (2018)
8. Central Bank of Russia: https://www.cbr.ru
9. Federal Agency for Sea and Inland Water Transport: http://www.morflot.ru
10. Russian Railway Magazine Partner.ru: http://www.rzd-partner.ru
11. Website of the Federal State Statistics Service of the Russian Federation: www.gks.ru
12. Website of the Ministry of Transport of the Russian Federation: www.mintrans.ru

Author Index

Printed in the United States
By Bookmasters